LONDON MATHEMATICAL SOCIETY LECTURE NOTE SERIES

Managing Editor: Professor M. Reid, Mathematics Institute, University of Warwick, Coventry CV4 7AL, United Kingdom

The titles below are available from booksellers, or from Cambridge University Press at http://www.cambridge.org/mathematics

London Mathematical Society Lecture Note Series: 393

Non-abelian Fundamental Groups and Iwasawa Theory

Edited by

JOHN COATES
University of Cambridge

MINHYONG KIM
University College London

FLORIAN POP
University of Pennsylvania

MOHAMED SAIDI
University of Exeter

PETER SCHNEIDER
Universität Münster

CAMBRIDGE
UNIVERSITY PRESS

CAMBRIDGE
UNIVERSITY PRESS

Shaftesbury Road, Cambridge CB2 8EA, United Kingdom

One Liberty Plaza, 20th Floor, New York, NY 10006, USA

477 Williamstown Road, Port Melbourne, VIC 3207, Australia

314–321, 3rd Floor, Plot 3, Splendor Forum, Jasola District Centre, New Delhi – 110025, India

103 Penang Road, #05–06/07, Visioncrest Commercial, Singapore 238467

Cambridge University Press is part of Cambridge University Press & Assessment,
a department of the University of Cambridge.

We share the University's mission to contribute to society through the pursuit of
education, learning and research at the highest international levels of excellence.

www.cambridge.org
Information on this title: www.cambridge.org/9781107648852

© Cambridge University Press & Assessment 2012

First published 2012

A catalogue record for this publication is available from the British Library

Library of Congress Cataloging-in-Publication data
Non-abelian fundamental groups and Iwasawa
theory / edited by John Coates . . . [et al.].
p.   cm. – (London Mathematical Society lecture note series ; 393)
ISBN 978-1-107-64885-2 (pbk.)
1. Iwasawa theory.   2. Non-Abelian groups.   I. Coates, J.   II. Title.   III. Series.
QA247.N56   2011
512.7′4–dc23
2011027955

ISBN   978-1-107-64885-2   Paperback

Contents

Contributors

Christophe Breuil *Bâtiment 425, C.N.R.S. et Université Paris-Sud, 91405 Orsay Cedex, France*

Kevin Buzzard *Department of Mathematics, Imperial College London, 180 Queen's Gate, London SW7 2AZ, UK*

Frank Calegari *Department of Mathematics, Northwestern University, 2033 Sheridan Road, Evanston, IL 60208-2730, USA*

J. Coates *DPMMS, Centre for Mathematical Sciences, Wilberforce Road, Cambridge CB3 0WB, UK*

Matthew Emerton *Department of Mathematics, Northwestern University, 2033 Sheridan Road, Evanston, IL 60208-2730, USA*

Mahesh Kakde *Department of Mathematics, University College London, Gower Street, London WC1E 6BT, UK*

Minhyong Kim *Department of Mathematics, University College London, Gower Street, London WC1E 6BT, UK*

Hiroaki Nakamura *Department of Mathematics, Okayama University, Okayama 700-8530, Japan*

Florian Pop *Department of Mathematics, University of Pennsylvania, 209 South 33rd Street, Philadelphia, PA 19104-6395, USA*

Mohamed Saïdi *College of Engineering, Mathematics, and Physical Sciences, University of Exeter, Harrison Building, North Park Road, Exeter EX4 4QF, UK*

R. Sujatha *School of Mathematics, Tata Institute of Fundamental Research, Homi Bhabha Road, Mumbai 400 005, India*

Zdzisław Wojtkowiak *Université de Nice-Sophia Antipolis, Dépt. of Math., Laboratoire Jean Alexandre Dieudonné, U.R.A. au C.N.R.S., No 168, Parc Valrose–B.P. N° 71, 06108 Nice Cedex 2, France*

Preface

In historical origins, serious topological input to arithmetic starts at the latest in the early years of the twentieth century with the foundations of class field theory by Takagi and Artin. Around the middle of the century, the rapid development of homological techniques in algebra led to an explosion of consequent activity in number theory and algebraic geometry, including the theory of coherent sheaves, the homological interpretation of Artin's reciprocity map, and Grothendieck's construction of arithmetic cohomology theories based upon the abstract notion of Grothendieck topologies. Then, in the 1980s, Grothendieck formulated his anabelian conjectures in the celebrated letter to Faltings, and brought to a hitherto-unexplored depth the interaction between topology and arithmetic. Therein came into focus the far-reaching vision that the perspective of non-abelian fundamental groups could lead to a fundamentally new understanding of deep arithmetic phenomena, including the arithmetic theory of moduli and Diophantine finiteness on hyperbolic curves. In the 1990s, profound (and perhaps unexpected) progress in Grothendieck's program was realized through the theorems of Nakamura, Tamagawa, Mochizuki. (See the contributions of Nakamura, Pop, and Saidi.)

Meanwhile, a certain amount of work in recent years linking fundamental groups to Diophantine geometry intimates deep and mysterious connections to the theory of motives and Iwasawa theory, notions usually associated to Diophantine problems involving exact formulae of which the most celebrated example is the conjecture of Birch and Swinnerton-Dyer (cf. Kim's article). In fact, the work thus far suggests that the still-unresolved section conjecture of Grothendieck, whereby maps from Galois groups of number fields to fundamental groups of hyperbolic arithmetic curves are all proposed to be of geometric origin, is exactly the sort of key problem that touches the core of all these areas of number theory and more. Therefore, the time seems right to encourage a much broader understanding of the arithmetic issues surrounding anabelian geometry and its ramifications.

While the overall importance of the theorems of anabelian geometry appears to be widely acknowledged, there is as yet not much specific knowledge within the arithmetic geometry community of its coherent body of concepts and philosophy, and of the new technology that yields actual results. It is our belief that a higher level of general awareness will lead to a genuine strengthening of our grasp of a wide range of Galois-theoretic and Diophantine phenomena, and hopefully to significant progress on the section conjecture. The goal then of this book is to present articles that contain the ideas and problems of anabelian geometry within the global context of mainstream arithmetic geometry, with strong emphasis on connections to non-commutative Iwasawa theory.

Broadly put, anabelian geometry and non-commutative Iwasawa theory deal in different ways with the lifting of homological and abelian ideas of arithmetic into the realm of the non-abelian, with connections to homotopy theory. The complementary nature of the two viewpoints is expressed in part by the exact sequence

$$(*) \qquad 0 \to \hat{\pi}_1(\bar{X}, b) \to \hat{\pi}_1(X, b) \to \mathrm{Gal}(\bar{\mathbb{Q}}/\mathbb{Q}) \to 0$$

associated to an algebraic curve X defined over the rationals, involving the geometric fundamental group $\hat{\pi}_1(\bar{X}, b)$ and the absolute Galois group $\mathrm{Gal}(\bar{\mathbb{Q}}/\mathbb{Q})$ of the rationals, both contributing in intricately inter-related ways to the structure of the arithmetic fundamental group $\hat{\pi}_1(X, b)$ in the middle. From this exact sequence, one extracts an induced outer action of $\mathrm{Gal}(\bar{\mathbb{Q}}/\mathbb{Q})$ on $\hat{\pi}_1(\bar{X}, b)$. When this action is viewed at the level of the abelianization $H_1(\bar{X}, \hat{\mathbb{Z}})$ of $\hat{\pi}_1(\bar{X}, b)$, it extends naturally to completed group algebras of quotient groups of $\mathrm{Gal}(\bar{\mathbb{Q}}/\mathbb{Q})$, which then become the central objects of study in non-commutative Iwasawa theory (cf. articles of Coates and Sujatha, and of Kakde).

Here, encoding the infinite tower of number fields represented by an infinite non-abelian quotient into a single group algebra is both convenient and crucial, and brings with it the intervention of refined tools that combine the structure theory of non-commutative algebras with cohomological aspects of Galois theory. In anabelian geometry, on the other hand, the main point is to study the first group $\hat{\pi}_1(\bar{X}, b)$ in its own right, without abelianizing, even as certain abelian or mildly non-abelian quotients or completions might mediate this study. Although the attendant action of $\mathrm{Gal}(\bar{\mathbb{Q}}/\mathbb{Q})$ is of key importance, it is safe to say that the overwhelmingly geometric nature of the techniques as well as the non-linearity of the action tend to obscure the role of fine invariants of $\mathrm{Gal}(\bar{\mathbb{Q}}/\mathbb{Q})$ such as might occur in Iwasawa theory. Thereby, with some oversimplification, one could describe both anabelian geometry and non-commutative Iwasawa theory within the single framework of the sequence (*), the difference lying only in the technical depths with which attention is drawn to either the kernel or the quotient, and in the degree of non-commutativity/non-linearity that is preserved in its study.

Having formulated thus the commonality and difference, the importance of coming up with an approach unified at both the philosophical and the technological level becomes quite obvious. As discussed in a number of different chapters of this book, there is a remarkable overlap between the anabelian and Iwasawa-theoretic viewpoints in the study of elliptic curves. On the other hand, when the geometric fundamental group is genuinely non-abelian, as happens for hyperbolic curves, it remains to develop the correct analogue of this interaction. In addition to progress on the section conjecture itself, it is hoped that an extended program of the Birch and Swinnerton-Dyer type for hyperbolic curves will result from investigation along these lines. Because these issues lie at the crossroads of important research in perhaps all of the main areas of arithmetic geometry, our intention is that the core scientific interaction of the book should be enriched further by surrounding essays (by Breuil, Buzzard, and Calegari and Emerton) that deal with one other key arena for non-abelian ideas and techniques, namely, the Langlands program.

The articles in this book were written by participants in the special programme on 'non-abelian fundamental groups in arithmetic geometry' that took place at the Isaac Newton Institute in the second half of the year 2009. In the course of the programme, we benefitted greatly from the advice of director David Wallace as well as the hospitality of the INI staff, especially Christine West, Esperanza de Felipe, and Mustapha Amrani. We are deeply indebted to their efficient help. Many thanks are due also to Diana Gillooly of Cambridge University Press, whose encouragement was critical to the inception of this book and whose sterling effort to bring together the disparate contributions into a single manuscript was essential to its completion.

Lectures on anabelian phenomena
in geometry and arithmetic

Florian Pop[a]

University of Pennsylvania

Part I. Introduction and motivation

The term "anabelian" was invented by Grothendieck, and a possible translation of it might be "beyond Abelian." The corresponding mathematical notion of "anabelian geometry" is vague as well, and roughly means that under certain "anabelian hypotheses" one has:

$$* \; * \; * \; \mathfrak{Arithmetic\ and\ Geometry\ are\ encoded\ in\ Galois\ Theory} \; * \; * \; *$$

It is our aim to try to explain the above assertion by presenting/explaining some results in this direction. For Grothendieck's writings concerning this, the reader should have a look at [G1], [G2].

A. First examples

(a) *Absolute Galois group and real closed fields*   Let K be an arbitrary field, K^{a} an algebraic closure, K^{s} the separable closure of K inside K^{a}, and finally $G_K = \mathrm{Aut}(K^{\mathrm{s}}|K) = \mathrm{Aut}(K^{\mathrm{a}}|K)$ the absolute Galois group of K. It is a celebrated well-known theorem by Artin–Schreier from the 1920s that asserts the following: *If G_K is a finite non-trivial group, then $G_K \cong G_{\mathbb{R}}$ and K is real closed.* In particular, $\mathrm{char}(K) = 0$, and $K^{\mathrm{a}} = K[\sqrt{-1}]$. Thus the non-triviality + finiteness of G_K imposes very strong restrictions on K. Nevertheless, the kind of restrictions imposed on K are not on the *isomorphism type of K* as a field, as there is a big variety of isomorphy types of real closed fields (and their classification up to isomorphism seems to be out of reach). The kind of restriction imposed on K is rather one concerning the algebraic behavior of K, namely that the algebraic geometry over K looks like the one over $\mathbb{R}$.

[a] This work was partially supported by NSF grant DMS-0801144.

Non-abelian Fundamental Groups and Iwasawa Theory, eds. John Coates, Minhyong Kim, Florian Pop, Mohamed Saïdi and Peter Schneider. Published by Cambridge University Press. ©Cambridge University Press 2012.

(b) *Fundamental group and topology of complex curves*   Let X be a smooth complete curve over an algebraically closed field of characteristic zero. Then, by using basic results about the structure of algebraic fundamental groups, it follows that the geometric fundamental group $\pi_1(X)$ of X is isomorphic – as a profinite group – to the profinite completion $\widehat{\Gamma}_g$ of the fundamental group Γ_g of the compact orientable topological surface of genus g. Hence $\pi_1(X)$ is the profinite group on $2g$ generators σ_i, τ_i $(1 \le i \le g)$ subject to the unique relation $\prod_i[\sigma_i, \tau_i] = 1$. In particular, the genus g of the curve X is encoded in $\pi_1(X)$. But, as above, the isomorphy type of the curve X, i.e., of the object under discussion, is not "seen" by its geometric fundamental group $\pi_1(X)$ (which in some sense corresponds to the absolute Galois group of the field K). Precisely, the restriction imposed by $\pi_1(X)$ on X is of topological nature (one on the complex points $X(\mathbb{C})$ of the curve).

B. Galois characterization of global fields

More than 40 years after the result of Artin–Schreier, it was Neukirch who realized (in the late 1960s) that there must be a p-adic variant of the Artin–Schreier theorem; and that such a result would have highly interesting consequences for the arithmetic of number fields (and more general, global fields). The situation is as follows. In the notations from (a) above, suppose that K is a field of algebraic numbers, i.e., $K \subset \mathbb{Q}^a \subset \mathbb{C}$. Then the Artin–Schreier theorem asserts that if G_K is finite and non-trivial, K is isomorphic to the field of real algebraic numbers $\mathbb{R}^{\mathrm{abs}} = \mathbb{R} \cap \mathbb{Q}^a$. This means that the only finite non-trivial subgroups of $G_\mathbb{Q}$ are the ones generated by the $G_\mathbb{Q}$-conjugates of the complex conjugation; in particular, all such subgroups have order 2, and their fixed fields are the conjugates of the field of real algebraic numbers. Now, Neukirch's idea was to understand the fields of algebraic numbers $K \subset \mathbb{Q}^a$ having absolute Galois group G_K isomorphic (as profinite group) to the absolute Galois group $G_{\mathbb{Q}_p}$ of the p-adics $\mathbb{Q}_p$. Note that $G_{\mathbb{Q}_p}$ is much more complicated than $G_\mathbb{R}$. It is nevertheless a topologically finitely generated field, and its structure is relatively well known, by the work of Jakovlev, Poitou, Jannsen–Wingberg, etc., see e.g. [J–W]. Finally, Neukirch proved the following surprising result, which in the case of subfields $K \subset \mathbb{Q}$ is the perfect p-adic analog of the theorem of Artin–Schreier.

Theorem (See e.g. Neukirch [N1])   *For fields of algebraic numbers $K, K' \subset \mathbb{Q}^a$ the following hold:*

(1) *Suppose that $G_K \cong G_{\mathbb{Q}_p}$. Then K is the decomposition field of some pro-*

longation of $||_p$ to $\mathbb{Q}^a$. *Or equivalently, K is $G_{\mathbb{Q}}$-conjugated to the field of algebraic p-adic numbers $\mathbb{Q}_p^{abs}$.*

(2) *Suppose that $G_{K'}$ is isomorphic to an open subgroup of $G_{\mathbb{Q}_p}$. Then there exists a unique $K \subset \mathbb{Q}^a$ as in (1) such that K' is a finite extension of K.*

The theorem above has the surprising consequence that an *isomorphism of Galois groups* of number fields gives rise functorially to an *arithmetical equivalence* of the number fields under discussion. The precise statement is as follows. For number fields K, let $\mathcal{P}(K)$ denote the set of their places. Let $\Phi : G_K \to G_L$ be an isomorphism of Galois groups of number fields. Then a consequence of the above theorem reads: Φ *maps the decomposition groups of the places of K isomorphically onto the decomposition places of L. This bijection respects the arithmetical invariants* $e(\mathfrak{p}|p)$, $f(\mathfrak{p}|p)$ *of the places* $\mathfrak{p}|p$, *thus defines an arithmetical equivalence:*

$$\varphi : \mathcal{P}(K) \to \mathcal{P}(L).$$

Finally, applying basic facts concerning arithmetical equivalence of number fields, one obtains the following. In the above context, suppose that $K|\mathbb{Q}$ is a Galois extension. Then $K \cong L$ as fields. Naturally, this isomorphism is a $\mathbb{Q}$-isomorphism. Since $K|\mathbb{Q}$ is a normal extension, it follows that $K = L$ when viewed as sub-extensions of fixed algebraic closure $\mathbb{Q}^a$. In particular, $G_K = G_L$ as subgroups of $G_{\mathbb{Q}}$. Thus the open normal subgroups of $G_{\mathbb{Q}}$ are *equivariant*, i.e., they are invariant under automorphisms of $G_{\mathbb{Q}}$. This led Neukirch to the following questions.

(1) Does $G_{\mathbb{Q}}$ have inner automorphisms only?
(2) Is every isomorphism $\Phi : G_K \to G_L$ as above defined by the conjugation by some element inside $G_{\mathbb{Q}}$?

Finally, the first peak in this development was reached at the beginning of the 1970s, with a *positive answer* to Question (1) by Ikeda [Ik] (and partial results by Komatsu), and the break through by Uchida [U1], [U2], [U3] (and unpublished notes by Iwasawa) showing that the answer to Question (2) is positive. Even more, the following holds.

Theorem *Let K and L be global fields. Then the following hold:*

(1) *If $G_K \cong G_L$ as profinite groups, $L \cong K$ as fields.*
(2) *More precisely, for every profinite group isomorphism $\Phi : G_K \to G_L$, there exists a unique field isomorphism $\phi : L^s \to K^s$ defining Φ, i.e., such that*

$$\Phi(g) = \phi^{-1} \circ g \circ \phi \quad \text{for all } g \in G_K.$$

In particular, $\phi(L) = K$. And therefore we have a bijection:

$$\text{Isom}_{\text{fields}}(L, K) \cong \text{Out}_{\text{prof.gr.}}(G_K, G_L).$$

This is indeed a very remarkable fact: the Galois theory of the global fields encodes the isomorphism type of such fields in a functorial way! This result is often called the *Galois characterization of global fields*.

We recall briefly the idea of the proof, as it is very instructive for future developments. First, recall that, by results of Tate and Shafarevich, we know that the virtual ℓ-cohomological dimension $\text{vcd}_\ell(K) := \text{vcd}(G_K)$ of a global field K is as follows, see e.g. Serre [S1], ch. II.

  (i) If K is a number field, then $\text{vcd}_\ell(K) = 2$ for all ℓ.
 (ii) If $\text{char}(K) = p > 0$, then $\text{vcd}_p(K) = 1$, and $\text{vcd}_\ell(K) = 2$ for $\ell \neq p$.

In particular, if $G_K \cong G_L$ then K and L have the same characteristic.

Case 1 $K, L \subset \mathbb{Q}^a$ are number fields.
Then the isomorphism $\Phi: G_K \to G_L$ defines an arithmetical equivalence of K and L. Therefore, K and L have the same normal hull M_0 over $\mathbb{Q}$ inside $\mathbb{Q}^a$; and moreover, for every finite normal sub-extension $M|\mathbb{Q}$ of $\mathbb{Q}^a$ which contains K and L one has: Φ maps G_M isomorphically onto itself, thus defines an isomorphism

$$\overline{\Phi}_M: \text{Gal}(M|K) \to \text{Gal}(M|L).$$

In order to conclude, one shows for a properly chosen *Abelian* extension $M_1|M$, every isomorphism $\overline{\Phi}_M$ that can be extended to an isomorphism $\overline{\Phi}_{M_1}$ can also be extended to an automorphism of $\text{Gal}(M|\mathbb{Q})$. Finally, one deduces from this that Φ can be extended to an automorphism of $G_\mathbb{Q}$, etc.

Note that the fact that the arithmetical equivalence of normal number fields implies their isomorphism relies on the Chebotarev density theorem, thus *analytical methods*. Until now we do not have a purely *algebraic proof* of that fact.

Case 2 K, L are global function fields over $\mathbb{F}_p$.
First recall that the space of all the non-trivial places $\mathcal{P}(K)$ of K is in a canonical bijection with the closed points of the unique complete smooth model $X \to \mathbb{F}_p$ of K. In particular, given an isomorphism $\Phi: G_K \to G_L$, the "arithmetical equivalence" of K and L, is just a bijection $X^0 \to Y^0$ from the closed points of X to the closed points of the complete smooth model $Y \to \mathbb{F}_p$ of L. And the problem is now to show that this abstract bijection comes from geometry. The way to do it is by using the class field theory of global function fields

as follows. First, one recovers the Frobenius elements at each place $\mathfrak{p}$ of K; and then the multiplicative group $K^\times$ by using Artin's reciprocity map; and finally the addition on $K = K^\times \cup \{0\}$. Since the recipe for recovering these objects is invariant under profinite group isomorphisms, it follows that $\Phi \colon G_K \to G_L$ defines a group isomorphism $\phi_K \colon K^\times \to L^\times$. Finally, one shows that ϕ_K respects the addition, by reducing it to the case $\phi_K(x + 1) = \phi_K(x) + 1$. Moreover, by performing this construction for all finite sub-extensions $K_1|K$ of $K^s|K$, and the corresponding sub-extensions $L_1|L$ of $L^s|L$ – which are finite as well, and by using the functoriality of the class field theory, one finally gets a field isomorphism $\phi \colon K^s \to L^s$ which defines Φ, i.e., $\Phi(g) = \phi \circ g \circ \phi^{-1}$ for all $g \in G_K$.

Part II. Grothendieck's anabelian geometry

The natural context in which the results above appear as the first prominent examples is *Grothendieck's anabelian geometry,* see [G1], [G2]. We will formulate Grothendieck's anabelian conjectures in a more general context later, after having presented the basic facts about étale fundamental groups. But it is easy and appropriate to formulate here the so-called *birational anabelian conjectures,* which involve only the usual absolute Galois group.

A. Warm-up: birational anabelian conjectures

The so-called birational anabelian conjectures place the results by Neukirch, Ikeda, Uchida, et al. – at least conjecturally – into a bigger picture. And in their most naive form, these conjectures assert that there should be a "Galois characterization" of the finitely generated infinite fields similar to that of the global fields; that is, if K and L are such fields and $G_K \cong G_L$, then K and L have finite purely inseparable extensions $K'|K$ and $L'|L$ such that $K' \cong L'$ as fields. (Note that the canonical projections $G_{K'} \to G_K$, $G_{L'} \to G_L$ are isomorphisms. Hence every isomorphism $G_K \cong G_L$ gives rise canonically to an isomorphism $G_{K'} \cong G_{L'}$. This is simply the translation of the fact that Galois theory "does not see" pure inseparable extensions.) To make a more precise conjecture, recall that for an arbitrary field K we denote by $K^i \subseteq K^a$ its maximal purely inseparable extension in some algebraic closure K^a of K. Thus if char$(K) = 0$, then $K^i = K$. Further, we say that two field homomorphisms $\phi, \psi \colon L \to K$ differ by an absolute Frobenius twist, if $\psi = \phi \circ \text{Frob}^n$ on L^i for some power Frob^n of the absolute Frobenius Frob. Finally, we identify G_K with G_{K^i} via the canonical projection $G_{K^i} \to G_K$, which is an isomorphism.

Birational anabelian conjectures

(1) *There exists a group theoretic recipe by which one can recover K^i from G_K for every finitely generated infinite field K. In particular, if for such fields K and L one has $G_K \cong G_L$, then $K^i \cong L^i$.*

(2) *Moreover, given such fields K and L, one has the following:*

- *Isom-form: Every isomorphism $\Phi \colon G_K \to G_L$ is defined by a field isomorphism $\phi \colon L^a \to K^a$ via $\Phi(g) = \phi^{-1} \circ g \circ \phi$ for $g \in G_K$, and ϕ is unique up to Frobenius twists. In particular, one has $\phi(L^i) = K^i$.*

- *Hom-form: Every open homomorphism $\Phi \colon G_K \to G_L$ is defined by a field embedding $\phi \colon L^a \hookrightarrow K^a$, and ϕ is unique up to Frobenius twists. In particular, one has $\phi(L^i) \subseteq K^i$.*

As in the case of global fields, the Isom-form of the birational anabelian conjecture is also called the *Galois characterization of the finitely generated infinite fields*. The main known facts are summarized below.

Theorem

(1) (See Pop [P2], [P3]) *There is a group theoretical recipe by which one can recover in a functorial way finitely generated infinite fields K from their absolute Galois groups G_K.*

   Moreover, this recipe works in such a way that it implies the Isom-form *of the birational anabelian conjecture, i.e., every isomorphism $\Phi \colon G_K \to G_L$ is defined by an isomorphism $\phi \colon L^a \to K^a$, and ϕ is unique up to Frobenius twists.*

(2) (See Mochizuki [Mzk3], theorem B) *The relative Hom-form of the birational anabelian conjecture is true in characteristic zero, which means the following: Given function fields K and L over $\mathbb{Q}$, every open $G_{\mathbb{Q}}$-homomorphism $\Phi \colon G_K \to G_L$ is defined by a unique field embedding $\phi \colon L^a \to K^a$, which in particular, maps L into K.*

We give here the sketch of the proof of the Isom-form. Mochizuki's Hom-form relies on his proof of the anabelian conjectures for curves over sub-p-adic fields, and we will say more about that later on.

The main steps of the proof are the following.

The first part of the proof develops a higher dimensional *Local Theory* which, roughly speaking, is a direct generalization of Neukirch's result above concerning the description of the places of global fields. Nevertheless, there are some difficulties with this generalization, because in higher dimensions the finitely generated fields do not have unique normal (or smooth) complete models. Recall that a model $X \to \mathbb{Z}$ for such a field K is by definition a separated, integral scheme of finite type over $\mathbb{Z}$ whose function field is K. We will consider only quasi-projective normal models, maybe satisfying some extra conditions, like regular, etc. In particular, if X is a model of K, then the Kronecker dimension $\dim(K)$ of K equals $\dim(X)$ as a scheme. One has the following.

- K is a global field if and only if every normal model X of K is an open of either $X_K := \operatorname{Spec} O_K$ if K is a number field, or of the unique complete smooth model $X_K \to \mathbb{F}_p$ of K if K is a global function field with $\operatorname{char}(K) = p$. Further, there exists a natural bijection between the *prime Weil divisors* of X_K and the non-archimedean places of K. The basic result by Neukirch [N1] can be interpreted as follows. First let us say that a closed subgroup $Z \subset G_K$ is a *divisorial like subgroup* if it is isomorphic to a decomposition group Z_q over some prime q of some global field L. Note that the structure of such groups as profinite groups is known; see, for example, Jannsen–Wingberg [J–W]. Then the decomposition groups over the places of K are the *maximal divisorial like subgroups* of G_K.

This gives then the group theoretic recipe for describing the prime Weil divisors of X_K in a functorial way.

- In general, i.e., if K is not necessarily a global field, there is a huge variety of normal complete models $X \to \mathbb{Z}$ of K. In particular, we cannot hope to obtain much information about a single specific model X of K, as in general there is no privileged model for K as in the global field case. (Well, maybe with the exception of arithmetical surfaces, where one could choose the minimal model, but this doesn't help much.) A way to avoid this is to consider – in a first approximation – the space of (Zariski) *prime divisors* $\mathcal{D}_K^1$ of K. This is, by definition, the set of all the discrete valuations v of K defined by the Weil prime divisors of all possible normal models $X \to \mathbb{Z}$ of K.

A prime divisor v of K is called geometrical if the residue field Kv of v has $\operatorname{char}(Kv) = \operatorname{char}(K)$, or equivalently, if v is trivial on the prime field of K, and arithmetical otherwise. Clearly, arithmetical prime divisors exist only if $\operatorname{char}(K) = 0$. If so, and v is defined by a Weil prime divisor X_1 of a normal model $X \to \mathbb{Z}$, then v is geometrical if and only if v is a "horizontal" divisor of $X \to \mathbb{Z}$.

For every prime divisor $v \in \mathcal{D}_K^1$ of K, let Z_v be the decomposition group of some prolongation v^s of v to K^s. We will call the totality of all the closed subgroups of the form Z_v the *divisorial subgroups* of G_K or of K. Finally, as above, a closed subgroup $Z \subset G_K$ is called a *divisorial like subgroup*, if it is isomorphic to a divisorial subgroup of a finitely generated field L with $\dim(L) = \dim(K)$. The main results of the Local Theory are as follows, see [P1].

(a) For a prime divisor v, the numerical data char(K), char(Kv), and dim(K) are group theoretically encoded in Z_v; in particular, whether v is geometric or not. Further, the inertia group $T_v \subset Z_v$ of $v^s | v$, and the canonical projection $\pi_v : Z_v \to G_{Kv}$ are also encoded group theoretically in Z_v. In particular, the residual absolute Galois group G_{Kv} at all the prime divisors v of K is group theoretically encoded in G_K.

(b) Every divisorial like subgroup $Z \subset G_K$ is contained in a unique divisorial subgroup Z_v of G_K. Thus the divisorial like subgroups of G_K are exactly the *maximal divisorial like subgroups* of G_K. And the space $\mathcal{D}_K^1$ is in bijection with the conjugacy classes of divisorial subgroups of G_K.

The results from the local theory above suggest that one should try to prove the birational anabelian conjecture by *induction on* dim(K). This is the idea for developing a *Global Theory* along the following lines.

For every field Ω and an abelian group A related to Ω, e.g., $A = \mathbb{Z}$ or $A = \mu_{\overline{\Omega}}$ the roots of unity in $\overline{\Omega}$, we consider the `prime to` char(Ω) `adic completion` of A denoted $\widehat{A}_\Omega := \varprojlim_m A/mA$, $(m, \text{char}(\Omega)) = 1$, or $\widehat{A}$ if Ω is clear from the context.

First, the Isom-form of the birational anabelian conjecture for global fields, i.e., dim(K) = 1, is known; and we think of it as the first induction step. Now suppose that dim(K) = $d > 1$. By the induction hypothesis, suppose that the Isom-form of the birational anabelian conjecture is true in dimension $< d$. Then one recovers the field K^i up to Frobenius twists from G_K along the following steps (from which it it will be clear what we mean by a "group theoretic recipe").

Step 1   Recover the cyclotomic character $\chi_K : G_K \to \widehat{\mathbb{Z}}^\times$ of G_K.
The recipe is as follows. Since dim(Kv) = dim(K) − 1 < d, the cyclotomic character χ_{Kv} is "known" for each prime divisor $v \in \mathcal{D}_K^1$. Thus

$$\chi_v : Z_v \xrightarrow{\pi_v} G_{Kv} \xrightarrow{\chi_{Kv}} \widehat{\mathbb{Z}}^\times$$

is known for all $v \in \mathcal{D}_K^1$. On the other hand, using the higher dimensional Chebotarev Density Theorem, see e.g., Serre [S3], it follows that ker(χ_K) is the closed subgroup of G_K generated by all the ker(χ_v), $v \in \mathcal{D}_K^1$. Thus χ_K is the unique character $\chi : G_K \to \widehat{\mathbb{Z}}^\times$ which coincides with χ_v on each Z_v.

Next let $\iota : \mathbb{T}_K \to \widehat{\mathbb{Z}}(1)$ be a fixed identification of $\widehat{\mathbb{Z}}(1)$ with the Tate G_K-module $\mathbb{T}_K = \varprojlim_m \mu_m$. Kummer theory gives:

$$K^\times \xrightarrow{\jmath_K} \widehat{K^\times} \xrightarrow{\hat{\delta}} \mathrm{H}^1(K, \widehat{\mathbb{Z}}(1)),$$

where $\widehat{K^\times}$ is the prime to char(K) adic completion of $K^\times$ and $\jmath_K$ is the completion homomorphism; and "functorial" means that performing this construction for finite extensions $M|K$ inside $K^a|K$, we get corresponding commutative "inclusion–restriction" diagrams (which we omit to write down here). An essential point to make here is that the completion morphism $\jmath_K \colon K^\times \to \widehat{K^\times}$ is injective. This follows by induction on $\dim(K)$ by using the following fact. Let $K|k$ be the function field of a geometrically irreducible complete curve $X \to k$. Then $K^\times/k^\times$ is the group of principal divisors of X, thus a free Abelian group. And for K a number field, one knows that $K^\times/\mu_K$ is a free Abelian group.

Step 2   Recover the geometric small sets of prime divisors.
We will say that a subset $\mathcal{D} \subset \mathcal{D}_K^1$ of prime divisors is *geometric* if there exists a quasi-projective normal model $X \to k$ of K such that $\mathcal{D} = \mathcal{D}_X$ is the set of prime divisors of K defined by the Weil prime divisors of X. Here, k is the field of constants of K. It is a quite technical point to show – by induction on the (absolute) transcendence degree $d = \mathrm{td}(K)$ – that the geometric sets of prime divisors can be recovered from G_K; see Pop [P3]. Next let $\mathcal{D} = \mathcal{D}_X$ be a geometric set of prime divisors. One has a canonical exact sequence

$$1 \to U_\mathcal{D} \to K^\times \to \mathrm{Div}(X) \to \mathfrak{Cl}(X) \to 0,$$

where $U_\mathcal{D}$ are the units in the ring of global sections on X, and the other notations are standard. Since the base field k is either finite or a number field, the Weil divisor class group $\mathfrak{Cl}(X)$ is finitely generated. Thus if X is "sufficiently small", then $\mathfrak{Cl}(X) = 0$. A geometric set of prime divisors $\mathcal{D} = \mathcal{D}_X$ will be called a *small geometric set* of prime divisors, if the adic completion $\widehat{\mathfrak{Cl}(X)}$ is trivial. One shows that the small geometric sets of prime divisors can be recovered from G_K, see loc. cit. In this process, one shows that the adic completion of the above exact sequence can be recovered from G_K too:

$$1 \to \widehat{U_\mathcal{D}} \to \widehat{K^\times} \to \widehat{\mathrm{Div}(X)} = \widehat{\oplus_v \mathbb{Z}} \to \widehat{\mathfrak{Cl}(X)} \to 0,$$

Step 3   Recover the multiplicative group $K^\times$ inside $\widehat{K^\times}$.
Let $\mathcal{D} = \mathcal{D}_X$ be a small geometric set of prime divisors of K. The resulting exact sequence defined above becomes $1 \to \widehat{U_\mathcal{D}} \to \widehat{K^\times} \to \widehat{\oplus_v \mathbb{Z}} \to 0$, as $\widehat{\mathfrak{Cl}(X)} = 0$. Next let $v \in \mathcal{D}$ be arbitrary. Then the group of global units $U_\mathcal{D}$ is contained in the group of v-units $O_v^\times$. Thus the $(\mathrm{mod}\,\mathfrak{m}_v)$ reduction $p_v \colon O_v^\times \to Kv^\times$ is defined on $U_\mathcal{D}$. Using some arguments involving Hilbertian fields, one shows that there exist "many" $v \in \mathcal{D}$ such that $U_\mathcal{D}$ as well as $\widehat{U_\mathcal{D}}$ are actually mapped isomorphically into $Kv^\times$, respectively $\widehat{Kv^\times}$; and moreover, that inside

$\widehat{Kv}^\times$ one has

(∗) $$p_v(U_\mathcal{D}) = \hat{p}_v(\widehat{U}_\mathcal{D}) \cap Kv^\times.$$

On the Galois theoretic side, the reduction map p_v is defined by the restriction coming from the inclusion $Z_v \hookrightarrow G_K$. And moreover, since $U_\mathcal{D}$ is contained in the v-units, it follows that, under the restriction map $\widehat{K} \to \mathrm{H}^1(Z_v, \widehat{\mathbb{Z}}(1))$, the image of $\widehat{U}_\mathcal{D}$ is contained in the image of $\inf(\pi_v) \colon \widehat{Kv}^\times \to \mathrm{H}^1(Z_v, \widehat{\mathbb{Z}}(1))$ defined by the canonical projection $\pi_v \colon Z_v \to G_{Kv}$.

Finally, the recipe to recover $K^\times$ inside $\widehat{K}^\times$ is as follows. First, for each small geometric set of prime divisors $\mathcal{D}$ and v as above, $U_\mathcal{D}$ is exactly the preimage of $\hat{p}_v(\widehat{U}_\mathcal{D}) \cap Kv^\times$, by assertion (∗) above. Since $K^\times = \cup_\mathcal{D} U_\mathcal{D}$, when $\mathcal{D}$ runs over smaller and smaller (small) geometric sets of prime divisors, we finally recover $K^\times$ inside $\widehat{K}^\times$.

Step 4   Define the addition in $K = K^\times \cup \{0\}$.
This is easily done using the induction hypothesis. Namely, let $x, y \in K^\times$ be given. Then $x+y = 0$ iff $x/y = -1$, and this fact is encoded in $K^\times$. Now suppose that $x + y \neq 0$. Then $x + y = z$ in K iff for all v such that x, y, z are all v-units one has: $p_v(x) + p_v(y) = p_v(z)$. On the other hand, this last fact is encoded in the field structure of G_{Kv}, which we already know.

Finally, in order to conclude the proof of the Isom-form of the birational anabelian conjecture, we proceed as follows. Let $\Phi \colon G_K \to G_L$ be an isomorphism of absolute Galois groups. Then the recipes for recovering the fields K and L are "identified" via Φ, and shows that the p-divisible hulls of K and L inside $\widehat{K}^\times \cong \widehat{L}^\times$ must be the same, where $p = \mathrm{char}(K)$. This finally leads to an isomorphism $\phi \colon L^{\mathrm{a}} \to K^{\mathrm{a}}$ which defines Φ. Its uniqueness up to Frobenius twists follows from the fact that, given two such field isomorphisms ϕ', ϕ'', then setting $\phi := {\phi'}^{-1} \circ \phi''$, we obtain an automorphism of K^{a} which commutes with G_K. And one checks that any such automorphism is a Frobenius twist.

B. Anabelian conjectures for curves

(a) *Étale fundamental groups*   Let X be a connected scheme endowed with a geometric base point $\overline{x}$. Recall that the étale fundamental group $\pi_1(X, \overline{x})$ of $(X, \overline{x})$ is the automorphism group of the fiber functor on the category of all the étale connected covers of X. The étale fundamental group is functorial in the following sense. Let connected schemes with geometric base points $(X, \overline{x})$ and $(Y, \overline{y})$ and a morphism $\phi \colon X \to Y$ be given such that $\overline{y} = \phi \circ \overline{x}$. Then ϕ gives rise to a morphism between the fiber functors $\mathcal{F}_{\bar{x}}$ and $\mathcal{F}_{\bar{y}}$, which induces a continuous morphism of profinite groups $\pi_1(\phi)\colon \pi_1(X,\bar{x}) \to \pi_1(Y,\bar{y})$ in the canonical way. In particular, setting $Y = X$ and $\bar{y}$ some geometric point of X, a "path" from between $\bar{x}$ and $\bar{y}$, gives rise to an inner automorphism of $\pi_1(X,\bar{x})$. In other words, $\pi_1(X,\bar{x})$ is *determined by X up to inner automorphisms*. (This means that the situation is completely parallel to the one in the case of the topological fundamental group.) A basic property of the étale fundamental group that it is invariant under *universal homeomorphisms,* hence under purely inseparable covers and Frobenius twists.

Next let $\mathcal{G}$ be the category of all profinite groups and outer continuous homomorphisms as morphisms. The objects of $\mathcal{G}$ are the profinite groups and, for given objects G and H, a $\mathcal{G}$-morphism from G to H is a set of the form $\mathrm{Inn}(H) \circ f$, where $f\colon G \to H$ is a morphism of profinite groups, and $\mathrm{Inn}(H)$ is the set of all the inner automorphisms of H. Clearly, if $f,g\colon G \to H$ differ by an inner automorphism of G, then $\mathrm{Inn}(H)\circ f = \mathrm{Inn}(H)\circ g$; thus they define the same $\mathcal{G}$-homomorphism from G to H. Further, $\mathrm{Inn}(H) \circ f$ is a $\mathcal{G}$-isomorphism if and only if $f\colon G \to H$ is an isomorphism of profinite groups.

Therefore, viewing the étale fundamental group π_1 as having values in $\mathcal{G}$ rather than in the category of profinite groups, the relevance of the geometric points $\bar{x}$ vanishes. Therefore, we will simply write $\pi_1(X)$ for the fundamental group of a connected scheme X.

In the same way, if S is a connected base scheme, and X is a connected S-scheme, then the structure morphism $\varphi_X\colon X \to S$ gives rise to an *augmentation morphism* $p_X\colon \pi_1(X) \to \pi_1(S)$. Thus the category $\mathfrak{Sch}_S$ of all the S-schemes is mapped by π_1 into the category $\mathcal{G}_S$ of all the $\pi_1(S)$-groups, i.e. the profinite groups G with an "augmentation" morphism $pr_G\colon G \to \pi_1(S)$.

Now let us consider the more specific situation when the base scheme S is a field, and the k-schemes X are geometrically connected. Denote by $\overline{X} = X\times_k k^s$ the base to the separable closure of k (in some fixed "universal field"), and remark that by the facts above one has an exact sequence of profinite groups of the form

$$1 \to \pi_1(\overline{X}) \to \pi_1(X) \to G_k \to 1.$$

In particular, we have a representation $\rho_X\colon G_k \to \mathrm{Out}(\pi_1(\overline{X})) = \mathrm{Aut}_{\mathcal{G}}(\pi_1(\overline{X}))$, which encodes most of the information carried by the exact sequence above. The group $\pi_1(\overline{X})$ is called the *algebraic* (or *geometric*) fundamental group of X. In general, little is known about $\pi_1(\overline{X})$, and in particular, even less about $\pi_1(X)$. Nevertheless, if X is a k-variety, and $k \subset \mathbb{C}$, then the base change to $\mathbb{C}$ gives a realization of $\pi_1(\overline{X})$ as the *profinite completion* of the topological fundamental group of $X^{\mathrm{an}} = X(\mathbb{C})$.

In terms of function fields, if $X \to k$ is geometrically integral and normal, one has the following. Let $k(X) \hookrightarrow k(\overline{X})$ be the function fields of $\overline{X} \to X$. Then the algebraic fundamental group $\pi_1(\overline{X})$ is (canonically) isomorphic to the Galois group of a maximal unramified Galois field extension $\mathcal{K}_{\overline{X}} \mid k(\overline{X})$.

Finally, we recall that $\pi_1(X)$ is a *birational invariant* in the case X is complete and regular. In other words, if X and X' are birationally equivalent complete regular k-varieties, then $\pi_1(X) \cong \pi_1(X')$ and $\pi_1(\overline{X}) \cong \pi_1(\overline{X}')$ canonically.

(b) *Étale fundamental groups of curves* Specializing even more, we turn our attention to curves, and give a short review of the basic known facts in this case. In this discussion we will suppose that X is a *smooth connected* curve, having a *smooth completion,* say X_0, over k. We denote $S = X_0 \backslash X$, and $\overline{S} = \overline{X}_0 \backslash \overline{X}$. We say that X is a (g, r) curve if X_0 has (geometric) genus g and $|\overline{S}| = r$. We say that X is a *hyperbolic curve* if its Euler–Poincaré characteristic $2 - 2g - r$ is negative. And we say that a curve X as above is *virtually hyperbolic* if it has an étale connected cover $X' \to X$ such that X' is a *hyperbolic curve* in the sense above. (Note that every étale cover $f \colon X' \to X$ as above is smooth and has a smooth completion which is a (g', r')-curve with $g \leq g'$ and $r' \leq r \deg(f)$ over some finite $k'|k$.)

In the above notation, let $X \to k$ be a (g, r) curve. Then a short list of the known facts about the algebraic fundamental group $\pi_1(\overline{X})$ is as follows. First, let $\Gamma_{g,r}$ be the fundamental group of the orientable compact topological surface of genus g with r punctures. Thus

$$\Gamma_{g,r} = \langle a_1, b_1, \ldots, a_g, b_g, c_1, \ldots, c_r \mid \textstyle\prod_i [a_i, b_i] \prod c_j = 1 \rangle$$

is the discrete group on $2g + r$ generators $a_1, b_1, \ldots, a_g, b_g, c_1, \ldots, c_r$ with the given unique relation. (The generators a_i, b_i, c_j have a precise interpretation as loops around the handles, respectively around the missing points.) In particular, if $r > 0$, then $\Gamma_{g,r}$ is the discrete free group on $2g + r - 1$ generators. It is well known that $\Gamma_{g,r}$ is *residually finite*, i.e., $\Gamma_{g,r}$ injects into its profinite completion:

$$\Gamma_{g,r} \hookrightarrow \widehat{\Gamma}_{g,r}.$$

Finally, given a fixed prime number p, respectively arbitrary prime numbers ℓ, we will denote by $\widehat{\Gamma}_{g,r} \to \widehat{\Gamma}'_{g,r}$ the maximal prime-p quotient of $\widehat{\Gamma}_{g,r}$, and by $\widehat{\Gamma}_{g,r} \to \widehat{\Gamma}^\ell_{g,r}$ the maximal pro-ℓ quotient of $\Gamma_{g,r}$.

Case 1   char$(k) = 0$.
Using the remark above, in the case $k \hookrightarrow \mathbb{C}$, it follows that $\pi_1(\overline{X}) \cong \widehat{\Gamma}_{g,r}$ via the base change $X \times_k \mathbb{C} \to \overline{X}$. If $\kappa(\overline{X})$ is the function field of $\overline{X}$, then $\pi_1(\overline{X})$ is the Galois group of a maximal Galois unramified field extension $\mathcal{K}_{k(\overline{X})} | k(\overline{X})$.

Moreover, the loops $c_j \in \Gamma_{g,r}$ around the missing points $x_i \in \overline{X}_0 \backslash \overline{X}$ are canonical generators of inertia groups T_{x_i} over these points in $\pi_1(\overline{X})$. In particular we have the following.

(a) X is a complete curve of genus g if and only if $\pi_1(\overline{X})$ has $2g$ generators a_i, b_i with the single relation $\prod_i [a_i, b_i] = 1$, provided X is not $\mathbb{A}_k^1$.

(b) X is of type (g, r) with $r > 0$ if and only if $\pi_1(\overline{X})$ is a profinite free group on $2g + r - 1$ generators, provided X is not k-isomorphic to $\mathbb{A}_k^1$ or $\mathbb{P}_k^1$.

Clearly, the dichotomy between the above subcases a) and b) can be as well deduced from the pro-ℓ maximal quotient $\pi_1^\ell(\overline{X})$ of $\pi_1(\overline{X})$, by simply replacing "profinite" by "pro-ℓ".

Further, the following conditions on X are equivalent:

(i) X is hyperbolic;

(ii) X is virtually hyperbolic;

(iii) $\pi_1(\overline{X})$ is non-Abelian, or equivalently, (iii)$^\ell$ $\pi_1^\ell(\overline{X})$ is non-Abelian.

Case 2   char$(k) > 0$.

First recall that the *tame fundamental group* $\pi_1^t(X)$ of X is the maximal quotient of $\pi_1(X)$ which classifies étale connected covers $X' \to X$ whose ramification above the missing points $x_i \in \overline{X}_0 \backslash \overline{X}$ is tame. We will denote by $\pi_1^t(\overline{X})$ the tame quotient of $\pi_1(\overline{X})$, and call it the *tame algebraic fundamental group* of X. Now the main technical tools used in understanding $\pi_1(\overline{X})$ and its tame quotient $\pi_1^t(\overline{X})$ are the following two facts.

Shafarevich's theorem   *In the context above, set* char$(k) = p > 0$, *and denote by $\pi_1^p(\overline{X})$ the maximal pro-p quotient of $\pi_1(\overline{X})$. Further let $r_{X_0} = \dim_{\mathbb{F}_p} \mathrm{Jac}_{\overline{X}_0}[p]$ denote the Hasse–Witt invariant of the complete curve $\overline{X}_0$. Then one has:*

(1) *If $X = X_0$, then $\pi_1^p(\overline{X})$ is a pro-p free group on $r_{X_0} \leq g$ generators.*

(2) *If X is affine, then $\pi_1^p(X)$ is a pro-p free group on $|k^a|$ generators.*

Let k be an arbitrary base field, and v a complete discrete valuation with valuation ring $R = R_v$ of k and residue field $kv = \kappa$. Let $X \to k$ be a smooth curve that has a smooth completion $X_0 \to k$. We will say that $X \to k$ *has good reduction* at v, if the following hold: $X_0 \to k$ has a smooth model $X_{0,R} \to R$ over R, and there exists an étale divisor $S_R \to R$ of $X_{0,R}$ such that the generic fiber of the complement $X_{0,R} \backslash S_R =: X_R \to R$ is X.

Now let $X \to k$ be a hyperbolic curve having good reduction at v. In the notations from above, let $X_s \to \kappa$ be the special fiber of $X_R \to R$. Then the canonical diagram of schemes

$$
\begin{array}{ccccc}
X & \hookrightarrow & X_R & \hookleftarrow & X_s \\
\downarrow & & \downarrow & & \downarrow \\
k & \hookrightarrow & R & \hookleftarrow & \kappa
\end{array}
$$

gives rise to a diagram of fundamental groups as follows:

$$
\begin{array}{ccccc}
\pi_1^t(X) & \to & \pi_1^t(X_R) & \leftarrow & \pi_1^t(X_s) \\
\downarrow & & \downarrow & & \downarrow \\
G_k & \to & G_k^t & \leftarrow & G_\kappa
\end{array}
$$

where $\pi_1^t(X_R)$ is the "tame fundamental group" of X_R, i.e., the maximal quotient of $\pi_1(X)$ classifying connected covers of X_R which have ramification only along S_R and the generic point of the special fiber, and this ramification is tame, and G_k^t is the Galois group of the maximal tamely ramified extension of k. The fundamental result concerning the fundamental groups in the diagram above is the following.

Grothendieck's specialization theorem *In the context above, let X be a smooth curve of type (g, r). Further let R^t be the extension of R to k^t, and $\overline{X}_R := X_R \times_R R^t$. Then one has the following:*

(1) *$\pi_1^t(\overline{X}) \to \pi_1^t(\overline{X}_R)$ is surjective, and $\pi_1^t(\overline{X}_R) \leftarrow \pi_1^t(\overline{X}_s)$ is an isomorphism. The resulting surjective homomorphism*

$$
\mathrm{sp}_v : \pi_1^t(\overline{X}) \to \pi_1^t(\overline{X}_s)
$$

   is called the specialization homomorphism *of tame fundamental groups at v. In particular, $\pi_1^t(\overline{X})$ is a quotient of $\widehat{\Gamma}_{g,r}$ in such a way that the generators c_j are mapped to inertia elements at the missing points $x_i \in \overline{X}_0 \backslash \overline{X}$.*

(2) *Further, let $\mathrm{char}(k) = p$, and denote by $\pi_1'(\overline{X})$ the maximal prime to p quotient of $\pi_1(\overline{X})$ (which then equals the maximal prime to p quotient of $\pi_1^t(\overline{X})$ too). Then $\pi_1'(\overline{X}) \cong \widehat{\Gamma}_{g,r}'$, and sp_v maps $\pi_1'(\overline{X})$ isomorphically onto $\pi_1'(\overline{X}_s)$. In particular, $\pi_1'(\overline{X})$ depends on (g, r) only.*

Combining Shafarevich's theorem and Grothendieck's specialization theorem above, we immediately see that the following facts and invariants of $X \to k$ are encoded in $\pi_1(\overline{X})$.

(a) If $\ell \neq p$, then $\pi_1^\ell(\overline{X}) \cong \widehat{\Gamma}_{g,r}^\ell$, and $\pi_1^p(\overline{X}) \ncong \widehat{\Gamma}_{g,r}^p$. Therefore, $p = \mathrm{char}(k)$ can be recovered from $\pi_1(\overline{X})$, provided X is not $\mathbb{P}_k^1$.

(a)^t The same is true correspondingly concerning the tame fundamental group $\pi_1^t(\overline{X})$, provided X is not isomorphic to $\mathbb{A}_k^1$ or $\mathbb{P}_k^1$.

(b) $X \to k$ is complete if and only if $\pi_1^p(\overline{X})$ is finitely generated.

(b)^t Correspondingly, $X \to k$ is complete if and only if $\pi_1^\ell(\overline{X})$ is not pro-ℓ-free, provided X is not isomorphic to $\mathbb{A}_k^1$.

(c) In particular, if X is complete, then $\pi_1^\ell(\overline{X})$ has $2g$ generators, thus g can be recovered from $\pi_1^\ell(\overline{X})$.

Finally, concerning the virtual hyperbolicity of X we have the following:

(d) In positive characteristic, every affine curve X is virtually hyperbolic.

Remark   Clearly, the applicability of Grothendieck's specialization theorem is limited by the fact that one would need *a priori* criteria for the good reduction of the given curve X at the (completions of k with respect to the) discrete valuations v of k. At least in the case of hyperbolic curves $X \to k$ such criteria do exist. The setting is as follows. Let $X \to k$ be a hyperbolic curve, and let v be a discrete valuation of k. Let $T_v \subseteq Z_v$ be the inertia, respectively the decomposition, groups of some prolongation of v to k^s. Recall the canonical projections $\pi_1(X) \to G_k$ and $\pi_1^t(X) \to G_k$ and the resulting Galois representations $\rho_X \colon G_k \to \mathrm{Out}(\pi_1(\overline{X}))$ and $\rho_X^t \colon G_k \to \mathrm{Out}(\pi_1^t(\overline{X}))$. Then one can characterize the fact that X has (potentially) good reduction at v as follows, see Oda [O] in the case of complete hyperbolic curves, and by Tamagawa [T1] in the case of arbitrary hyperbolic curves.

In the above notations, $X \to k$ has good reduction at v if and only if the representation ρ_X^t is trivial on T_v.

The concrete picture of how to apply the above remark in studying fundamental groups of hyperbolic curves $X \to k$ over either finitely generated infinite base field k or finitely generated fields over some fixed base field k_0 is as follows. Let $X \to k$ be a smooth curve of type (g, r). Further let S be a smooth model of k over $\mathbb{Z}$, if k is a finitely generated field, respectively over the base k_0 otherwise. For every closed point $s \in S$, there exists a discrete valuation v_s whose valuation ring R_s dominates the local ring $O_{S,s}$, and having residue field $\kappa_{v_s} = \kappa(s)$. Let us choose such a valuation v_s. Then in the previous notations, X has good reduction at s if and only if ρ_X^t is trivial on the inertia group T_s over the point s. Note that by the uniqueness of the smooth model $X_{R_s} \to R_s$ – in the case it does exist – the existence of such a good reduction *does not depend on the concrete valuation* v_s used. One should also remark here, that in the context above, $X \to k$ has good reduction on a Zariski open subset of S. This follows e.g., from the Jacobian criterion for smoothness.

Finally, we now come to announcing Grothendieck's anabelian conjectures for curves and the section conjectures.

Let $\mathcal{P}$ be a property defined for some category of schemes X. We will say that the property $\mathcal{P}$ is an *anabelian property*, if it is encoded in $\pi_1(X)$ in a group theoretical way, or in other words, if $\mathcal{P}$ can be recovered by a group theoretic recipe from $\pi_1(X)$. In particular, if X has the property $\mathcal{P}$, and $\pi_1(X) \cong \pi_1(Y)$, then Y has the property $\mathcal{P}$.

Examples
(a) In the category of all the fields K, the property *"K is real closed"* is anabelian. This is the theorem of Artin–Schreier from above.
(b) In the category of all the smooth k-curves X which are not isomorphic to $\mathbb{A}_k^1$, the property: *"X is complete and has genus g"* is anabelian. This follows from the structure theorems for the fundamental group of complete curves as discussed above.

We will say that a scheme X is *anabelian* if the isomorphy type of X up to some natural transformations, which are not encoded in Galois theory, can be recovered *group theoretically* from $\pi_1(X)$ in a functorial way; or equivalently, if there exists a *group theoretic recipe* to recover the isomorphy type of X, up to the natural transformations in discussion, from $\pi_1(X)$. Typical examples of such "natural transformations" that are not seen by Galois theory are the radicial covers and the birational equivalence of complete regular schemes. Concretely, for k-varieties $X \to k$ with char$(k) = p > 0$, there are two typical radicial covers. First the maximal purely inseparable cover $X^i \to k^i$. And second, the Frobenius twists $X(n) \to k$ of $X \to k$ and/or $X^i(n) \to k^i$ of $X^i \to k^i$ obtained by acting by Frob^n "on the coefficients": $X(n) := X \times_{\mathrm{Frob}^n} k \to k$. In the same way, if $X \to k$ and $Y \to k$ are complete regular k varieties which are birationally equivalent, then $\pi_1(X)$ and $\pi_1(Y)$ are canonically isomorphic, but X and Y might be very different.

A good set of examples of anabelian schemes are the *finitely generated infinite fields*, as we have seen in the previous section. Given such a field K, one has $\pi_1(\mathrm{Spec}\, K) = G_K$, and by the birational anabelian conjectures we know that K can be recovered from $\pi_1(K)$ in a functorial way, up to pure inseparable extensions and Frobenius twists.

Anabelian conjecture for curves (absolute form)

(1) *Let $X \to k$ be a virtually hyperbolic curve over a finitely generated base field k. Then X is anabelian in the sense that the isomorphism type of X can be recovered from $\pi_1(X)$ up to purely inseparable covers and Frobenius twists.*

(2) *Moreover, given such curves $X \to k$ and $Y \to l$, one has the following:*

- Isom-form: *Every isomorphism* $\Phi \colon \pi_1(X) \to \pi_1(Y)$ *is defined by an isomorphism* $\phi \colon X^i \to Y^i$, *and* ϕ *is unique up to Frobenius twists.*
- Hom-form: *Every open homomorphism* $\Phi \colon \pi_1(X) \to \pi_1(Y)$ *is defined by a dominant morphism* $\phi \colon X^i \to Y^i$, *and* ϕ *is unique up to Frobenius twists.*

One could also consider a relative form of the above conjecture as follows.

Anabelian conjecture for curves (relative form)

(1) *Let* $X \to k$ *be a virtually hyperbolic curve over a finitely generated base field* k. *Then* $X \to k$ *is anabelian in the sense that* $X \to k$ *can be recovered from* $\pi_1(X) \to G_k$ *up to purely inseparable covers and Frobenius twists.*

(2) *Moreover, given such curves* $X \to k$ *and* $Y \to k$, *one has the following:*

- Isom-form: *Every* G_k-*isomorphism* $\Phi \colon \pi_1(X) \to \pi_1(Y)$ *is defined by a unique* k^i-*isomorphism* $\phi \colon X^i(n) \to Y^i$ *for some n-twist.*
- Hom-form: *Every open* G_k-*homomorphism* $\Phi \colon \pi_1(X) \to \pi_1(Y)$ *is defined by a unique dominant* k^i-*morphism* $\phi \colon X^i(n) \to Y^i$ *of some n-twist.*

Concerning **higher dimensional anabelian conjectures** there are only vague ideas. There are some obvious necessary conditions which higher dimensional varieties X have to satisfy in order to be anabelian (like being of general type, being $K(\pi_1)$, etc.). Also, easy counterexamples show that one cannot expect a naive Hom-form of the conjectures. See Grothendieck [G2], and Ihara–Nakamura [I–N], Mochizuki [Mzk3], [Mzk4] for more about this.

Remark (Standard reduction technique)   Before going into the details concerning the known facts about the anabelian conjectures for curves, let us set the technical frame for a fact used several times below. Let $X \to k$ be a smooth curve over the field k. Suppose that k is either a finitely generated infinite field, or a function field over some base field k_0. Let $S \to \mathbb{Z}$, respectively $S \to k_0$ be a smooth model of k.

Next let $X \to k$ be a hyperbolic curve, say with smooth completion X_0. Let $\pi_1(\overline{X}) \to G_k$, respectively $\pi_1^t(\overline{X}) \to G_k$ be the corresponding canonical projections. Then choosing for each closed point $s \in S$ a discrete valuation v_s which dominates the local ring of s, we have the Oda–Tamagawa criterion (mentioned above) for deciding whether $X \to k$ has good reduction at s. We also know, that $X \to k$ has good reduction on a Zariski open subset of S. In particular, the Oda–Tamagawa criterion is a group theoretic criterion for describing the Zariski open subset of S on which $X \to k$ has good reduction. Moreover, if s is a point of good reduction of $X \to k$, then Grothendieck's specialization theorem for π_1^t gives a commutative diagram of the form:

$$
\begin{array}{ccc}
\pi_1^t(X_{k_v}) & \xrightarrow{\text{sp}_s} & \pi_1^t(X_s) \\
\downarrow & & \downarrow \\
Z_v & \xrightarrow{\text{pr}_v} & G_{\kappa(s)}
\end{array}
$$

where $k_v \subset k^s$ is the decomposition field over v defining Z_v inside G_k. Therefore, given a point $s \in S$, one can recover the fact that X has good reduction at s, and if this the case, also the canonical projection $\pi_1^t(X_s) \to G_{\kappa(s)}$ from the following data: $\pi_1^t(X) \to G_k$ endowed with a decomposition group Z_v above a discrete valuation v whose valuation ring R_v dominates the local ring $O_{S,s}$.

We conclude that the set of points $s \in S$ of good reduction of $X \to k$ as well as the canonical projections $\pi_1^t(X_s) \to G_{\kappa(s)}$ at such points *can be recovered from* $\pi_1^t(X) \to G_k$ if we endow G_k with decomposition groups over some discrete valuations v_s dominating $O_{S,s}$ (all closed points $s \in S$)

(I) Tamagawa's results concerning affine hyperbolic curves In this subsection we will sketch a proof of the following result by Akio Tamagawa concerning affine hyperbolic curves.

Theorem (See Tamagawa [T1])

(1) *There exists a group theoretic recipe by which one can recover an affine smooth connected curve X defined over a finite field from $\pi_1(X)$. Moreover, if X is hyperbolic, then this recipe recovers X from $\pi_1^t(X)$.*

   Further, the absolute and the relative Isom-form of the anabelian conjecture for curves holds for affine curves over finite fields; and its tame form holds for affine hyperbolic curves over finite fields.

(2) *There exists a group theoretic recipe by which one can recover affine hyperbolic curves $X \to k$ defined over finitely generated fields k of characteristic zero from $\pi_1(X)$.*

   Further, the absolute and the relative Isom-form of the anabelian conjecture for curves holds for affine hyperbolic curves over finitely generated fields of characteristic zero.

The strategy of the proof is as follows.

First consider the case when the base field k is finitely generated and has $\text{char}(k) = 0$. We claim that the canonical exact sequence

$$
(*) \qquad\qquad 1 \to \pi_1(\overline{X}) \to \pi_1(X) \to G_k \to 1,
$$

is encoded in $\pi_1(X)$. Indeed, recall that the algebraic fundamental group $\pi_1(\overline{X})$

is a finitely generated normal subgroup of $\pi_1(X)$. Therefore, since G_k has no proper finitely generated normal subgroups, see e.g. [F–J], ch.16, proposition 16.11.6, it follows that $\pi_1(\overline{X})$ is the unique maximal finitely generated normal subgroup of $\pi_1(X)$. Thus the exact sequence above can be recovered from $\pi_1(X)$. Further, by either using the characterization of the geometric inertia elements in $\pi_1(X)$ given by Nakamura [Na1], or by using specialization techniques, one finally recovers the projection $\pi_1(X) \rightarrow \pi_1(X_0)$, where X_0 is the completion of X.

After having recovered the exact sequence (∗) above, one reduces the case of hyperbolic affine curves over finitely generated fields of characteristic zero to the π_1^t-case of affine hyperbolic curves over finite fields. This is done by using the "standard reduction technique" mentioned above.

Tamagawa also shows that the absolute form of the Isom-conjecture and the relative one are roughly speaking equivalent (using the birational anabelian conjecture described at the beginning of Part II).

We now turn our attention to the case of affine curves over finite fields, respectively the π_1^t-case of hyperbolic curves over finite fields. Tamagawa's approach is a tremendous refinement of Uchida's strategy to tackle the birational case, i.e., to prove the birational anabelian conjecture for global function fields. (Naturally, since $\pi_1(X)$ seems to encode much less information than the absolute Galois group $G_{k(X)}$, things might/should be much more intricate in the case of curves.) A rough approximation of Tamagawa's proof is as follows. Let $X \rightarrow k$ be an affine smooth geometrically connected curve, where k is a finite field with char$(k) = p$. As usual let X_0 be the smooth completion of X. Thus we have surjective canonical projections

$$\pi_1(X) \rightarrow \pi_1(X_0) \rightarrow G_k \rightarrow 1$$

and correspondingly for the tame fundamental groups.

The first part of the proof consists in developing a *Local Theory*, which as in the birational case, will give a description of the closed points $x \in X_0$ in terms of the conjugacy classes of the decomposition groups $Z_x \subset \pi_1(X)$ above each closed point $x \in X$.

The steps for doing this are as follows.

Step 1   Recovering the several arithmetical invariants.
Here Tamagawa shows that the canonical projections $\pi_1(X) \rightarrow G_k$, thus $\pi_1(\overline{X})$, as well as $\pi_1(\overline{X}) \rightarrow \pi_1^t(\overline{X})$ are encoded in $\pi_1(X)$. Further, by a combinatorial argument he recovers the Frobenius element $\varphi_k \in G_k$. In particular, one gets the *cyclotomic character* of $\pi_1(X)$, and so one knows the ℓ-adic cohomology

of $\pi_1(X)$, as well as the Galois action of G_k on the ℓ-adic Galois cohomology groups of $\pi_1(\overline{X})$.

The next essential remark is that, after replacing X by some "sufficiently general" finite étale cover $Y \to X$, the completion $Y_0 \to k$ of Y is itself hyperbolic. In particular, the ℓ-adic Galois cohomology groups $H^i(\pi_1(\overline{Y_0}), \mathbb{Z}_\ell(r))$ of $\pi_1(\overline{Y_0})$ are the same as the ℓ-adic étale cohomology $H^i(\overline{Y_0}, \mathbb{Z}_\ell(r))$ of $\overline{Y_0}$. Thus by the remarks above, one can recover the ℓ-adic cohomology of $\overline{Y_0}$ for every étale cover $Y \to X$ having a hyperbolic completion Y_0.

This is a fundamental observation in Tamagawa's approach, as it can be used in order to tackle the following problem.

Which sections of the canonical projection $\mathrm{pr}_X \colon \pi_1(X) \to G_k$ *are defined by points* $x \in X_0(k)$ *in the way as described in the section conjecture?*

We remark that since $G_k \cong \widehat{\mathbb{Z}}$ is profinite free on one generator, one cannot expect that all such sections are defined by points as asked by the section conjecture. (Indeed, there are uncountably many such conjugacy classes of sections, thus too "many" to be defined by points, even if X_0 has no k-rational points.)

Here is Tamagawa's answer. Let $s \colon G_k \to \pi_1(X)$ be a given section. For every open neighborhood $U \subset \pi_1(X)$ of $s(G_k)$, we denote by $X_U \to X$ the finite étale cover of X classified by U. First, since U projects onto G_k, the curve $X_U \to k$ is geometrically connected. Further, we have in tautological way: $\pi_1(X_U) = U$, and $\overline{U} := U \cap \pi_1(\overline{X}) = \pi_1(\overline{X_U})$. Let $X_{U,0}$ be the smooth completion of X_U. Then by Step (1), the canonical projection $\pi_1(X_U) \to \pi_1(X_{U,0})$ can be recovered from $U = \pi_1(X_U)$, thus from $\pi_1(X)$ endowed with the section $s \colon G \to \pi_1(X)$.

Now we remark that for U sufficiently small, the complete curve $X_{U,0}$ is hyperbolic, both in the case where X is affine, or if X was hyperbolic and we were working with $\pi_1^t(X)$. We set $U_0 := \pi_1(X_{U,0})$ and view it as quotient of $\pi_1(X_U)$, and $\overline{U}_0 = \pi_1(\overline{X_{U,0}})$. Since $X_{U,0}$ is complete and hyperbolic, the ℓ-adic cohomology group $H^i_{\mathrm{et}}(\overline{X_{U_0}}, \mathbb{Z}_\ell(1))$ equals the Galois cohomology group $H^i(\pi_1(\overline{X_{U_0}}), \mathbb{Z}_\ell(1))$, thus the cohomology group $H^i(\overline{U}_0, \mathbb{Z}_\ell(1))$ for all i.

Finally, since the Frobenius element $\varphi_k \in G_k$ is known, by applying the Lefschetz Trace Formula, we recover the *number of k-rational points* of X_{U_0}:

$$|X_{U_0}(k)| = \Sigma_{i=0}^2 (-1)^i \mathrm{Tr}(\varphi_k) | H^i(\overline{U}_0, \mathbb{Z}_\ell(1))$$

In this way we obtain the technical input for the following.

Proposition *Let* $s \colon G_k \to \pi_1(X)$ *be a section of* $\mathrm{pr}_X \colon \pi_1(X) \to G_k$. *Then* s *is defined by a point of* $X_0(k)$ *if and only if for every open sufficiently small neighborhood* U *of* $s(G_k)$ *as above, one has* $X_{U_0}(k) \neq \varnothing$.

Step 2   Recovering the decomposition groups Z_x over closed points.

This is done by using the proposition above. Actually, using Artin's reciprocity law, one shows that in the case of a *complete hyperbolic curve*, like the X_{U_0} above, the set $X_{U_0}(k)$ is in bijection with the conjugacy classes of sections s defining points. And this gives a recipe to recover the points $X_0(k)$ which come from points in $X_{U_0}(k)$ for some U as above; thus finally for recovering all the points in $X_0(k)$. By replacing k by finite extension $l \mid k$, one recovers in a functorial way $X(l)$ too, etc. Thus finally one recovers the closed points x of X_0 as being in bijection with the conjugacy classes of decomposition groups $Z_x \subset \pi_1(X)$. Correspondingly the same is done for π_1^t-case.

The second part of the proof is to develop a *Global Theory,* as done by Uchida in the birational case. Naturally, $\pi_1(X)$ endowed with all the decomposition groups over the closed points of X_0 carries much less information than $G_{k(X)}$ endowed with all the decomposition groups Z_v over the places v of $k(X)$.

Step 3   Recovering the multiplicative group $k(X)^{\times}$ together with the valuations $v_x \colon k(X) \to \mathbb{Z}$.

Since the Frobenius elements $\varphi_x \in Z_x$ are known for closed points $x \in X_0$, by applying global class field theory as in the birational case, one gets the multiplicative group $k(X)^{\times}$ together with the valuations $v_x \colon k(X)^{\times} \to \mathbb{Z}$.

Step 4   Recovering the addition on $k(X)^{\times} \cup \{0\}$.

This is much more difficult than that in the birational case. And here is where the hypothesis that X is affine is used. Namely, if $x \in X_0 \backslash X$ is any point "at infinity", then from a decomposition group Z_x over x, one finally can recover the evaluation map

$$\mathfrak{p}_x \colon k(X) \to k^{\mathrm{a}} \cup \infty.$$

One proceeds by applying the following.

Proposition   *Let $X_0 \to \kappa$ be a complete smooth curve over an algebraically closed field κ. Suppose that the multiplicative group $k(X_0)^{\times}$ together with the valuations $v_x \colon k(X_0)^{\times} \to \mathbb{Z}$ at closed points $x \in X_0$, and the evaluation of the functions at at least three k-points x_0, x_1, x_{∞} of X, are known. Then from these data the structure field of $k(X_0)$ can be recovered.*

In order to conclude the proof of Tamagawa's theorem above over finite fields, we remark that all closed points of X_0 were recovered from $\pi_1(X)$ via the decomposition groups above them. In particular, such a point lies in X if and only if the inertia group above such a point is trivial. This gives the recipe to identify X inside X_0. Thus we finally have a group theoretic recipe for recovering the curve $X \to k$ from its fundamental group $\pi_1(X)$.

Finally, the functoriality of the recipe for recovering X shows that any isomorphism of fundamental groups $\Phi \colon \pi_1(X) \to \pi_1(Y)$ is defined by an isomorphism of function fields $\phi \colon k(X_0) \to k(Y_0)$ which induces an isomorphism of
schemes $X \to Y$. The uniqueness of ϕ up to Frobenius twists follows the same
pattern as in the birational case, but using the fact that the center of $\pi_1(X)$ is
trivial in the cases under discussion.

(II) Mochizuki's results over sub-p-adic fields In this subsection we discuss
briefly some of Mochizuki's results concerning hyperbolic curves in characteristic zero and applications of these results to other anabelian questions.

The first such result was announced by Mochizuki shortly after Tamagawa's
theorem discussed above. The result deals with hyperbolic curves over finitely
generated fields of characteristic zero, and more or less extends the corresponding result by Tamagawa to *complete hyperbolic curves*. The proof relies heav-
ily on Tamagawa's theorem, but Mochizuki's strategy for the proof goes beyond Tamagawa's approach.

Theorem (See Mochizuki [Mzk1]) *The hyperbolic curves over finitely generated fields of characteristic zero are anabelian. Further, both the relative and
the absolute Isom-form of the anabelian conjecture for hyperbolic curves over
such fields hold.*

We indicate briefly the idea of the proof. Let $X \to k$ be a hyperbolic curve
over a finitely generated field k of characteristic zero. Proceeding as Tamagawa
did in the case of affine hyperbolic curves, we can recover the exact sequence

$$1 \to \pi_1(\overline{X}) \to \pi_1(X) \to G_k \to 1,$$

and also the projection $\pi_1(X) \to \pi_1(X_0)$, where X_0 is the completion of X. In
this way one reduces the question to the case of *complete hyperbolic curves*.

Thus let $X \to k$ be a complete hyperbolic curve over some finitely generated
field k of characteristic zero. The idea of Mochizuki is to reduce the problem in
this case to the π_1^t-case of affine hyperbolic curves over finite fields and then use
Tamagawa's theorem for affine hyperbolic curves over finite fields. In order to
do that, Mochizuki uses log-schemes and log-fundamental groups. In essence
one does the following. In the context above, recall the setting explained in the
"standard reduction technique". In the notations from there, let v_s be a discrete
valuation of k with valuation ring R_s dominating some closed point $s \in S$ such
that the residue field equals $\kappa(s)$, thus finite. Let $p = \text{char}(\kappa(s))$, thus $\kappa(s)$ is
finite over $\mathbb{F}_p$. In the case $X \to k$ has good reduction at s – and this is the case
on a Zariski open subset of S, in particular if p is big enough, the special fiber
$X_s \to \kappa(s)$ of $X_{R_s} \to R_s$ at s is a complete hyperbolic curve. Thus one <u>cannot</u>

apply and use Tamagawa's theorem in order to recover $X_s \to \kappa(s)$ from $\pi_1^t(X_s)$ (even if the projection $\pi_1^t(X_s) \to G_{\kappa_v}$ is known). Nevertheless, the fact that X has good reduction at v is encoded in the canonical exact sequence $\pi_1(X) \to G_k$ endowed with a decomposition group $Z_{v_s} \subset G_k$ above v_s by Oda's criterion for good reduction of hyperbolic complete curves.

In the above notations, suppose that $X_{R_s} \to R_s$ is smooth. Let us consider a finite Galois étale cover $Y^{(p)} \to X$ such that its geometric part $\overline{Y}^{(p)} \to \overline{X}$ is the maximal p-elementary Abelian cover of $\overline{X}$. After enlarging k, we can eventually suppose that $Y^{(p)} \to k$ is geometrically connected, and that $\mathrm{Aut}_X(Y^{(p)})$ is defined over k. Under this hypothesis, one has

$$\deg(Y^{(p)} \to X) = p^{2g_X} \quad (g_X \text{ is the genus of } X).$$

On the other hand, considering the maximal geometric p-elementary Abelian cover $Z^{(p)} \to X_s$, and recalling that $r_{X_s} \le g_{X_s} = g_X$ denotes the Hasse–Witt invariant of X_s, we see that

$$\deg(Z^{(p)} \to X_s) = p^{r_{X_v}} \le p^{g_X}.$$

We conclude that $Y^{(p)} \to k$ does not have potentially good reduction. Moreover, for p getting larger, the special fiber $Y_s^{(p)} \to \kappa_s$ of the stable model of $Y^{(p)} \to k$ (which is defined over some finite extension $l|k$ and corresponding extensions R_w of R_{v_s}, etc.) has "many" double points. We set $Y_s^{(p)} = \cup_i Y_i$, where Y_i are the irreducible components of Y_s. For each Y_i, let U_i be the smooth part of Y_s inside Y_i. Since Y_s is connected, it follows that each $U_i \to \kappa_w$ is an affine hyperbolic curve over the finite field κ_w.

Now it is part of the theory of log-fundamental groups that the tame fundamental group $\pi_1^t(U_i)$ can be recovered from $\pi_1(Y^{(p)}) \to G_l$. Thus applying Tamagawa's theorem for the π_1^t-case of affine hyperbolic curves over finite fields, we can recover $U_i \to \kappa_w$ in a functorial way from $\pi_1^t(U_i)$. And finally, one can recover $Y_s^{(p)} \to \kappa_w$, and $X_s \to \kappa_v$ as well.

One concludes by using the standard reduction/globalization techniques.

The result above by Mochizuki is the precursor of his much stronger result concerning hyperbolic curves over sub-p-adic fields as explained below. First let us introduce Mochizuki's notations. A *sub-p-adic field* k is any field which can be embedded into some function field over $\mathbb{Q}_p$. Let k be a sub-p-adic field, and let $X \to k$ be a geometrically connected scheme over k. Consider the exact sequence of fundamental groups

$$1 \to \pi_1(\overline{X}) \to \pi_1(X) \to G_k \to 1.$$

We denote $\Delta_X := \pi^p(\overline{X})$ the maximal pro-p quotient of $\pi_1(\overline{X})$, and remark that the kernel N of the map $\pi_1(\overline{X}) \to \Delta(X)$ is a characteristic subgroup of $\pi_1(\overline{X})$, i.e., it is invariant under all automorphisms of $\pi_1(\overline{X})$. In particular, N is invariant under the conjugation in $\pi_1(X)$. Thus N is a normal subgroup in $\pi_1(X)$ too. We will set $\Pi_X = \pi_1(X)/N$. Therefore, the above exact sequence gives rise to a canonical exact sequence of fundamental groups:

$$1 \to \Delta_X = \pi_1^p(\overline{X}) \to \Pi_X = \pi_1(X)/N \to G_k \to 1.$$

With these notations, the main result by Mochizuki can be stated as follows.

Theorem (See Mochizuki [Mzk3]) *Let $Y \to k$ be a geometrically integral hyperbolic curve over a sub-p-adic field. Then Y can be recovered from the canonical projection $\Pi_Y \to G_k$.*

Moreover, this recipe is functorial in such a way that it implies the following Hom-*form of the relative anabelian conjecture for curves. Let $X \to k$ be a geometrically integral smooth variety. Then every open G_k-homomorphism $\Phi\colon \Pi_X \to \Pi_Y$ is defined in a functorial way by a unique dominant k-morphism $\phi\colon X \to Y$.*

The main tools used by Mochizuki are the p-adic Tate–Hodge theory and Faltings' theory of almost étale morphisms. The proof is very technical and difficult to follow for non-experts (maybe even for experts!). I will nevertheless try to summarize here the main points in the proof (which are, though, more intricate and complex than I might suggest here...). I should also mention that in this case we do not have a recipe for recovering $X \to k$ from $\pi_X \to G_k$ that is as explicit as in the previous cases. The main difficulty in this respect lies in not having an *explicit local theory* as in the previous case. In particular, and unfortunately, we do not yet have a way of describing $X(k)$, i.e., we do not have any kind of an answer to the section conjecture so far.

Coming back to the proof of the theorem above, the first observation is that via more or less standard specialization techniques, the problem is reduced to the following case: $k|\mathbb{Q}_p$ is a finite extension, and $X \to k$ is a smooth hyperbolic curve, and $Y \to k$ is a complete hyperbolic curve. And finally replacing X by the étale cover classified by the image of Φ, one can suppose that Φ is surjective.

One should remark that a further reduction step to the case where X is complete is not at all trivial, and it is one of the facts which complicates things a lot. Naturally, by using the canonical projection $\Pi_{k(X)} \to \Pi_X$, one might reformulate the problem above correspondingly, and ask whether every surjective G_k-morphism $\Phi\colon \Pi_{k(X)} \to \Pi_Y$ is defined by a unique morphism of function fields $k(Y) \hookrightarrow k(X)$ over k. We will nevertheless take this last reduction step for granted, and only outline the proof of the following.

Let $X \to k$ and $Y \to k$ be complete hyperbolic curves, where $k|\mathbb{Q}_p$ is finite. Then every surjective G_k-homomorphism $\Phi \colon \Pi_X \to \Pi_Y$ is defined by a unique k-morphism $\phi \colon X \to Y$ in a functorial way.

The *Local Theory* in this case is as follows. As mentioned above, one has no clue at all how to recover $X(k)$ from the given data $\Pi_X \to G_k$, and the situation is no better if we replace Π_X by the full étale fundamental group $\pi_1(X)$. Nevertheless, Mochizuki develops another kind of a "Local Theory", which fortunately does the right job for the problem. The kind of points one can recover are as follows. Let R be the valuation ring of k, and let $X_R \to R$ be a semi-stable model of $X \to k$ (such models exist after enlarging the base field k). Let $(X_i)_i$ be the irreducible components of the special fiber $X_s \to \kappa$ of $X_R \to R$. If $\eta_i \in X_R$ is the generic point of X_i, then the local ring O_{X_R, η_i} is a discrete valuation ring of $k(X)$ dominating R, and the residue field $\kappa(\eta_i)$ is the function field of $X_i \to \kappa$. Let us call such points η_i *arithmetical points* of X (arising for the several models X_R).

Next let (L, v) be a discrete complete valued field over k such that the valuation of L prolongs the p-adic valuation of k, and the residue field $Lv | kv$ is a function field in one variable. Remark that the completion of $k(X)$ with respect to the valuation $v := v_{\eta_i}$ defined by an arithmetical point is actually such a discrete complete valued field over k. We denote for short $\mathcal{H}_L^\Omega = H^1(G_{Lk^a}, \widehat{O}_{L^a}(1))/(\text{torsion})$, where $\widehat{O}_{L^a}$ is the completion of the valuation ring of L^a (similar to the completion $\mathbb{C}_p$ of the algebraic closure of $\mathbb{Q}_p^a$).

We will say that a G_k-homomorphism $\Phi_X \colon G_L \to \Pi_X$ is *non-degenerate*, if the induced map on the p-adic cohomology

$$H^1(\Delta_X, \mathbb{Z}_p(1)) \xrightarrow{\text{infl}_{\Phi_X}} H^1(G_L, \widehat{O}_{L^a}(1)) \xrightarrow{\text{can}} \mathcal{H}_L^\Omega$$

is non-trivial. Now the main technical points of the proof are as follows.

(1) In the context above, let $\Phi \colon \Pi_X \to \Pi_Y$ be an open G_k-homomorphism. Then there exists a non-degenerate G_k-homomorphism $\Phi_X \colon G_L \to \Pi_X$ such that the composition $\Phi_Y := \Phi \circ \Phi_X$ is a non-degenerate homomorphism. Thinking of the Local Theory from the birational case, this assertion here corresponds more or less to the characterization of arithmetical prime divisors.

(2) Every non-degenerate G_k-morphism $\Phi_X \colon G_L \to \Pi_X$ as above is of geometrical nature: given such a Φ_X, there exists a unique *L-rational point* $\psi_{\Phi_X} \colon L \to X$ defining Φ_X *in a functorial way*.

(3) In particular, for Φ and Φ_X as at (1) above, there exist L-rational points $\psi_{\Phi_X} \colon L \to X$ and $\psi_{\Phi_Y} \colon L \to Y$ defining the non-degenerate morphisms Φ_X and $\Phi_Y = \Phi \circ \Phi_X$ in a functorial way.

Finally, Mochizuki's *Global Theory* is a very nice application of the p-adic

Hodge–Tate theory and of Faltings' theory of almost étale morphisms. The idea is as follows.

First, by the p-adic Hodge–Tate theory, the sheaf of global differentials on X, say $D_X := \mathrm{H}^0(X, \Omega_X)$, can be recovered from the action of G_k on the $\mathbb{C}_p$-twists with the p-adic cohomology $\mathrm{H}^i_{\mathrm{et}}(\overline{X}, \mathbb{Z}_p(j))$ of $\overline{X}$. On the other hand, since X is a complete hyperbolic curve over a field of characteristic zero, thus $\neq p$, the p-adic cohomology $\mathrm{H}^i_{\mathrm{et}}(\overline{X}, \mathbb{Z}_p(j))$ is the same as the Galois cohomology of Δ_X, thus known.

Let us denote $D^i_X = \mathrm{H}^0(X, \Omega^{\otimes i}_X)$, and let $\mathcal{R}^i_X := \ker(D^{\otimes i}_X \to D^i_X)$ be the space of ith homogeneous relations in D^i_X. If X is not a hyperelliptic curve (which we can suppose after replacing X by a properly chosen étale cover whose geometric part is p-elementary Abelian), then the system of all D^i_X completely defines X. Equivalently, the system of all data $\mathcal{R}^i_X \subset D^{\otimes i}_X$ completely defines X.

Second, let $\Phi \colon \Pi_X \to \Pi_Y$ be a surjective G_k-homomorphism. Then Φ induces in a functorial way a morphism of k vector spaces $\iota_\Phi \colon D_Y \to D_X$; thus also morphisms of k vector space $\iota^{\otimes i}_\Phi \colon D^{\otimes i}_Y \to D^{\otimes i}_X$ for each $i \geq 1$. And by the general non-sense concerning the canonical embedding, if each $\iota^{\otimes i}_\Phi$ "respects the relations", i.e., it maps $\mathcal{R}^i_Y$ into $\mathcal{R}^i_X$, then ι_Φ is defined by some dominant k-morphism $\phi \colon X \to Y$ in the canonical way.

Finally, in order to check that ι_Φ does indeed respect the relations, one uses the Local Theory and Faltings' theory of almost étale morphisms. Choose a non-degenerate morphism $\Phi_X \colon G_L \to \Pi_X$ as at (3) above. Let Ω_L denote the p-adically continuous k-differentials of L, and let Ω^i_L be its powers. Since ψ_{Φ_X} is a non-degenerate point of X, the differential $d_X := d(\jmath_{\Phi_X}) \colon D_X \to \Omega_L$ of $\jmath_{\Phi_X} \colon L \to X$ and its powers $d^i_X \colon D^i_X \hookrightarrow \Omega^i_L$ are embeddings. Thus in order to check that ι_Φ respects the relations, it is sufficient to check that this is the case for the composition

$$\iota^{\otimes i}_L \colon D^{\otimes i}_Y \to D^{\otimes i}_X \hookrightarrow \Omega^{\otimes i}_L .$$

On the other hand, the composition of the map $\iota^{\otimes i}_L$ with $\Omega^{\otimes i}_L \to \Omega^i_L$ is exactly the canonical map $D^{\otimes i}_Y \to D^i_Y \hookrightarrow \Omega^i_L$ defined via the non-degenerate morphism $\Phi_Y = \Phi \circ \Phi_X$ and the resulting point $\psi_{\Phi_Y} \colon L \to \Pi_Y$. This concludes the proof.

Remarks (1) First, theorem A$'$ of Mochizuki [Mzk3] shows that *truncated-Π_X versions* of the assertion of the main result above are also valid. Namely, one can replace Δ_X by its central series quotient $\Delta^{(n)}_X$, and consequently Π_X by the corresponding quotient $\Pi^{(n)}_X$ which fits into the exact sequence

$$1 \to \Delta^{(n)}_X \to \Pi^{(n)}_X \to G_k \to 1.$$

Let $n \geq 3$. Then given an open G_k-homomorphism $\Phi^{(n+2)} \colon \Pi_Y^{(n+2)} \to \Pi_X^{(n+2)}$ there exists a unique dominant k-homomorphism $Y \to X$ such that the canonical open morphism $\Pi_Y^{(n)} \to \Pi_X^{(n)}$ induced by ϕ coincides with $\Phi^{(n)}$ on $\Pi_Y^{(n)}$.

(2) Using specialization techniques, Mochizuki proves the relative Hom-form of the birational anabelian conjecture for finitely generated fields over sub-p-adic fields k as follows.

Theorem (Mochizuki [Mzk3], theorem B) *Let $K|k$, $L|k$ be regular function fields. Then every open G_k-homomorphism $\Phi \colon G_K \to G_L$ is defined functorially by a unique k-embedding of fields $L \to K$.*

I would like to remark that by using techniques developed in order to prove pro-ℓ birational type results, see Part III of these notes, one can sharpen the above result and show the following.

Theorem (Corry–Pop [C–P]) *Every open Π_k-homomorphism $\Phi \colon \Pi_K \to \Pi_L$ is defined by a unique k-embedding $L \to K$ in a functorial way.*

(3) Mochizuki also shows that *hyperbolically fibered surfaces* are anabelian. And moreover, the Isom-form of an anabelian conjecture for fibered surfaces is true. Here, a hyperbolically fibered surface X is the complement of an étale divisor in a smooth proper family $\tilde{X} \to X_1$ of hyperbolic complete curves over a hyperbolic base curve X_1. The result is as follows.

Theorem (Mochizuki [Mzk3], theorem D) *Let $Y \to k$ and $X \to k$ be geometrically integral hyperbolically fibered surfaces over a sub-p-adic field k. Then every G_k-isomorphism $\Phi \colon \pi_1(Y) \to \pi_1(X)$ is defined by a unique k-isomorphism $\phi \colon Y \to X$ in a functorial way.*

Note that in the theorem above the *full fundamental group* π_1 is needed. It is maybe useful to remark that a naive Hom-form of the above theorem is not true. Indeed, let k be an infinite base field. Then by using general hyperplane arguments, one can show that for every smooth quasi-projective k-variety $X \subseteq \mathbb{P}^N$, there exist smooth k-curves $Y \subseteq X$ obtained from $X \to k$ by intersections with general hyperplanes such that the canonical map $\pi_1^t(Y) \to \pi_1^t(X)$ is surjective. In a second step, one can realize $\pi_1^t(Y)$ in many ways as quotients of fundamental groups $\pi_1^t(Z) \to \pi_1^t(Y)$ for several smooth k-varieties (which can be chosen to be projective, if X is complete), e.g., $Z = Y \times \cdots \times Y$ finitely many times. Finally, the composition

$$\pi_1^t(Z) \to \pi_1^t(Y) \to \pi_1^t(X)$$

is a surjective G_k-morphism, but by its construction, it does not originate from a dominant k-rational map.

(III) Jakob Stix's results concerning hyperbolic curves in positive characteristic The results of Stix deal with *hyperbolic non-constant curves* over finitely generated infinite fields k of positive characteristic (but also apply to such fields of characteristic zero, where the results are already known). Recall that given a curve $X \to k$ with char$(k) = p > 0$, one says that X is *potentially isotrivial* if there exists a finite étale cover $X' \to X$ such that X' is defined over a finite field. One can show that, if X is hyperbolic, then X is potentially isotrivial if and only if there exists a finite field extension $k'|k$ such that the base change $X' = X \times_k k'$ is defined over a finite field. Further, recall that for a curve $X \to k$ as above, we denote by $X^i \to k^i$ the maximal purely inseparable cover of $X \to k$. And for every integer n we denote by $X^i(n) \to k^i$ the relative Frobenius n-twist.

Theorem (Stix [St1], [St2]) *Let $X \to k$ be a non potentially isotrivial hyperbolic curve over a finitely generated infinite field k with char$(k) = p > 0$. Then one can recover $X^i \to k^i$ from $\pi_1^t(X) \to G_k$ in a functorial way.*

Moreover, the relative Isom-form of the anabelian conjecture for hyperbolic curves over k is true in the following sense. Let $Y \to k$ be some hyperbolic curve, and let a G_k-isomorphism $\Phi: \pi_1^t(X) \to \pi_1^t(Y)$ be given. Then there exists a unique n and a k^i-isomorphism $\phi: X^i(n) \to Y^i$ defining Φ.

The strategy of proof is as follows.

Let k be a finitely generated infinite field, and $X \to k$ a hyperbolic curve over k. In the notations from the "standard reduction technique", let $X_S \to S$ be a smooth surjective family of hyperbolic curves whose generic fiber is $X \to k$. The idea is as follows.

Case 1 $X \to k$ is an affine hyperbolic curve.
By shrinking S if necessary, we can suppose that $X \to k$ has good reduction at all closed points $s \in S$. From $\pi_1^t(X) \to G_k$ one recovers the local projections $\Phi_s: \pi_1^t(X_s) \to G_{\kappa(s)}$ for all closed points $s \in S$. By Tamagawa's theorem, we can recover the isomorphy type of $X_s^i \to \kappa(s)$ *up to Frobenius twists*. In particular, let $\Phi: \pi_1^t(X) \to \pi_1^t(Y)$ be a G_k-isomorphism, where $Y \to k$ is some hyperbolic curve over k. By the "standard specialization technique", we obtain $\kappa(s)$-isomorphisms of some relative Frobenius twists of the special fibers, say $\phi_s: X_s^i(n_s) \to Y_s^i$ defining Φ_s. Unfortunately, the usual globalization techniques work only under the hypothesis the Frobenius twists n_s are constant, say equal to n, on a non-empty open of S (and then they are constant on the

whole S). If this is the case, then the local isomorphisms ϕ_s indeed originate from a unique global k^i-isomorphism $\phi: X^i(n) \to Y^i$, which defines the given G_k-isomorphism $\Phi: \pi_1^t(X) \to \pi_1^t(X)$.

Here is the way Stix shows that the exponents n_s are indeed constant. First, by replacing X by a properly chosen tame étale cover, we can suppose that the smooth completion X_0 of X is hyperbolic too. Next we fix some $m > 2$ relatively prime to $p = \mathrm{char}(k)$, and replace k by its finite extension over which the m-torsion of Jac_{X_0} becomes rational. And we choose an m-level structure on X_0 by fixing an isomorphism

$$\varphi_{X,m}: {}_m\mathrm{Jac}_{X_0} = \pi_1^{ab}(\overline{X}_0)/m \to (\mathbb{Z}/m)^{2g}.$$

Then X_0 endowed with $\varphi_{X,m}$ is classified by a k-rational point $\psi_X: k \to \mathcal{M}_g[m]$. Moreover, using the "standard reduction technique", the level structure $\varphi_{X,m}$ gives rise via the specialization homomorphisms $\mathrm{sp}_s: \pi_1(\overline{X}_0) \to \pi_1(\overline{X}_{0,s})$ canonically to level structures

$$\varphi_{X_s,m}: {}_m\mathrm{Jac}_{X_{0,s}} = \pi_1^{ab}(\overline{X}_{0,s})/m \to (\mathbb{Z}/m)^{2g}.$$

This happens in such a way that $\psi_X: k \to \mathcal{M}_g[m]$ defined above becomes the generic fiber of a morphism $\psi_{X_S}: S \to \mathcal{M}_g[m]$ whose special fibers classify the curves $X_s \to \kappa(s)$ endowed with the level structures $\varphi_{X_s,m}$.

Now let us come back to the G_k-isomorphism $\Phi: \pi_1^t(X) \to \pi_1^t(Y)$. Clearly, Φ transports the m-level structure $\varphi_{X,m}$ of X_0 to an m-level structure $\varphi_{Y,m}$ for Y_0. And the local $G_{\kappa(s)}$-isomorphisms $\Phi_s: \pi_1^t(X_s) \to \pi_1^t(Y_s)$ transport the m-level structures $\varphi_{X_s,m}$ to m-level structures $\varphi_{Y_s,m}$ which are compatible with the specialization morphisms $\mathrm{sp}_s: \pi_1(\overline{Y}_0) \to \pi_1(\overline{Y}_{0,s})$.

Now suppose that there exist some exponents n_s and $\kappa(s)$-isomorphisms $\phi_s: X_s^i(n_s) \to Y_s^i$ which define the G_κ-isomorphisms $\Phi_s: \pi_1^t(X_s) \to \pi_1^t(Y_s)$. Then ϕ_s prolongs to an $\kappa(s)$-isomorphism $\phi_{0,s}: X_{0,s}(n_s) \to Y_{0,s}$. Next, remark that $X_{0,s}$ and its relative Frobenius twists endowed with the same m-level structure $\varphi_{X_s,m} = \varphi_{X_s(n_s),m}$ factor through the same closed point of $\mathcal{M}_g[m]$. Thus we have: The classifying morphisms $\psi_{X_S}: S \to \mathcal{M}_g[n]$ for $X_{0,s}$ and $\psi_{Y_S}: S \to \mathcal{M}_g[n]$ for $Y_{0,s}$ defined above coincide (topologically) on the closed points $s \in S$.

In order to conclude, Stix proves the following.

Proposition (Stix [St1]) *Let S and M be irreducible $\mathbb{Z}$-varieties. Let $f, g: S \to M$ be two morphisms which coincide topologically on the closed points of S. Suppose that $f \neq g$. Then S is defined over $\mathbb{F}_p$ for some p, and f and g differ by a power of Frobenius, which is unique if f is not constant.*

Thus applying the proposition above we conclude that the classifying morphisms ϕ_X and ϕ_Y differ by a power Frob^n of Frobenius. In particular, fiber-wise the same is the case. From this one finally deduces that $\Phi\colon \pi_1^t(X) \to \pi_1^t(Y)$ is defined by some k-isomorphisms $\phi\colon X^i(n) \to Y^i$ for some integer n.

This completes the proof of the case when X is an affine hyperbolic curve.

Case 2 $X \to k$ is a complete hyperbolic curve.

Let us try to mimic Mochizuki's strategy from the case of complete hyperbolic curves over finitely generated fields of characteristic zero. Then we run immediately into the following difficulty. If k has positive characteristic $p > 0$, there is no obvious reason that some finite properly chosen étale (Galois) covers $X' \to X$ have bad reduction at points $s \in S$ where X has good reduction. Note that the "trick" used by Mochizuki in [Mzk2] in the case $\mathrm{char}(k) = 0$ definitely does not work in positive characteristic. (This follows from Grothendieck's specialization theorem. Let $X_{R_s} \to R_s$ be a complete smooth curve, and let $X' \to X$ be an étale Galois cover whose geometric part has degree prime to p. Then X' has potentially good reduction.)

In order to avoid this difficulty, one can nevertheless use the Raynaud, Pop–Saidi, Tamagawa theorem (see Part III of these notes). A consequence of that result is the following: Let a closed point $s \in S$ be given. Then there exists a finite étale cover $X^{(s)} \to X$ whose geometric part is a cyclic étale cover of $\overline{X}$ of degree prime to p having this property: any finite étale cover $X' \to X^{(s)}$ whose geometric part factors through the maximal p-elementary étale cover of $\overline{X}^{(s)}$ *does not have potentially good reduction*. With this input, Stix uses the theory of log-étale fundamental groups in order to conclude the proof in the same style as Mochizuki [Mzk1], but using the methods developed to treat the case of affine hyperbolic curves.

(IV) Mochizuki's cuspidalization results and applications   As we have seen so far, the information about "points" and how this information is encoded in Galois theory plays a crucial role in the strategies to tackle anabelian type assertions, both in the birational and in the curve case. The difficulty of proving the anabelian conjecture for complete curves lies exactly in the impossibility of having obvious candidates for "points" which are encoded in the $\pi_1(X)$, because the inertia groups at closed points $x \in X$ are all trivial in $\pi_1(X)$. Mochizuki [Mzk5] had the idea how to detect tame inertia type information at closed points $x \in X$ in a functorial way together with the Galois action of the Frobenius on them. He calls this procedure `cuspidalization`. Here is a short introduction to this topic. Let X be a proper hyperbolic curve over a field k which is either finite or a finite extension of $\mathbb{Q}_p$ for some $p > 0$. Further, let

$U \subseteq X$ be an open sub-scheme of X. Then one has a canonical surjective projection $\pi_1(U) \to \pi_1(X)$ which is compatible with the projections $\pi_1(U) \to G_k$ and $\pi_1(X)$, hence defines a surjective map of the geometric fundamental groups $\pi_1(\overline{U}) \to \pi_1(\overline{X})$.

(1) *Abelian cuspidalization* Let $\pi_1^{\mathrm{c.ab}}(\overline{U})$ be the maximal quotient of $\pi_1(\overline{U})$ which is a central prime to p extension of $\pi_1(\overline{X})$, i.e., $\pi_1^{\mathrm{c.ab}}(\overline{U})$ is the maximal quotient of $\pi_1(\overline{U})$ such that the kernel of $\pi_1(\overline{U}) \to \pi_1^{\mathrm{c.ab}}(\overline{U})$ is contained in the kernel of $\pi_1(\overline{U}) \to \pi_1(\overline{X})$ and the kernel of $\pi_1^{\mathrm{c.ab}}(\overline{U}) \to \pi_1(\overline{X})$ is central in $\pi_1^{\mathrm{c.ab}}(\overline{U})$ and has order prime to the characteristic. We notice that if $x \in X \backslash U$ is a "cuspidal point" of U, i.e., a closed point of X which does not lie in U, and $T_x^{\mathrm{t}} \subset \pi_1(\overline{U})$ is the tame part of an inertia group above x, then T_x^{t} is mapped isomorphically onto its image $T_x^{\mathrm{c.ab}} \subset \pi_1^{\mathrm{c.ab}}(\overline{U})$. More precisely, one has for all cuspidal points x of U the following: $T_x^{\mathrm{t}} \cong \widehat{\mathbb{Z}}' \cong T_x^{\mathrm{c.ab}}$ and the groups $T_x^{\mathrm{c.ab}}$ generate the kernel of $\pi_1^{\mathrm{c.ab}}(\overline{U}) \to \pi_1(\overline{X})$ with a unique relation. Finally, we notice that the kernel of $\pi_1(\overline{U}) \to \pi_1^{\mathrm{c.ab}}(\overline{U})$ is a normal subgroup in $\pi_1(U)$, thus $\pi_1^{\mathrm{c.ab}}(\overline{U})$ fits into an exact sequence of the form $1 \to \pi_1^{\mathrm{c.ab}}(\overline{U}) \to \pi_1^{\mathrm{c.ab}}(U) \to G_k \to 1$, and $\pi_1(U) \to \pi_1(X)$ gives rise to a canonical surjective projection $\pi_1^{\mathrm{c.ab}}(U) \to \pi_1(X)$.

(2) *Pro-ℓ cuspidalization* In the above context, let $\ell \neq \mathrm{char}$ be a fixed prime number. Then one can consider as above the $\pi_1^{\mathrm{c.}\ell}(\overline{U})$ be the maximal quotient of $\pi_1(\overline{U})$ which is a pro-ℓ extension of $\pi_1(\overline{X})$, i.e., $\pi_1^{\mathrm{c.}\ell}(\overline{U})$ is the maximal quotient of $\pi_1(\overline{U})$ such that the kernel of $\pi_1(\overline{U}) \to \pi_1^{\mathrm{c.}\ell}(\overline{U})$ is contained in the kernel of $\pi_1(\overline{U}) \to \pi_1(\overline{X})$, and the kernel of $\pi_1^{\mathrm{c.}\ell}(\overline{U}) \to \pi_1(\overline{X})$ is a pro-ℓ group. As above, we notice that if $x \in X \backslash U$ is a "cuspidal point" of U, and $T_x^{\ell} \subset \pi_1(\overline{U})$ is a Sylow ℓ-group of a tame inertia group above x, then T_x^{ℓ} is mapped isomorphically onto its image $T_x^{\mathrm{c.}\ell} \subset \pi_1^{\mathrm{c.}\ell}(\overline{U})$. More precisely, one has for all cuspidal points x of U the following: $T_x^{\mathrm{t}} \cong \mathbb{Z}_{\ell} \cong T_x^{\mathrm{c.}\ell}$, and there are properly chosen groups T_x^{ℓ} such that their images $T_x^{\mathrm{c.}\ell}$ generate the kernel of $\pi_1^{\mathrm{c.}\ell}(\overline{U}) \to \pi_1(\overline{X})$ with a unique relation. As above, the kernel of $\pi_1(\overline{U}) \to \pi_1^{\mathrm{c.}\ell}(\overline{U})$ is a normal subgroup in $\pi_1(U)$, thus $\pi_1^{\mathrm{c.}\ell}(\overline{U})$ fits into an exact sequence of the form $1 \to \pi_1^{\mathrm{c.}\ell}(\overline{U}) \to \pi_1^{\mathrm{c.}\ell}(U) \to G_k \to 1$, and $\pi_1(U) \to \pi_1(X)$ gives rise to a canonical surjective projection $\pi_1^{\mathrm{c.ab}}(U) \to \pi_1(X)$.

Theorem (See Mochizuki [Mzk5])

(1) *Let $U \subset X$ be a Zariski open sub-scheme as above. Then there are group theoretical recipes to recover the canonical projection $\pi_1^{\mathrm{c.ab}}(U) \to \pi_1(X)$ from $\pi_1(X)$, and the same holds for the projection $\pi_1^{\mathrm{c.}\ell}(U_x) \to \pi_1(X)$ for U_x of the form $U_x := X \backslash \{x\}$ with $x \in X$ a closed point of X.*

(2) *Moreover, the group theoretical recipes above are invariant under isomor-*
    phisms. Precisely, let $Y \rightarrow l$ be a hyperbolic curve with l either finite or a
    finite extension of some $\mathbb{Q}_q$ for some prime number q such that one has an
    isomorphism $\Phi \colon \pi_1(X) \rightarrow \pi_1(Y)$. Then the following hold:
    (a) *Φ is compatible with the projections $\pi_1(X) \rightarrow G_k$, $\pi_1(Y) \rightarrow G_l$ thus de-*
        fines an isomorphism $G_k \rightarrow G_l$, and k and l have the same (residual)
        characteristic.
    (b) *For every $U \subseteq X$ there exists $V \subseteq Y$ such that $\pi_1(X) \rightarrow \pi_1(Y)$ can*
        be lifted to an isomorphism $\pi_1^{c.ab}(U) \rightarrow \pi_1^{c.ab}(V)$ which maps inertia
        isomorphically onto inertia. A similar more technical assertion holds
        for $\pi_1^{c,\ell}(U_x)$.

As applications, Mochizuki [Mzk5] showed that the Isom-form of the anabelian conjecture for hyperbolic *complete* curves over finite fields holds, thus
extending the results by Stix mentioned previously, and thus reproving the
Isom-form of the anabelian conjecture for complete curves in general via Stix's
specialization result.

Nevertheless, meanwhile there is a much stronger result concerning the
Isom-form of the anabelian conjecture for hyperbolic (complete) curves by
Saidi–Tamagawa [S–T], which is based on Mochizuki's cuspidalization, and is
stated as follows. Let Σ be any non-empty set of rational prime numbers. Then
for a geometrically integral curve $X \rightarrow k$ over some base field k, we denote
by $\pi_1^{\Sigma}(\overline{X})$ the pro-Σ completion of the geometric fundamental group of X. Note
that the kernel of $\pi_1(\overline{X}) \rightarrow \pi_1^{\Sigma}(\overline{X})$ is characteristic in $\pi_1(\overline{X})$, thus this kernel is
normal in $\pi_1(X)$. In particular, $\pi_1^{\Sigma}(\overline{X})$ fits canonically into an exact sequence

$$1 \rightarrow \pi_1^{\Sigma}(\overline{X}) \rightarrow \pi_1^{(\Sigma)}(X) \rightarrow G_k \rightarrow 1,$$

and we will say that $\pi_1^{(\Sigma)}(X)$ is the `geometrically pro-`Σ fundamental group
of $X \rightarrow k$. Note that the geometrically pro-Σ fundamental group $\pi_1^{(\Sigma)}$ is one of
the several possible `geometrically pro-`C completions $\pi_1^{(C)}$ of $\pi_1(X)$.

Theorem (See Saidi–Tamagawa [S–T]) *Let Σ, Ξ be sets of rational prime
numbers, with Σ consisting of all but maybe finitely many prime numbers. Let
X and Y be hyperbolic curves over finite fields, and $\pi_1^{(\Sigma)}(X)$ and $\pi_1^{(\Xi)}(Y)$ be
their geometrically pro-Σ, respectively pro-Ξ, fundamental groups. Then every
isomorphism $\Phi \colon \pi_1^{(\Sigma)}(X) \rightarrow \pi_1^{(\Xi)}(Y)$ originates from geometry as predicted by
the anabelian conjectures for curves, and moreover, if such an isomorphism Φ
exists, then $\Sigma = \Xi$.*

Further very exciting developments resulting from Mochizuki's cuspidalization theory are applications to the *absolute Isom-form* of the anabelian conjec-

tures over sub-p-adic fields, in particular over finite extensions $k|\mathbb{Q}_p$ of $\mathbb{Q}_p$. See Mochizuki [Mzk6], etc., for more about this.

C. The section conjectures

Let $X_0 \to k$ be an arbitrary irreducible k-variety, and $X \subset X_0$ an open k-subvariety. Let $x \in X_0$ be a regular k^i-rational point of X_0. Then choosing a system of regular parameters $(t_1, \ldots, t_d)$ at x, we can construct – by the standard procedure – a valuation v_x of the function field $k(X)$ of X with value group $v_x(K^\times) = \mathbb{Z}^d$ ordered lexicographically, and residue field $k(X)v_x = \kappa(x)$, thus a subfield of k^i. Let v be a prolongation of v_x to $k(X)^s$, and let $T_v \subset Z_v$ be the inertia, respectively decomposition group, of $v|v_x$ in G_K. By general valuation theory, see e.g., Kuhlmann–Pank–Roquette [K–P–R], one has: T_v *has complements* G_v *in* Z_v. And clearly, since $k(X)v_x = \kappa(x) \subset k^i$, under the canonical exact sequence

$$1 \to T_v \to Z_v \to G_{k(X)v_x} = G_k \to 1,$$

every complement G_v is mapped isomorphically onto $G_k = G_{k^i}$. Therefore, the canonical projection

$$(*) \qquad \mathrm{pr}_{k(X)} \colon G_{k(X)} \to G_k$$

has sections $s_v \colon G_k \to G_v \subset G_{k(X)}$ constructed as shown above.

Moreover, let us recall that, under the canonical projection $G_{k(X)} \to \pi_1(X)$, the decomposition group Z_v is mapped onto the decomposition group Z_x of v in $\pi_1(X)$, and T_v is mapped onto the inertia group T_x of v in $\pi_1(X)$. And finally, any complement G_v of T_v is mapped isomorphically onto a complement G_x of T_x in Z_x. Clearly $G_x \to G_k$ isomorphically, thus

$$(**) \qquad \mathrm{pr}_X \colon \pi_1(X) \to G_k$$

has a section $s_x \colon G_k \to G_x \subset \pi_1(X)$ defined via the k^i-rational point $x \in X(k^i)$. One has the following possibilities.

(a) Suppose that $x \in X$. Then $T_x = \{1\}$, as the étale covers of X are not ramified over x. Therefore, $Z_x = G_x$. And in this case the sections of pr_X of the form above build a full conjugacy class of sections.

(b) Next let $x \in (X_0 \backslash X)(k^i)$. Then $T_x \neq \{1\}$ and $G_x \neq Z_x$ in general. Therefore, for a given k^i-rational point x, there might exist several complements G_x of T_x in Z_x, thus a "bouquet" of sections of $\mathrm{pr}_X \colon \pi_1(X) \to G_k$, *which are not conjugate* inside $\pi_1(X)$. Such sections are called *sections at infinity* (for the variety X), and the "bouquet" of conjugacy classes of sections is the non-commutative continuous cohomology pointed set $H^1_{cont}(G_k, T_x)$ defined via the split exact sequence $1 \to T_x \longrightarrow Z_x \overset{pr_x}{\longrightarrow} G_k \to 1$.

Nevertheless, in the case X is a curve and $\mathrm{char}(k) = 0$, one has: T_x is an abelian group and $T_x \cong \widehat{\mathbb{Z}}(1)$ as G_k-modules. In particular, the space of sections $H^1_{cont}(G_k, T_x)$ is $H^1_{cont}(G_k, T_x) \cong \widehat{k^\times}$ by Kummer theory, where the last group is the adic completion of the multiplicative group $k^\times$ of the field k.

Grothendieck's section conjecture asserts roughly that under certain "anabelian hypotheses" all the sections of pr_x arise in the way described above. Precisely, recall that a curve $X \to k$ is non-isotrivial, if it has no finite cover which is defined over a finite field.

Section conjectures *Let k be a finitely generated infinite field and $X \to k$ be a hyperbolic curve which is non-isotrivial. Let $X_0 \supseteq X$ be the smooth completion of X. In the notations from above, the following hold:*

(1) Birational SC: *The sections of $pr_{k(X)}$ arise from k^i-rational points of X_0 as indicated above.*

(2) Curve SC: *The sections of pr_X arise from k^i-rational points of X_0 of X as indicated above.*

We notice that the Curve SC implies the Birational SC. Indeed, the Birational SC can be easily derived from the Curve SC by starting with the complete geometrically integral smooth curve X_0 and taking limits over a system $X_i \subset X_0$ of Zariski open neighborhoods of the generic point $\eta \in X_0$. But the Birational SC can be formulated for other "interesting" Galois field extensions $\tilde{K}|K$ of $K := k(X_0)$ as follows. First, recall that there exists a canonical bijection between the k-places v of $K|k$ and the closed points x of X_0, by interpreting each such closed point as a Weil prime divisor of X_0. If v and x correspond to each other, then the corresponding residue fields are equal: $\kappa(x) = \kappa(v)$. Hence x is a k^i-rational point if and only if v is k^i-rational place of $K|k$. Let $\tilde{K}|K$ be some Galois extension, and let $\tilde{G}_K := \mathrm{Gal}(\tilde{K}|K)$ denote the Galois group of $\tilde{K}|K$. Further let $\tilde{k} := \overline{k} \cap \tilde{K}$ be the "constants" of $\tilde{K}$, and set $\tilde{G}_k := \mathrm{Gal}(\tilde{k}|k)$. Hence one has a canonical exact sequence

$$1 \to \mathrm{Gal}(\tilde{K}|K\tilde{k}) \longrightarrow \mathrm{Gal}(\tilde{K}|K) \overset{\tilde{p}_K}{\longrightarrow} \mathrm{Gal}(\tilde{k}|k) \to 1 \, .$$

For a k^i-rational point x of X and its k^i-rational place v of K, let $\tilde{v}$ be a prolongation to $\tilde{K}$, and $T_{\tilde{v}} \subseteq Z_{\tilde{v}}$ be the inertia, respectively decomposition, groups of $\tilde{v}|v$, and $G_{\tilde{v}} := \mathrm{Aut}(\tilde{K}\tilde{v}\,|\,Kv)$ the residual automorphism group. By general Hilbert decomposition theory one has a canonical exact sequence

$$(*) \qquad\qquad\qquad 1 \to T_{\tilde{v}} \to Z_{\tilde{v}} \to G_{\tilde{v}} \to 1 \, .$$

We remark that in general $\tilde{k} \subset \tilde{K}\tilde{v}$ can be a strict inclusion, hence in particular, the canonical projection $\tilde{G}_v \to \tilde{G}_k$ is not an isomorphism, even if v is a k^i-rational place of $K|k$. Further, the above exact sequence $(*)$ is not necessarily split. The conclusion is that in general a k^i-rational point x of X does not necessarily give rise to a section of $\tilde{p}_K : \mathrm{Gal}(\tilde{K}|K) \to \mathrm{Gal}(\tilde{k}|k)$.

Nevertheless, if $\tilde{k} = \bar{k}$ is an algebraic closure of k, then $\tilde{K}\tilde{v} = \bar{k}$, and if v is a k^i-rational place of $K|k$, then $G_{\tilde{v}} = G_k$, and the exact sequence $(*)$ is split.

Thus if $\tilde{k} = \bar{k}$, every k^i-rational point x of X gives rise via its k^i-rational place v to a "bouquet" of conjugacy classes of sections $s_x : G_k \to \mathrm{Gal}(\tilde{K}|K)$ of the projection $\tilde{p}_K : \mathrm{Gal}(\tilde{K}|K) \to G_k$ which consists of the conjugacy classes in the non-commutative continuous cohomology pointed set $\mathrm{H}^1_{\mathrm{cont}}(G_k, T_{\tilde{v}})$ defined via the split exact sequence $(*)$ above.

There are several variants of the the above conjectures, from which we mention here the following.

(Birational) p-adic section conjecture *The above (birational) section conjecture holds over a base field k which is a finite extension $k|\mathbb{Q}_p$ of $\mathbb{Q}_p$.*

It seems that the initial intention of Grothendieck concerning the Curve SC was to make it part of a strategy for proving the Mordell conjecture, now Faltings' theorem (which had not yet been proved when Grothendieck made the conjectures). Unfortunately, at present the precise relation between the Curve SC and (an effective) Mordell conjecture is not yet clear. However, there has been recent progress on this question by Minhyong Kim in [K1], where a new proof of Siegel's theorem (on the finiteness of integer points on affine curves of genus one) is given using motivic fundamental group techniques; see also [K2], [K3]. In Kim [K4] he extends these ideas to curves X of arbitrary genus (defined over number fields k). Using pro-linear completions of the geometric fundamental group of X (precisely: pro-unipotent completions to suitably chosen primes), one attaches to X a `Selmer variety`, which is a "non-commutative group like" object (commutative in the case of curves of genus one) that carries information about the sections s of the canonical projection $\mathrm{pr}_X : \pi_1(X) \to G_k$. Conjecturally, the Selmer variety has nice properties, and Minhyong Kim designs in [K4] an algorithm (of p-adic nature) which – under the conjectural properties of the Selmer varieties – produces the rational points of the curve under discussion, and more impressively, the effectiveness of the algorithm is guaranteed by the validity of the section conjecture. This is a very promising strategy for finally proving an *effective Mordell conjecture* using Grothendieck's section conjecture – at least in the case where k is a number field. This moves the Curve SC to the center of a very intensive research effort.

The first known examples of curves over number fields that satisfy the section conjecture were probably given in Stix [St3] and later by Harari–Szamuely [H–Sz], and Stix [St4]. More recently, Hain [Hn] proved the Curve SC for the generic curve of genus $g \geq 5$. Unfortunately, all these examples are *no sections examples,* in the sense that there are no sections of pr_X, and hence no rational points.

Concerning the Curve SC over number fields, I would also like to mention the following. First, Nakamura's [Na1] result concerning the Galois characterization of points at infinity of affine curves over number fields (and more general) in terms of fundamental groups. Second, the very recent negative result by Hoshi [Ho], which asserts that the `geometrically pro-`p form of the conjecture does not hold over $k := \mathbb{Q}(\zeta_p)$ for p a regular prime and X_0 the Fermat curve given by $X_0^p + X_1^p + X_2^p \subset \mathbb{P}_k^2$.

Nevertheless, the Curve SC is wide open, and so is the weaker Birational SC. One of the strongest – unfortunately conditional – results concerning the Birational SC is the following.

Theorem (See Esnault–Wittenberg [E–W]) *Let $K = k(X)$ be the function field of a complete curve X over a number field k, and suppose that the Tate–Shafarevich group of the Jaconian of X is finite. Let $\mathcal{K}$ be the maximal abelian extension of $K\bar{k}$, and suppose that the canonical projection $\mathrm{pr}_{\mathcal{K}} \colon \mathrm{Gal}(\mathcal{K}|K) \to G_k$ has a section $s_{\mathcal{K}}$. Then the index of X, i.e., the g.c.d. of the degrees $d_x := [\kappa(x) : k]$ of all the closed points $x \in X$ is 1.*

The p-adic section conjectures have recently moved into the focus of several investigations, as pieces of evidence for the p-adic section conjecture emerged in recent years. Among the results are the following, see Koenigsmann [Ko3] and Pop [P8].

Theorem *Let $k|\mathbb{Q}_p$ be a finite extension, and $K|k$ an arbitrary regular field extension. Then for every section $s \colon G_k \to G_K$ of the canonical projection $\mathrm{pr}_K \colon G_K \to G_k$ one has that the fixed field $K^{(s)}$ of $s(G_k)$ in K^{a} is p-adically closed. Moreover, if v_s is the valuation of $K^{(s)}$ defining it as a p-adically closed field, then $Kv_s = k$.*

In particular, if $K = k(X)$ is the function field of a complete smooth k-variety, then every section $s \colon G_k \to G_{k(X)}$ is defined by a k-rational point $x_s \in X(k)$. The point x_s is exactly the center of v_s on the complete k-variety X.

This is so far the best unconditional result concerning the (birational) section conjecture we have. But it is not at all clear how to "globalize" such p-

adic results in order to get the birational section conjecture over number fields. There is nevertheless an unexpected sharpening of the p-adic Birational SC as follows: Let $k|\mathbb{Q}_p$ be a finite field extension which contains the pth roots of unity. Then in the notations from above, let $K' \subset K''$ be the maximal elementary $\mathbb{Z}/p$-abelian, respectively $\mathbb{Z}/p$-metabelian extensions of K, and pr: $\mathrm{Gal}(K'|K) \to \mathrm{Gal}(k'|k)$ the canonical projection. We say that a section of pr is `liftable`, if it can be lifted to a section of pr: $\mathrm{Gal}(K''|K) \to \mathrm{Gal}(k''|k)$.

Theorem (See Pop [P9]) *In the above notations, there is a canonical bijection between the "bouquets" of liftable sections of* $\mathrm{Gal}(K'|K) \to \mathrm{Gal}(k'|k)$ *and the k-rational points $X(k)$ of X, where X is a complete smooth curve with $K = k(X)$.*

Concerning the p-adic Curve SC, there are several partial results, such as Mochizuki's [Mzk5] concerning `cuspidal sections` of curves which cover $\mathbb{P}^1 \backslash \{0, 1, \infty\}$ as well as Mochizuki's [Mzk6], etc., concerning – among other things – the relation between the p-adic Curve SC and the absolute anabelian conjecture for curves; and "conditional results" proved by Saidi [Sa2], where so-called `good sections` are defined and it is shown that they originate indeed from points, using techniques inspired by the (proof of the) above result of Pop [P9]. (In all these stories, Mochizuki's cuspidalization plays heavily into the game.) But so far, it seems that the strongest unconditional result towards a proof of the p-adic Curve SC seems to be the following. Let $k|\mathbb{Q}_p$ be a finite field extension, X/k be a hyperbolic curve, and $K := k(X)$. Further let $\tilde{K} = k(\tilde{X})$ be the function field of the pro-étale universal cover of X, thus $\mathrm{Gal}(\tilde{K}|K) = \pi_1(X)$.

Theorem (See Pop–Stix [P–St]) *For every section $s: G_k \to \pi_1(X)$ of the canonical projection $\mathrm{pr}_X: \pi_1(X) \to G_k$ there exists a valuation w of $k(X)$ and a prolongation $\tilde{w}$ to $\tilde{K}|K$ such that $\mathrm{im}(s) \subset Z_{\tilde{w}}$, where $Z_{\tilde{w}}$ is the decomposition group of $\tilde{w}|w$.*

One of the main technical result in the proof is Stix' theorem [St4] asserting that the existence of a section of the canonical projection $\mathrm{pr}_X: \pi_1(X) \to G_k$ implies that the `index` of X is a power of p. (Recall that the index of a curve is the g.c.d. of the degrees of all its closed points. In particular, if the curve has a rational point, than the index of the curve is 1.) This makes it possible to apply machinery similar to that used in the birational case, namely the Tate, Roquette, Lichtenbaum Local-Global Principle for $\mathrm{Br}(X)$ and its generalization by Pop [P8], and reduce the problem to a fixed point question about actions of groups on graphs. One concludes by using a concrete structural assertion about the log-étale fundamental group of X as developed by Mochizuki [Mzk1] and Stix [St5]. Finally, Tamagawa [T6] is used to make the assertion more precise.

Part III. Beyond the arithmetical action

It is/was a widespread belief that the reason for the existence of anabelian schemes is strong interaction between the arithmetic and a rich algebraic fundamental group, and that this interaction makes étale fundamental groups so rigid, that the only way isomorphisms, respectively open homomorphisms, can occur is the geometrical one. (That is to say, morphisms between étale fundamental groups which do not have a reason to exist, do not exist indeed...)

On the other hand, some developments from the 1990s showed evidence for very strong anabelian phenomena for curves and higher dimensional varieties over *algebraically closed fields,* thus in a total absence of a Galois action of the base field. We mention here the following.

(a) *Bogomolov's birational anabelian program* (see [Bo])   Let ℓ be a fixed rational prime number. For algebraically closed base fields k of characteristic $\neq \ell$, we consider integral k-varieties $X \to k$, with function field $K := k(X)$. It turns out that there is a major difference between the cases $\dim(X) = 1$ and $\dim(X) > 1$. Indeed, if $\dim(X) = 1$, then the absolute Galois group G_K is profinite free on $|k|$ generators. This is the so-called *geometric case of a conjecture of Shafarevich,* proved by Harbater [Ha2], and Pop [Po]. On the other hand, if $d = \dim(X) > 1$, then G_K is very complicated (in particular, having $\mathrm{cd}_\ell G_K = d$, etc.).

The guess of Bogomolov [Bo] is that if $\dim(X) > 1$, then the Galois group G_K should be so rich as to encode the birational class of X up to purely inseparable covers and Frobenius twists. More precisely, Bogomolov proposes and gives evidence for the following. In the context above, let $G_K(\ell)$ be the maximal pro-ℓ quotient of G_K, i.e., $G_K(\ell)$ is the Galois group of the maximal Galois pro-ℓ sub-extension $K(\ell)$ of $K^s | K$. Further let $G_K^{(n)}$ denote the nth ℓ^∞-factor in the central series of $G_K(\ell)$, and let $K^{(n)}$ be the corresponding fixed fields inside $K(\ell)$. Hence setting $\Pi_K^{(n)} := G_K(\ell)/G^{(n)}$ we have: $\Pi_K := \Pi_K^{(2)}$ is the Galois group of the maximal prol-ℓ abelian sub-extension $K^{(2)} = K^{\ell,\mathrm{ab}}$ of $K(\ell)$, and $\Pi_K^c := \Pi_K^{(3)}$ is the Galois group of the maximal pro-ℓ abelian-by-central sub-extension $K^{(3)}$ of $K(\ell)$. Now Bogomolov's [Bo] claim is that in fact the isomorphy type of the function $K|k$ is encoded in Π_K^c. (Note that in Bogomolov [Bo] the notation for $\Pi_K^c := G_K(\ell)/G_K^{(3)}$ is PGal_K^c.) The starting point in this development was Bogomolov's theory of *liftable commuting pairs,* see a result mentioned below concerning this.

(b) *Tamagawa's theorem concerning* $\mathbb{P}^1_{k_0} \setminus \{0, 1, \infty, x_1, \ldots, x_n\}$ In the mid-1990s Tamagawa gave evidence for the fact that some curves over the algebraic closure $k_0 = \overline{\mathbb{F}}_p$ are weakly anabelian, i.e., their isomorphy type as a scheme can be recovered from π_1 or even π_1^t. The first precursor of this fact is Tamagawa's result that, given a smooth curve $X \to k$, the type (g, r) of the curve is encoded in the algebraic fundamental group $\pi_1(\overline{X})$; and moreover, the canonical projections $\pi_1(\overline{X}) \to \pi_1^t(\overline{X}) \to \pi_1(X_0)$ are encoded in the algebraic fundamental group of X. This answered a question raised by Harbater. And finally, Tamagawa showed the following.

Theorem (See Tamagawa [T2])   *Let* $U = \mathbb{P}^1_{k_0} \setminus \{0, 1, \infty, x_1, \ldots, x_n\}$ *be an affine open. Then the isomorphy type of* U *as a scheme can be recovered from* $\pi_1^t(U)$. *Moreover, if* X *is any other curve over some algebraically closed field* k, *and* $\pi_1(X) \cong \pi_1(U)$, *then* $k = k_0$, *and* $X \cong U$ *as schemes.*

The kind of results above show that one can expect anabelian phenomena over algebraically closed base fields, thus in a complete absence of arithmetical Galois action. This kind of anabelian phenomena go *beyond Grothendieck's anabelian geometry*. Here is a short list of the kind of such anabelian results.

A. Small Galois groups and valuations

Let ℓ be a fixed prime number. We consider fields K of characteristic $\neq \ell$, such that $\mu_\ell \subset K$. We denote by $K(\ell)$ the maximal Galois pro-ℓ extension of K in some fixed algebraic closure K^a of K, and denote by $G_K(\ell)$ the Galois group of $K(\ell)|K$. In order to avoid complications arising from orderings in case $\ell = 2$, we will also suppose that $\mu_4 \subset K$ if $\ell = 2$.

In the above context, let v be a non-trivial valuation of $K(\ell)$ such that value group vK is not ℓ-divisible and the residue field Kv has characteristic $\neq \ell$. Let $V_v \subseteq T_v \subseteq Z_v$ be respectively the ramification, the inertia, and the decomposition groups of v in $G_K(\ell)$. Then by the Hilbert decomposition theory for valuations one has, see e.g., [BOU]: $V_v = \{1\}$, as $\text{char}(Kv) \neq \ell$. Thus $T_v = Z_v/V_v$ is an Abelian pro-ℓ group. Further, $K(\ell)v = (Kv)(\ell)$, and one has the canonical exact sequence:

$$1 \to T_v \to Z_v \to G_{Kv}(\ell) \to 1.$$

Finally, $vK(\ell)$ is the ℓ-divisible hull of vK. And denoting by $\widehat{vK}$ the ℓ-adic completion of vK, there is an isomorphism of G_{Kv}-modules $T_v \cong \text{Hom}(\widehat{vK}, \mathbb{T}_\ell)$ where $\mathbb{T}_\ell = \varprojlim_m \mu_{\ell^m}$ is the Tate module of Kv. This reduces the problem of describing Z_v to that of describing $Kv(\ell)$. But the essential observation here is that T_v is a non-trivial Abelian normal subgroup of Z_v.

The following result is based on work by Ware [W] if $\ell = 2$, and Koenigsmann [Ko1] if $\ell \neq 2$, see also Efrat [Ef1], [Ef2]. It is the best possible converse to the above description of Z_v.

Theorem (Engler–Koenigsmann [E–K]) *In the above notations let Z be a closed non-procyclic subgroup of $G_K(\ell)$ having a non-trivial Abelian normal subgroup T. Then there exists a valuation w of $K(\ell)$ with the following properties:*

(1) $Z \subseteq Z_w$ and $T \subseteq T_w$.
(2) *The residue field Kw has* $\mathrm{char}(Kw) \neq \ell$.

The proof of the theorem above is based an a fine analysis of the multiplicative structure of fields with *very small pro-ℓ Galois group*. We will say that K has a very small pro-ℓ Galois group if $K(\ell)$ is non pro-cyclic but fits into an exact sequence of the form $0 \to \mathbb{Z}_\ell \to K(\ell) \to \mathbb{Z}_\ell \to 0$. In such a case one simply can write down the valuation ring of a valuation w on K satisfying the properties (i), (ii) above, see loc. cit. The rest is just valuation theory techniques.

The above assertion concerning fields with very small pro-ℓ Galois group is somehow parallel to the Bogomolov's theory of liftable commuting pairs mentioned above, first mentioned in Bogomolov [Bo], and finally proved in [B–T1].

Theorem (Bogomolov–Tschinkel [B–T1]) *Suppose that K contains an algebraically closed subfield. Let $\Gamma \cong \mathbb{Z}_\ell \times \mathbb{Z}_\ell$ be a closed subgroup of Π_K having an abelian preimage under $\Pi_K^c \to \Pi_K$. Then there exists a non-trivial valuation w of K such that denoting by $T_w \subseteq Z_w \subseteq \Pi_K$ the inertia/decomposition group of w in Π_K, the following hold:*

(1) $\Gamma \subseteq Z_w$ and $\Gamma \cap T_w \neq 1$.
(2) *The residue field Kw has* $\mathrm{char}(Kw) \neq \ell$.

The proof relies on a very ingenious idea of Bogomolov to compare maps between affine geometries and projective geometries. The two kind of geometries arise as follows. First, by Kummer theory one has a canonical map

$$K^\times \xrightarrow{\jmath} \mathrm{Hom}_{\mathrm{cont}}(\Pi_K, \mathbb{Z}_\ell),$$

which is trivial on $k^\times$, as k is algebraically closed. This allows us to interpret $\jmath$ as a map from the projectivization $K^\times/k^\times$ of the k-vector space $(K, +)$ to the "affine" space on the right, which is $\mathrm{Hom}_{\mathrm{cont}}(\Pi_K, \mathbb{Z}_\ell)$, or even $\mathrm{Hom}_{\mathrm{cont}}(\Pi_K, \mathbb{F}_\ell)$. And in particular, if Π_K is very small, then on the right we do really have an affine geometry. Finally, since such maps between projective and affine geometries are of very special shape, Bogomolov–Tschinkel show that a liftable commuting subgroup $\Gamma \cong \mathbb{Z}_\ell \times \mathbb{Z}_\ell$ of Π_K must contain an element σ which – by duality – defines a *flag function* on $K^\times$. Strictly speaking, this means that σ is an inertia element to a valuation w with the claimed properties.

It is interesting to remark that, as a by-product of the theory of very small pro-ℓ Galois groups, one obtains a p-adic analog of the Artin–Schreier theorem for the Galois characterization of the real closed fields. The result is as follows.

Theorem (See Pop [P8], Koenigsmann [Ko1], Efrat [Ef1])   *Let K be a field having G_K isomorphic to some open subgroup of $G_{\mathbb{Q}_p}$. Then K is p-adically closed, i.e., K is Henselian with respect to a valuation v having divisible value group, and residue field Kv contained and relatively algebraically closed in some finite extension $k|\mathbb{Q}_p$ of $\mathbb{Q}_p$.*

The proof of this assertion has two main parts: In a first approximation, one uses the techniques mentioned above as developed by Ware, Koeningsmann, Efrat in order to show that the fixed field of every Sylow ℓ-group of G_K carries a Henselian valuation. Using *valuation* theoretical and *Galois* theoretical techniques, one proves that actually K itself carries a non-trivial Henselian valuation v having residual characterisitc char$(Kv) > 0$. One concludes by applying an *arithmetical* type result by Pop [P8], theorem E9, and showing that the valuation is actually a p-adic valuation.

B. Variation of fundamental groups in families of curves

As we saw at the beginning of Part III, one might/should expect strong anabelian phenomena for curves (maybe even more general varieties) over algebraically closed fields of positive characteristic. Maybe a good hint in that direction is the fact that *we do not have a description of the algebraic fundamental group of any potentially hyperbolic curve.* Indeed, if we can recover X from its fundamental group in a functorial way, then the fundamental group of X must encode "moduli" of X, thus information of a completely different nature than crude profinite group theory. See Tamagawa [T3], [T4] for more about this conjectural world.

Now let me explain the best result we have so far for the variation of the geometric fundamental groups in families of complete curves. Recall that, for a complete smooth connected curve X of genus $g \geq 2$ over a field of characteristic 0 one has $\pi_1(\overline{X}) \cong \widehat{\Pi}_g$, thus $\pi_1(\overline{X})$ depends on g only. As mentioned above, in positive characteristic $\pi_1(\overline{X})$ is unknown, and it *depends on the isomorphy type* of $\overline{X}$. By Grothendieck's specialization theorem, $\pi_1(\overline{X})$ is a quotient of

$\widehat{\Gamma}_g$, thus it is topologically finitely generated. In particular, $\pi_1(\overline{X})$ is completely determined by its set of their finite quotients. (Terminology: $\pi_1(\overline{X})$ is a *Pfaffian group*.)

Let $\mathcal{M}_g \to \mathbb{F}_p$ be the coarse moduli space of proper and smooth curves of genus g in characteristic p. One knows that $\mathcal{M}_g$ is a quasi-projective and geometrically irreducible variety. And if k is an algebraically closed field of characteristic p, then $\mathcal{M}_g(k)$ classifies the isomorphism classes of curves of genus g over k. For $\overline{x} \in \mathcal{M}_g(k)$ let $C_{\overline{x}} \to k$ be a curve classified by $\overline{x}$, and let $x \in \mathcal{M}_g$ be such that $\overline{x} \colon k \to \mathcal{M}_g$ factors through x. We set

$$\pi_1(x) := \pi_1(C_{\overline{x}}),$$

and remark that the structure of $\pi_1(x)$ as a profinite group depends on x only, and not on the concrete geometric point $\overline{x} \in \mathcal{M}_g(k)$ used to define it. In particular, the fundamental group functor gives rise to a map as follows:

$$\pi_1 \colon \mathcal{M}_g \to \mathcal{G}, \quad x \to \pi_1(x).$$

To finish our preparation we notice that for points $x, y \in \mathcal{M}_g$ such that x is a specialization of y, by Grothendieck's specialization theorem there exists a surjective continuous homomorphism $\mathrm{sp} \colon \pi_1(y) \to \pi_1(x)$. In particular, if η is the generic point of $\mathcal{M}_g$, then C_η is the generic curve of genus g; and every point x of $\mathcal{M}_g$ is a specialization of η. For every $x \in \mathcal{M}_g$, there is a surjective homomorphism $\mathrm{sp}_x \colon \pi_1(\eta) \to \pi_1(x)$ which is determined up to Galois-conjugacy by the choice of the local ring of x in the algebraic closure of $\kappa(\eta)$. For every $x \in \mathcal{M}_g$ we fix such a map once for all; in particular, if x is a specialization of y, there exists a specialization homomorphism $\mathrm{sp}_{y,x} \colon \pi_1(y) \to \pi_1(x)$ such that $\mathrm{sp}_{y,x} \circ \mathrm{sp}_y = \mathrm{sp}_x$.

Theorem (Raynaud [R2], Pop–Saidi [P–S], Tamagawa [T5])   *For all points $s \neq x$ in $\mathcal{M}_g$ with s closed and specialization of x, the specialization homomorphism* $\mathrm{sp}_{x,s} \colon \pi_1(x) \to \pi_1(s)$ *is not an isomorphism.*

More precisely, there exist cyclic étale covers of X_x of order prime to p, which do not have good reduction under the specialization $x \mapsto s$.

This gives an answer to a question raised by Harbater.

Corollary   *There is no non-empty open subset $U \subset \mathcal{M}_g$ such that the isomorphy type of the geometric fundamental group $\pi_1(x)$ is constant on U.*

Concerning the proof of the above theorem: in the case $g = 2$ the above theorem was proved by Raynaud, by introducing a new kind of *theta divisor* (now called the Raynaud theta divisor). Using this tool, he showed that given a projective curve $X \to k_0$ of genus 2, there exist only finitely many curves $X' \to$

k_0 with $\pi_1(X) \cong \pi_1(X')$; see Raynaud [R2]. Around the same time Pop–Saidi proposed a way of generalizing Raynaud's result to all genera, by combining the theory of Raynaud's theta divisor with the results by Hrushovski [Hr] on the geometric case of the Manin–Mumford conjecture as follows. First suppose that $g = 2$. Then for points $x_0 \neq x_1$ in $\mathcal{M}_g$ such that x_0 is a specialization of x_1, it turns out that Raynaud's result follows from: *If x_0 is a closed point, then* sp_{x_1,x_0} *is not an isomorphism.* Pop–Saidi showed in [P–S] that this is the case for arbitrary genera $g > 1$, provided x_0 has some special properties, see loc.cit. Finally, Tamagawa [T5] elaborating on the method proposed in [P–S] showed that sp_{x_1,x_0} is not an isomorphism, provided $x \neq x_0$ and x_0 is a closed point.

C. Pro-ℓ abelian-by-central birational anabelian geometry

First, I want to make precise the assertion one expects to prove under Bogo-molov's [Bo] hypotheses mentioned at the beginning of Part III and to report on some recent results/developments. Recall that for a fixed prime number ℓ, we denote by $G_K(\ell) \rightarrow \Pi_K^c \rightarrow \Pi_K$ the Galois groups of a maximal pro-ℓ Galois extension of $K(\ell)|K$, respectively of the maximal abelian-by-central, respectively abelian, sub-extensions $K^{(3)}|K$ and $K^{(2)}|K$ of $K(\ell)|K$. In the case $K = k(X)$ is the function field of a variety over k, the more precise conjecture one can/should make is the following.

Conjecture AbC *In the above notations, there exists a group theoretical recipe by which one can recover in a functorial way the isomorphy type of $K|k$ from Π_K^c, provided $\mathrm{td}(K|k) > 1$. Moreover, if $L|l$ is a further function field over an algebraically closed field l, and $\Phi: \Pi_K^c \rightarrow \Pi_L^c$ is an isomorphism, then $L^i \cong K^i$.*

A sketch of a strategy to tackle Conjecture AbC from pro-ℓ Galois information, at least in the case that k is an algebraic closure of a finite field, was presented in Pop [P3]. It has as starting point the following idea. Let $\widehat{K^\times}$ be the ℓ-adic completion of the multiplicative group $K^\times$ of $K|k$.[1] Since the cyclotomic character of K is trivial, one can identify the ℓ-adic Tate module $\mathbb{T}_{K,\ell}$ of K with $\mathbb{Z}_\ell$ (non-canonically), and let $\iota_K: \mathbb{T}_{K,\ell} \rightarrow \mathbb{Z}_\ell$ be a fixed identification. Via Kummer theory, one has isomorphisms of ℓ-adically complete groups:

$$\widehat{K^\times} = \mathrm{Hom}_{\mathrm{cont}}(\Pi_K, \mathbb{T}_{K,\ell}) \xrightarrow{\iota_K} \mathrm{Hom}_{\mathrm{cont}}(\Pi_K, \mathbb{Z}_\ell),$$

i.e., $\widehat{K^\times}$ can be recovered from Π_K, hence as well from Π_K^c via the canonical projection $\Pi_K^c \rightarrow \Pi_K$. On the other hand, since $k^\times$ is divisible, $\widehat{K^\times}$ equals the

---

[1] Recall that for an abelian group A, its ℓ-adic completion is $\widehat{A} := \varprojlim_e A/\ell^e$.

ℓ-adic completion of the free abelian group $K^\times/k^\times$. Now the idea of recovering $K|k$ is as follows.

(a) First, give a recipe to recover the image $j_K(K^\times) = K^\times/k^\times$ of the ℓ-adic completion functor $j_K\colon K^\times \to K^\times/k^\times \subset \widehat{K^\times}$ inside the "known" ℓ-adically complete group $\widehat{K^\times} = \mathrm{Hom}_{\mathrm{cont}}(\Pi_K, \mathbb{Z}_\ell)$.

(b) Second, interpreting $K^\times/k^\times =: \mathcal{P}(K)$ as the projectivization of the (infinite) dimensional k-vector space $(K, +)$, give a recipe to recover the projective lines $\mathfrak{l}_{x,y} := (kx + ky)^\times/k^\times$ inside $\mathcal{P}(K)$, where $x, y \in K$ are k-linearly independent.

(c) Third, apply the *Fundamental Theorem of Projective Geometries*, see e.g., Artin [Ar], and deduce that $K|k$ can be recovered from $\mathcal{P}(K)$ endowed with all the lines $\mathfrak{l}_{x,y}$.

(d) Finally, show that the recipes above are functorial, i.e., they are invariant under isomorphisms of profinite groups $\Pi_K \to \Pi_L$ which are abelianizations of isomorphisms $\Pi_K^c \to \Pi_L^c$. In particular, such isomorphisms $\Pi_K \to \Pi_L$ originate actually from geometry.

The strategy from Pop [P3] for tackling problems (a), (b), (c) and (d) above is in principle similar to the strategies (initiated by Neukirch and Uchida) for tackling Grothendieck's anabelian conjectures. It has two main parts, namely a *local theory*, in which one has to recover the prime divisors v of $K|k$ as well as the so-called generalized prime divisors of $K|k$; and a *global theory*, in which one recovers $\mathcal{P}(K) = K^\times/k^\times$ inside $\widehat{K^\times}$ and finally the so-called rational projections $\Pi_K \to \Pi_{k(t)}$ defined by "generic" functions $t \in K$, i.e., functions having the property that $k(t)$ is relatively algebraically closed in K. See the introduction to Pop [P5] for more about this. And a quite surprising new, but very basic aspect of the above strategy is the following fact (see Pop [P10], introduction, for definitions).

Inertia versus Frobenius   *The set of all the (tame) inertia elements in G_K is topologically closed in G_K. Further, the tame quasi-divisorial inertia elements are dense in the set of all the tame inertia elements.*

This is in contrast to the behavior of the set of all the *Frobenius elements* $\mathfrak{Frob}_X$ of models X of finitely generated fields $\mathcal{K}$, because by the Chebotarev density theorem, $\mathfrak{Frob}_X$ are dense in the absolute Galois group $G_{\mathcal{K}}$.

A first application of the above strategy was to prove a stronger form of the *full pro-ℓ* Conjecture AbC in the case k is an algebraic closure of a finite field, see Pop [P4]. Finally, the above strategy led to a positive answer to Conjecture AbC in the case k is an algebraic closure a finite field as follows. First,

Bogomolov–Tschinkel [B–T2] proved that in the case k is an algebraic closure of a finite field, the function fields $K|k$ of transcendence degree $\mathrm{td}(K|k) = 2$ can be recovered from Π_K^c as predicted by Conjecture AbC. The strongest assertion concerning the case k is an algebraic closure of a finite field was proved in Pop [P6] and is the following.

Theorem (See Pop [P6])   *In the above notations, the following hold:*

(1) *There exists a group theoretical recipe by which one can recover the transcendence degree* $\mathrm{td}(K|k) = \dim(X)$ *of $K|k$ and the fact that k is an algebraic closure of a finite field from Π_K^c. Further suppose that $\mathrm{td}(K|k) > 1$ and that k is an algebraic closure of a finite field. Then there exists a group theoretical recipe which recovers $K|k$ from Π_K^c in a functorial way.*

(2) *The recipes above are invariant under isomorphisms as follows. Let $L|k$ be a further function field with l algebraically closed, and $\Phi \colon \Pi_K \to \Pi_L$ be a continuous isomorphism which can be lifted to an abstract isomorphism $\Pi_K^c \to \Pi_L^c$. Then there exists $\epsilon \in \mathbb{Z}_\ell^\times$ and an isomorphism of field extensions $\phi \colon L^i|l \to K^i|k$ such that $\epsilon{\cdot}\Phi$ is induced via Kummer theory by ϕ. Moreover, ϕ is unique up to Frobenius twists and ϵ is unique up to multiplication by p-powers, where p is the characteristic.*

The theorem above is a far reaching extension of Grothendieck's birational conjecture in positive characteristic, as it implies Grothendieck's birational conjecture if $\mathrm{td}(K) > 1$; moreover, it implies the *geometrically pro-ℓ* form of Grothendieck's birational anabelian conjecture. But I should mention right away that the arithmetical pro-ℓ form of Grothendieck's birational conjecture is completely open at this moment.

D. The Ihara/Oda–Matsumoto conjecture

The Ihara question from the 1980s, which in the 1990s became a conjecture by Oda–Matsumoto, for short I/OM, is about giving a topological/combinatorial description of the absolute Galois group of the rational numbers. Let me explain the question/conjecture I/OM in detail.

Let $k_0 \subset \mathbb{C}$ be a fixed finitely generated field, e.g., a number field, and k the algebraic closure of k_0 inside $\mathbb{C}$. For every geometrically integral k_0-variety X, let $\overline{X} := X \times_{k_0} k$ be the base change to k, and let $\overline{\pi}(X) := \pi_1(\overline{X}, *)$ denote the algebraic fundamental group of X, where $*$ is a geometric point of X (which we will not mention any more in order to simplify notation) defined by some fixed algebraic closure $\overline{k}|k$, thus defining the absolute Galois group G_{k_0} of k_0 as well. Via the exact sequence of étale fundamental groups

$$1 \to \bar{\pi}(X) \to \pi_1(X) \to G_{k_0} \to 1,$$

one gets a representation $\rho_X \colon G_{k_0} \to \mathrm{Out}(\bar{\pi}(X))$. By the functoriality of the étale fundamental group functor, the collection of all the representations $(\rho_X)_X$, is compatible with k_0-morphisms of geometrically connected k_0-varieties.

In particular, for every small category $\mathcal{V}$ of geometrically integral varieties over k_0, the `algebraic fundamental group functor` $\bar{\pi} \colon \mathcal{V} \to \mathcal{G}$, $X \mapsto \bar{\pi}(X)$, of $\mathcal{V}$ gives rise to a representation

$$\iota_{\mathcal{V}} \colon G_{k_0} \to \mathrm{Aut}(\bar{\pi}_{\mathcal{V}}),$$

where $\mathrm{Aut}(\bar{\pi}_{\mathcal{V}})$ is the automorphism group of $\bar{\pi}_{\mathcal{V}}$. In down to earth terms, the elements $\sigma \in \mathrm{Aut}(\bar{\pi}_{\mathcal{V}})$ are the families $\sigma = (\sigma_X)_{X \in \mathcal{V}}$, $\sigma_X \in \mathrm{Out}(\bar{\pi}_1(X))$, which are compatible with $\bar{\pi}_1(f) \colon \bar{\pi}_1(X) \to \bar{\pi}_1(Y)$ for all morphisms $f \colon X \to Y$ in $\mathcal{V}$.

Recall that $\bar{\pi}(X)$ is nothing but the profinite completion of the topological fundamental group $\pi_1^{\mathrm{top}}(X(\mathbb{C}), *)$ of the "good" topological space $X(\mathbb{C})$, thus $\bar{\pi}(X)$ is an object of topological/combinatorial nature. Following Grothendieck, one should give a new description of G_{k_0} by finding categories $\mathcal{V}$ for which $\mathrm{Aut}(\bar{\pi}_{\mathcal{V}})$ has a "nice" description, and $\iota_{\mathcal{V}}$ is an isomorphism. If so, then via the isomorphism $\iota_{\mathcal{V}}$, we would have a new non-tautological description of G_{k_0}. For $k_0 = \mathbb{Q}$, Grothendieck suggested to study $\bar{\pi}_{\mathcal{V}}$, for $\mathcal{V}$ sub-categories of the *Teichmüller modular tower* $\mathcal{T}$ of all the moduli spaces $M_{g,n}$. For instance, if $\mathcal{V}_0 = \{M_{04}, M_{05}\}$ is endowed with "connecting homomorphisms," $\mathrm{Aut}(\bar{\pi}_{\mathcal{V}_0})$ is the famous *Grothendieck–Teichmüller group* $\widehat{GT}$, which was intensively studied first by Drinfel'd [Dr], Ihara [I1], [I2], [I3], Deligne [De], and lately by several others, e.g. Hain–Matsumoto [H–M], Harbater–Schneps [H–Sch], Ihara–Matsumoto [I–M], Lochak–Schneps [L–Sch], Nakamura–Schneps [N–Sch], and many others.

We know little about the nature of $\iota_{\mathcal{V}}$ in general, and the situation is quite mysterious. Concerning injectivity, Drinfel'd remarked that using Belyi's theorem [Be] it follows that $\iota_{\mathcal{V}}$ is injective, provided $C := \mathbb{P}^1 \backslash \{0, 1, \infty\}$ lies in $\mathcal{V}$. Further, Voevodsky showed that the same is true if $C \in \mathcal{V}$ with X some affine open of a curve of genus one. It is conjectured though that $\iota_{\mathcal{V}}$ should be injective as soon as $\mathcal{V}$ contains at least one hyperbolic curve (and it seems that Hoshi–Mochizuki have announced a proof of this fact), but so far the following is the best result one has in this direction.

Theorem (See Matsumoto [Ma]) *Suppose that $\mathcal{V}$ contains some affine hyperbolic curve C. Then the representation $\iota_{\mathcal{V}} \colon G_{k_0} \to \mathrm{Aut}(\bar{\pi}_{\mathcal{V}})$ is injective.*

The question about the surjectivity of the representation ι_{r_V} is of a completely different nature and less understood, and Ihara asked in the 1980s whether ι_{r_V} is onto, thus an isomorphism, provided $k_0 = \mathbb{Q}$ and $\mathcal{V}$ is the category of all the k_0-varieties. Finally, based on some "motivic evidence," Oda–Matsumoto conjectured in the 1990s that the answer to Ihara's question should be positive. The author showed (unpublished), that the answer to I/OM is positive – for more general fields, thus giving a positive answer to I/OM. See also André [An], where the author defines `tempered fundamental groups`, introduces a p-adic tempered variant $\widehat{GT}_p$ of $\widehat{GT}$, and (re)proves the I/OM over p-adic fields.

We now formulate the **pro-ℓ abelian-by-central I/OM**, which implies the usual I/OM. Let $\overline{\pi}(X) \longrightarrow \Pi_X^c \longrightarrow \Pi_X$ be the pro-ℓ abelian-by central, respectively pro-ℓ abelian, quotients of $\overline{\pi}$. Since the kernels of the above group homomorphisms are characteristic subgroups of $\overline{\pi}(X)$ and Π_X^c, respectively, the above homomorphisms give rise to homomorphisms $\mathrm{Out}(\overline{\pi}(X)) \to \mathrm{Out}(\Pi_X^c) \to \mathrm{Out}(\Pi_X)$. Further, $\mathbb{Z}_\ell^\times$ acts by multiplication on Π_X and this multiplication can be lifted to an action of $\mathbb{Z}_\ell^\times$ on Π_X^c. For every category $\mathcal{V}$ as above, one has a corresponding morphism $\overline{\pi}_{\mathcal{V}} \to \Pi_{\mathcal{V}}^c \to \Pi_{\mathcal{V}}$ of functors, which by the discussion above gives rise to a group homomorphism $\mathrm{Aut}(\overline{\pi}_{\mathcal{V}}) \to \mathrm{Aut}(\Pi_{\mathcal{V}}^c) \to \mathrm{Aut}(\Pi_{\mathcal{V}})$. Let $\mathrm{Aut}^c(\Pi_{\mathcal{V}})$ be the image of $\mathrm{Aut}(\Pi_{\mathcal{V}}^c)$ in $\mathrm{Aut}(\Pi_{\mathcal{V}})$ modulo the action of $\mathbb{Z}_\ell^\times$. In other words, the elements of $\mathrm{Aut}^c(\Pi_{\mathcal{V}})$ are exactly the $\mathbb{Z}_\ell^\times$ equivalence classes of automorphisms of $\Pi_{\mathcal{V}}$ which can be lifted to automorphisms of $\Pi_{\mathcal{V}}^c$. As above, we get a group homomorphism

$$\iota_{\mathcal{V}}^c \colon G_{k_0} \to \mathrm{Aut}^c(\Pi_{\mathcal{V}}).$$

Pro-ℓ abelian-by-central I/OM   *The homomorphism $\iota_{\mathcal{V}}^c$ is an isomorphism in the case $k_0 = \mathbb{Q}$ and $\mathcal{V}$ is the category of all the geometrically integral $\mathbb{Q}$-varieties.*

We will actually prove a stronger/more precise assertion than the above pro-ℓ abelian-by-central I/OM which evolves as follows. Let $U_0 := \mathbb{P}^1 \setminus \{0, 1, \infty\} \times k_0$ be the `tripod` over k_0 (terminology by Mochizuki–Hoshi). For every geometrically integral k-variety X, and a basis $\{U_i\}_i$ of Zariski (affine) open neighborhoods of the generic point η_X of X, we consider the category $\mathcal{V}_X = \{U_i\}_i \cup \{U_0\}$ with morphisms the inclusions $U_{i''} \hookrightarrow U_{i'}$ and the dominant k-morphisms $\varphi_i \colon U_i \to U_0$. In particular, $\mathrm{Aut}(\mathcal{V}_X) = 1$ is trivial, and thus the autmorphism group of $\Pi_{\mathcal{V}_X}$ are the systems $(\sigma_i)_i$ with $\sigma_i \in \mathrm{Aut}^c(\Pi_{U_i})$ and $\sigma_0 \in \mathrm{Aut}^c(\Pi_{U_0})$

such that the following diagrams are commutative, when defined:

$$
\begin{array}{ccc}
\Pi_{U_j} & \xrightarrow{\sigma_j} & \Pi_{U_j} \\
\downarrow & & \downarrow \\
\Pi_{U_i} & \xrightarrow{\sigma_i} & \Pi_{U_i}
\end{array}
\qquad\qquad
\begin{array}{ccc}
\Pi_{U_i} & \xrightarrow{\sigma_i} & \Pi_{U_i} \\
\downarrow{\scriptstyle p_i} & & \downarrow{\scriptstyle p_i} \\
\Pi_{U_0} & \xrightarrow{\sigma_0} & \Pi_{U_0}
\end{array}
$$

Note that every k_0 embedding $k_0(t) \hookrightarrow k_0(X)$ originates from some dominant k_0 morphism $\varphi_j : U_i \to U_0$ for U_j sufficiently small. Therefore, every $(\sigma_i)_i$ gives rise to an automorphism $\sigma \in \mathrm{Aut}^c(\Pi_K)$ which is compatible with the projections $p_\iota : \Pi_K \to \Pi_{U_0}$ defined by all embeddings $\iota : k_0(t) \hookrightarrow k_0(X)$, i.e., one has:

$$
p_\iota \circ \sigma = \sigma_0 \circ p_\iota.
$$

This suggests that a possible way to prove the pro-ℓ abelian-by-central I/OM is to prove its *birational form* for each category $\mathcal{V}_X$ with $\dim(X)$ sufficiently large, which is the stronger assertion that every $\sigma \in \mathrm{Aut}^c(\Pi_K)$ which is compatible with all the projections p_ι originates from G_{k_0}.

Theorem (See Pop [P7]) *In the above notation, the following holds. Let k_0 be arbitrary and $\dim(X) > 1$. Then the canonical representation*

$$
\iota^c_{\mathcal{V}_X} : G_{k_0} \to \mathrm{Aut}^c(\Pi_{\mathcal{V}_X})
$$

is an isomorphism. In particular, the pro-ℓ abelian-by-central I/OM holds for the category $\mathcal{V}_X$, thus the classical I/OM holds too.

Some major open questions/problems

Q1 Let k be an algebraically closed field of positive characteristic. Can one recover $\mathrm{td}(k)$ from $\pi_1(\mathbb{A}^1_k)$?

Q2 For k as above, give a non tautological description of of $\pi_1(\mathbb{A}^1_k)$.

Q3 For k as above, give a non tautological description of $\pi_1(X)$ and/or $\pi_1^t(X)$ for some hyperbolic curve $X \to k$.

Q4 Give an algebraic proof of the fact that $\pi_1(\mathbb{P}^1_{\mathbb{C}} \backslash \{0, 1, \infty\})$ is generated by inertia elements c_0, c_1, c_∞ over $0, 1, \infty$ with a single relation $c_0 c_1 c_\infty = 1$.

Q5 Give a proof of the Hom-form of Grothendieck's birational anabelian conjecture in positive characteristic.

Q6 Prove the geometrically pro-ℓ Isom-form and/or the Hom-form of the anabelian conjecture for curves in positive characteristic.

Q7 Let $x_1 \neq x_0$ in $\mathcal{M}_g \to \mathbb{F}_p$ be such that x_0 is a specialization of x_1. Show that $\mathrm{sp}_{x_1 x_0} : \pi_1(x_1) \to \pi_1(x_0)$ is not an isomorphism.

Q8 Prove the section conjecture, say over number fields and/or p-adic fields.
Q9 Prove the AbC Conjecture over arbitrary algebraically closed base fields k.

Bibliography

[An] André, Y., *On a geometric description of* $\mathrm{Gal}(\overline{\mathbb{Q}}_p|\mathbb{Q})$ *and a p-adic avatar of* $\widehat{GT}$, Duke Math. J. **119** (2003), 1–39.

[Ar] Artin, E., Geometric Algebra, Interscience Publishers, New York 1957.

[Be] Belyi, G. V., *On Galois extensions of a maximal cyclotomic field,* Mathematics USSR Izvestija, Vol. **14** (1980), no. 2, 247–256. (Original in Russian: Izvestiya Akademii Nauk SSSR, vol. **14** (1979), no. 2, 269–276.)

[Bo] Bogomolov, F. A., *On two conjectures in birational algebraic geometry,* in Algebraic geometry and analytic geometry, ICM-90 Satellite Conference Proceedings, ed. A. Fujiki et al., Springer Verlag, Tokyo 1991.

[B–T1] Bogomolov, F. A., and Tschinkel, Y. *Commuting elements in Galois groups of function fields,* in: Motives, Polylogs and Hodge Theory, International Press 2002, 75–120.

[B–T2] ——, *Reconstruction of function fields,* Geometric and Functional Analysis **18** (2008), 400–462.

[BOU] Bourbaki, *Algèbre commutative,* Hermann, Paris 1964.

[C–P] Corry, S. and Pop, F., *Pro-p hom-form of the birational anabelian conjecture over sub-p-adic fields,* Journal reine angew. Math. **628** (2009), 121-128.

[De] Deligne, P., *Le groupe fondamental de la droite projective moins trois points,* in: Galois groups over **Q**, Math. Sci. Res. Inst. Publ. **16**, 79–297, Springer 1989.

[Dr] Drinfeld, V. G., *On quasi-triangular quasi-Hopf algebras and on a group that is closely connected with* $\mathrm{Gal}(\overline{\mathbb{Q}}/\mathbb{Q})$ (Russian), Algebra i Analiz **2**, no. 4 (1990), 149–181; translation in: Leningrad Math. J. **2**, no. 4 (1991), 829–860.

[Ef1] Efrat, I,, *Construction of valuations from K-theory,* Mathematical Research Letters 6 (1999), 335–344.

[Ef2] ——, *Recovering higher global and local fields from Galois groups— an algebraic approach,* Invitation to higher local fields (Münster, 1999), 273–279 (electronic), Geom., Topol. Monogr. **3**, Geom. Topol. Publ., Coventry, 2000

[E–K] Engler, A. J. and Koenigsmann, J., *Abelian subgroups of pro-p Galois groups,* Trans. AMS **350** (1998), no. 6, 2473–2485.

[E–W] Esnault, H. and Wittenberg, O., *On abelian birational sections,* J. AMS, **23** (2010), 713–724.

[F1] Faltings, G., *p-adic Hodge theory,* Journal AMS **1** (1988), 255-299.

[F2] ——, *Hodge–Tate structures and modular forms,* Math. Annalen **278** (1987), 133–149.

[F3] ——, *Curves and their fundamental groups [following Grothendieck, Tamagawa, Mochizuki],* Séminaire Bourbaki, Vol 1997-98, Exposé 840, Mars 1998.

[F–J] Fried, M. D. and Jarden, M., *Field arithmetic,* in: Ergebnisse der Mathematik und ihre Grenzgebiete, 3. Folge, Vol 11, Springer Verlag 2004.

[G1] Grothendieck, A., *Letter to Faltings,* June 1983. See [GGA].

[G2] ——, *Esquisse d'un programme,* 1984. See [GGA].

[GGA] *Geometric Galois Actions I,* LMS LNS Vol. **242**, eds. L. Schneps and P. Lochak, Cambridge University Press 1998.

[Hn] Hain, R., *Rational points of universal curves,* preprint, January 2010; see `arXiv:mathNT/1001.5008v1`.

[H–M] Hain, R. and Matsumoto, M., *Tannakian fundamental groups associated to Galois groups,* in: Galois Groups and Fundamental Groups, ed. L. Schneps, MSRI Pub. Series **41**, 2003, pp.183–216.

[H–Sz] Harari, D. and Szamuely, T., *Galois sections for abelianized fundamental groups,* and *Appendix* by E. V. Flynn, Math. Annalen **344** (2009), 779–800.

[Ha1] Harbater, D., *Abhyankar's conjecture on Galois groups over curves,* Invent. Math. **117** (1994), 1–25.

[Ha2] ——, *Fundamental groups of curves in characteristic p,* in: Proceedings of the ICM (Zürich, 1994), Birkhauser, Basel, 1995, 656–666.

[H–Sch] Harbater, D. and Schneps, L., *Fundamental groups of moduli and the Grothendieck–Teichmüller group,* Trans. AMS, Vol. **352** (2000), 3117-3148.

[Ho] Hoshi, Y., *Existence of non-geometric pro-p Galois sections of hyperbolic curves,* preprint, January 2010, RIMS preprint # 1689.

[Hr] Hrushovski, E., *The Mordell–Lang conjecture for function fields,* Journal AMS **9** (1996), 667–690.

[I1] Ihara, Y., *On Galois represent. arising from towers of covers of* $\mathbb{P}^1 \setminus \{0, 1, \infty\}$, Invent. Math. **86** (1986), 427–459.

[I2] ——, *Braids, Galois groups, and some arithmetic functions,* Proceedings of the ICM'90, Vol. I, II, Math. Soc. Japan, Tokyo, 1991, pp.99–120.

[I3] ——, *On beta and gamma functions associated with the Grothendieck–Teichmüller group* II, J. reine angew. Math. **527** (2000), 1–11.

[I–M] Ihara, Y. and Matsumoto, M., *On Galois actions on profinite completions of braid groups,* in: Recent developments in the inverse Galois problem (Seattle, WA, 1993), 173–200, Contemp. Math. **186** AMS Providence, RI, 1995.

[I–N] Ihara, Y., and Nakamura, H., *Some illustrative examples for anabelian geometry in high dimensions,* in: Geometric Galois Actions I, 127–138, LMS **242**, Cambridge University Press, Cambridge, 1997.

[Ik] Ikeda, M., *Completeness of the absolute Galois group of the rational number field,* J. reine angew. Math. **291** (1977), 1–22.

[J–W] Jannsen, U., and Wingberg, K., *Die Struktur der absoluten Galois-gruppe p-adischer Zahlkörper,* Invent. Math. **70** (1982/83), no. 1, 71–98.

[J] de Jong, A. J., *Families of curves and alterations,* Annales de l'institute Fourier, **47** (1997), pp. 599–621.

[Ka] Kato, K., *Logarithmic structures of Fontaine–Illusie,* Proceedings of the First JAMI Conference, Johns Hopkins Univ. Press (1990), 191–224.

[K–L] Katz, N., and Lang, S. *Finiteness theorems in geometric class field theory,* with an Appendix by K. Ribet, Enseign. Math. **27** (1981), 285–319.

[K–P–R] Kuhlmann, F.-V., Pank, M., and Roquette, P., *Immediate and purely wild extensions of valued fields,* Manuscripta Math. **55** (1986), 39–67.

[K1] Kim, M., *The motivic fundamental group of* $\mathbb{P}\setminus\{0, 1, \infty\}$ *and the theorem of Siegel,* Inventiones Math. **161** (2005), 629–656.

[K2] ——, *The unipotent Albanese map and Selmer varieties for curves,* Publ. Res. Inst. Math. Sci. **45** (2009), 89–133.

[K3] ——, *Massey products for elliptic curves of rank one,* J. AMS **23** (2010), 725–747.

[K4] ——, *A remark on fundamental groups and effective Diophantine methods for hyperbolic curves,* in: Number theory, analysis and geometry – In memory of Serge Lang, eds: Goldfeld, Jorgenson, Jones, Ramakrishnan, Ribet, Tate; Springer-Verlag 2010.

[Ko1] Koenigsmann J., *From p-rigid elements to valuations (with a Galois-characterization of p-adic fields). With an appendix by Florian Pop,* J. Reine Angew. Math. **465** (1995), 165–182.

[Ko2] ——, *Solvable absolute Galois groups are metabelian,* Invent. Math. **144** (2001), no. 1, 1–22.

[Ko3] ——, *On the 'section conjecture' in anabelian geometry,* J. reine angew.
       Math. **588** (2005), 221–235.

 [Ko] Koch, H., Die Galoissche Theorie der *p*-Erweiterungen, Math. Mono-
       graphien **10**, Berlin 1970.

 [La] Lang, S., Algebra, Springer-Verlag 2001.

[L–T] Lang, S. and Tate, J., Principal homogeneous spaces over Abelian va-
       rieties, Am. J. Math. **80** (1958), 659–684.

[LS1] The Grothendieck Theory of Dessins d'Enfants, ed. Leila Schneps,
       LMS LNS **200**, Cambridge Univ Press, 1994.

[LS2] Around Grothendieck's Esquisse d'un Programme, eds. Schneps &
       Lochak, LMS LNS **242**, Cambridge University Press, 1997.

[L–Sch] Lochak, P. and Schneps, L., *A cohomological interpretation of the
       Grothendieck-Teichmüller group. Appendix by C. Scheiderer,* Invent.
       Math. **127** (1997), 571–600.

 [Ma] Matsumoto, M., *Galois representations on profinite braid groups on
       curves,* J. reine angew. Math. **474** (1996), 169–219.

[Mzk1] Mochizuki, Sh., *The profinite Grothendieck conjecture for closed hy-
       perbolic curves over number fields,* J. Math. Sci. Univ Tokyo **3** (1966),
       571–627.

[Mzk2] ——, *A version of the Grothendieck conjecture for p-adic local fields,*
       Internat. J. Math. no. 4, **8** (1997), 499–506.

[Mzk3] ——, *The local pro-p Grothendieck conjecture for hyperbolic curves,*
       Invent. Math. **138** (1999), 319-423.

[Mzk4] ——, *The absolute anabelian geometry of hyperbolic curves,* Galois
       theory and modular forms, 77–122, Dev. Math. **11**, Kluwer Acad.
       Publ., Boston, MA, 2004.

[Mzk5] ——, *Absolute anabelian cuspidalizations of proper hyperbolic curves.,*
       J. Math. Kyoto Univ. **47** (2007), 451–539.

[Mzk6] ——, *Topics in absolute anabelian geometry II: Decomposition groups
       and endomorphisms,* RIMS Preprint **1625**, March 2008.

  [N] Nagata, M., *A theorem on valuation rings and its applications,* Nagoya
       Math. J. **29** (1967), 85–91.

[Na1] Nakamura, H., *Galois rigidity of the étale fundamental groups of punc-
       tured projective lines,* J. reine angew. Math. **411** (1990) 205–216.

[Na2] ——, *Galois rigidity of algebraic mappings into some hyperbolic va-
       rieties,* Int. J. Math. **4** (1993), 421–438.

[N–Sch] Nakamura, H. and Schneps, L., *On a subgroup of the Grothendieck–
       Teichmüller group acting on the profinite Teichmüller modular group,*
       Invent. Math. **141** (2000), 503–560.

[N1] Neukirch, J., *Über eine algebraische Kennzeichnung der Henselkörper,* J. reine angew. Math. **231** (1968), 75–81.

[N2] ——, *Kennzeichnung der p-adischen und endlichen algebraischen Zahlkörper,* Inventiones math. **6** (1969), 269–314.

[N3] ——, *Kennzeichnung der endlich-algebraischen Zahlkörper durch die Galoisgruppe der maximal auflösbaren, Erweiterungen,* J. für Math. **238** (1969), 135–147.

 [O] Oda, T., *A note on ramification of the Galois representation of the fundamental group of an algebraic curve I,* J. Number Theory (1990) 225–228.

[Pa] Parshin, A. N., *Finiteness Theorems and Hyperbolic Manifolds,* in: The Grothendieck Festschrift III, ed P. Cartier et al., PM Series vol. 88, Birkhäuser, Boston Basel Berlin 1990.

[Po] Pop, F., *Étale Galois covers of affine, smooth curves,* Invent. Math. **120** (1995), 555–578.

[P1] ——, *On Grothendieck's conjecture of birational anabelian geometry,* Ann. of Math. **139** (1994), 145–182.

[P2] ——, *On Grothendieck's conjecture of birational anabelian geometry* II, preprint, 1995

[P3] ——, *MSRI talk notes,* fall 1999. See http://www.msri.org/publications/ln/msri/1999/gactions/pop/1/index.html,

[P4] ——, *Pro-ℓ birational anabelian geometry over alg. closed fields* I, manuscript, Bonn 2003. See http://arxiv.org/PS_cache/math/pdf/0307/0307076

[P5] ——, *Recovering fields from their decomposition graphs,* in: Number theory, analysis and geometry – In memory of Serge Lang, Springer special volume 2010; eds: Goldfeld, Jorgenson, Jones, Ramakrishnan, Ribet, Tate.

[P6] ——, *On the birational anabelian program initiated by Bogomolov I,* (to appear).

[P7] ——, *On I/OM,* Manuscript 2010.

[P8] ——, *Galoissche Kennzeichnung p-adisch abgeschlossener Körper,* dissertation, Heidelberg 1986.

[P9] ——, *On the birational p-adic section conjecture,* Compositio Math. **146** (2010), 621–637.

[P10] ——, *Inertia elements versus Frobenius elements,* Math. Annalen **348** (2010), 1005–1017.

[P–S] Pop, F., and Saidi, M. *On the specialization homomorphism of fundamental groups of curves in positive characteristic,* in: Galois groups and fundamental groups (ed. L. Schneps), MSRI Pub. Series **41**, 2003, pp.107–118.

[P–St] Pop, F., and Stix, J., *Arithmetic in the fundamental group of a p-adic curve,* manuscript, Cambridge–Heidelberg 2009.

[R1] Raynaud, M., *Revêtements des courbes en caractéristique p > 0 et ordinarité,* Compositio Math. **123** (2000), no. 1, 73–88.

[R2] ——, *Sur le groupe fondamental d'une courbe complète en caractéristique p > 0,* in: Arithmetic fundamental groups and noncommutative algebra (Berkeley, CA, 1999), 335–351, Proc. Sympos. Pure Math., 70, AMS, Providence, RI, 2002.

[Ro] Roquette, P., *Some tendencies in contemporary algebra,* in: Perspectives in Math. Anniversary of Oberwolfach 1984, Basel 1984, 393–422.

[Sa1] Saidi, M., *Revêtements modérés et groupe fondamental de graphes de groupes,* Compositio Math. **107** (1997), 319–338.

[Sa2] ——, *Good sections of arithmetic fundamental groups,* manuscript 2009.

[Sa3] ——, *Around the Grothendieck anabelian section conjecture.* See `arXiv:1010.1314v2[math.AG]`

[S–T] Saidi, M. and Tamagawa, A, *A prime-to-p version of the Grothendieck anabelian conjecture for hyperbolic curves over finite fields of characteristic p > 0,* Publ. Res. Inst. Math. Sci. **45** (2009), 135–186.

[S1] Serre, J.-P., Cohomologie Galoisienne, 5th ed., LNM 5, Springer Verlag, Berlin, 1994

[S2] ——, *Zeta and L functions,* in: Arithmetical algebraic geometry (Proc. conf. Purdue Univ., 1963), pp.82–92; Harper & Row, New York 1965.

[S3] ——, *Corps locaux,* Hermann, Paris 1962.

[Sp] Spiess, M., *An arithmetic proof of Pop's theorem concerning Galois groups of function fields over number fields,* J. reine angew. Math. **478** (1996), 107–126.

[St1] Stix, J., *Affine anabelian curves in positive characteristic,* Compositio Math. **134** (2002), no. 1, 75–85

[St2] ——, *Projective anabelian curves in positive characteristic and descent theory for log-étale covers,* dissertation, Bonner Mathematische Schriften **354**, Universität Bonn, Math. Inst. Bonn, 2002; see `www.mathi.uni-heidelberg.de/\~{}stix`

[St3] ——, *On the period-index problem in light of the section conjecture,* American J. Math. **132** (2010), 157–180.

[St4] ——, *The Brauer–Manin obstruction for sections of the fundamental group,* preprint, Cambridge–Heidelberg, Oct. 2009; see arXiv: mathAG/0910.5009v1.

[St5] ——, *A general Seifert–Van Kampen theorem for algebraic fundamental groups,* Publications of RIMS **42** (2006), 763–786.

[Sz] Szamuely, T., *Groupes de Galois de corpes de type finit [d'après Pop],* Astérisque **294** (2004), 403–431.

[T1] Tamagawa, A., *The Grothendieck conjecture for affine curves,* Compositio Math. **109** (1997), 135-194.

[T2] ——, *On the fundamental groups of curves over algebraically closed fields of characteristic* > 0, Internat. Math. Res. Notices 1999, no. 16, 853–873.

[T3] ——, *On the tame fundamental groups of curves over algebraically closed fields of characteristic* > 0, in: Galois groups and fundamental groups, 47–105, MSRI Publ., **41**, Cambridge Univ. Press, Cambridge, 2003.

[T4] ——, *Fundamental groups and geometry of curves in positive characteristic,* in: Arithmetic fundamental groups and noncommutative algebra (Berkeley, CA, 1999), 297–333, Proc. Sympos. Pure Math., 70, AMS, Providence, RI, 2002.

[T5] ——, *Finiteness of isomorphism classes of curves in positive characteristic with prescribed fundamental groups,* J. Algebraic Geom. **13** (2004), no. 4, 675–724.

[T6] ——, *Resolution of non-singularities of families of curves,* Publ. Res. Inst. Math. Sci. **40** (2004), 1291–1336.

[U1] Uchida, K., *Isomorphisms of Galois groups of algebraic function fields,* Ann. of Math. **106** (1977), 589–598.

[U2] ——, *Isomorphisms of Galois groups of solvably closed Galois extensions,* Tôhoku Math. J. **31** (1979), 359–362.

[U3] ——, *Homomorphisms of Galois groups of solvably closed Galois extensions,* Journal Math. Soc. Japan **33**, No.4, 1981.

[Ta1] Tate, J., *Relations between K_2 and Galois cohomology,* Invent. Math. **36** (1976), 257–274.

[Ta2] ——, *p-divisible groups,* Proceeding of a conference on local fields, Driebergen, Springer Verlag 1969, 158–183.

[W] Ware, R., *Valuation rings and rigid elements in fields,* Can. J. Math. **33** (1981), 1338–1355.

On Galois rigidity of fundamental groups of algebraic curves

Hiroaki Nakamura

Okayama University

English translation of [31] (1989)

§1. Conjecture and result Let U be an (absolutely irreducible, nonsingular) algebraic curve defined over a number field k. It is well known that the etale fundamental group $\pi_1(U)$ is naturally regarded as a group extension of the absolute Galois group $G_k := \mathrm{Gal}(\bar{k}/k)$ by a finitely generated topological group. We shall consider a question of how much the equivalence class of this group extension depends on the isomorphism class of the issued curve.

First, let us recall that the etale fundamental group of the algebraic curve U/k is a profinite topological group defined as the projective limit of finite groups as follows:

$$\pi_1(U) := \varprojlim_{Y} \mathrm{Aut}_U(Y),$$

where Y runs over the projective system of the connected finite etale Galois covers of U. If we restrict the projective system to a subsystem consisting of the covers of the form $\{Y = U \otimes K \mid K/k : \text{finite Galois extension}\}$ and note that $\mathrm{Aut}_U(U \otimes K) \cong \mathrm{Gal}(K/k)$, then we obtain a canonical surjective homomorphism

$$p_{U/k} : \pi_1(U) \longrightarrow G_k(:= \mathrm{Gal}(\bar{k}/k)).$$

We will treat $\pi_1(U)$ as an object associated with the "augmentation map" $p_{U/k}$ onto G_k. The kernel (= $\ker p_{U/k}$) is isomorphic to $\pi_1(U \otimes \bar{k})$, which in this article will often be denoted simply by π_1. In fact, we may present it by generators and relations as a topological group:

Note 1

Non-abelian Fundamental Groups and Iwasawa Theory, eds. John Coates, Minhyong Kim, Florian Pop, Mohamed Saïdi and Peter Schneider. Published by Cambridge University Press. ©Cambridge University Press 2012.

$$(\Pi) \quad \pi_1 = \pi_1(U \otimes \bar{k}) = \left\langle \begin{matrix} \alpha_1,\ldots,\alpha_g \\ \beta_1,\ldots,\beta_g \\ \gamma_1,\ldots,\gamma_n \end{matrix} \;\middle|\; \begin{matrix} \alpha_1\beta_1\alpha_1^{-1}\beta_1^{-1}\cdots\alpha_g\beta_g\alpha_g^{-1}\beta_g^{-1}\cdot \\ \cdot\gamma_1\cdots\gamma_n = 1 \end{matrix} \right\rangle_{\text{top}}.$$

Here, α_i, β_i and γ_j $(1 \le i \le g, 1 \le j \le n)$ are taken as suitable loops at a base point on the associated complex curve $U_{\mathbb{C}}^{\text{an}}$ with U, which is assumed to be a Riemann surface of genus g with n punctures. The suffix *top* designates the profinite completion. Note here that in our issued group extension

$$1 \longrightarrow \pi_1 \longrightarrow \pi_1(U) \longrightarrow G_k \longrightarrow 1 \quad \text{(exact)},$$

the kernel part π_1 is a group whose isomorphism class is determined only by the topological type (g, n), hence may be written as $\pi_1 = \Pi_{g,n}$. In other words, whenever a topological space of type (g, n) gets a structure of an algebraic curve over k, it gives rise to a group extension of $\mathrm{Gal}(\bar{k}/k)$ by $\Pi_{g,n}$.

Conjecture (Part of Grothendieck's fundamental conjecture of "anabelian" al-   <u>Note 2</u>
gebraic geometry [2])   When $\Pi_{g,n}$ is non-abelian, namely, $(g, n) \ne (0, 0), (0, 1)$, $(0, 2), (1, 0)$, the above correspondence

$$\text{algebraic curve } U/k \rightsquigarrow \text{group extension } \pi_1(U)/G_k$$

is faithful.

In this note, we would like to report the following result.

Theorem   Conjecture is true for $(g, n) = (0, n)$ $(n \ge 3)$ and $(1, 1)$.

§2. Finiteness for π_1 modulo π_1''   As a non-abelian profinite group $\pi_1 = \Pi_{g,n}$ is so large, a basic approach to the above problem would be to look at suitable quotients of it. For any class of finite groups C, set

$$J_C := \bigcap_{G \in C} \bigcap_{f \in \mathrm{Hom}(\pi_1, G)} \ker(f).$$

Noting that J_C is a characteristic subgroup of π_1 determined by the class C, we obtain an exact sequence

$$(1) \qquad\qquad 1 \longrightarrow \pi_1/J_C \longrightarrow \pi_1(U)/J_C \longrightarrow G_k \longrightarrow 1.$$

(Here, π_1/J_C is so called the maximal pro-C quotient of the topological group π_1. In a particular case when $C = \{$all l-groups$\}$ for a fixed rational prime l, π_1/J_C (written $\pi_1^{\text{pro-}l}$) is called the pro-l fundamental group, and the Galois representation $G_k \to \mathrm{Out}(\pi_1^{\mathrm{pro}\text{-}l})$ associated with the above exact sequence have been studied in depth by Y. Ihara and other authors (cf. [3])).

For two algebraic curves U/k and U'/k, if there is a commutative diagram of profinite groups

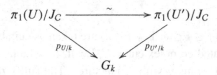

with the horizontal arrow being an isomorphism, then U and U' are called π_1-equivalent modulo J_C and written

$$\pi_1(U)/J_C \cong_{G_k} \pi_1(U')/J_C.$$

In the case C being the class of all finite groups, we find $J_C = \{1\}$. In this case, we just say that they are π_1-equivalent. It is obvious that

(2) $\pi_1(U) \cong_{G_k} \pi_1(U') \Rightarrow \pi_1(U)/J_C \cong_{G_k} \pi_1(U')/J_C.$

Let us first consider the case $C = \{$abelian groups$\}$. In this case, $J_C = \overline{[\pi_1, \pi_1]} = \pi_1'$ is the (closure of the) commutator subgroup of π_1, and $\pi_1/\pi_1' = \pi_1^{\mathrm{ab}}$ can be identified with the etale homology group $H_1(U \otimes \bar{k}, \hat{\mathbb{Z}})$. It then follows from (2) that

$$\pi_1(U) \cong_{G_k} \pi_1(U') \Longrightarrow H_1(U \otimes \bar{k}) \cong H_1(U' \otimes \bar{k}).$$

$$\langle \text{as } G_k\text{-modules} \rangle$$

<u>Note 3</u> When U has genus ≥ 1, a rough application of the finiteness theorem of Shafarevich–Faltings implies finiteness of U'/k that are π_1-equivalent over k to a fixed U/k. However, when U has genus 0, it is impossible to deduce such finiteness only from H_1. In fact, suppose Λ is a finite subset of $\mathbf{P}^1(k)$ with cardinality $n(\geq 4)$ and let $U = \mathbf{P}_k^1 - \Lambda$. Then, one should ask how the equivalence class of the group extension

$$1 \longrightarrow \underset{\substack{\wr\| \\ \hat{\mathbb{Z}}^{\oplus n-1}}}{H_1} \longrightarrow \pi_1(U)/\pi_1' \longrightarrow G_k \longrightarrow 1$$

(3)

varies according to the relative position of the point set Λ on $\mathbf{P}^1$. But one finds that (3) has no more information on the cardinality of Λ, immediately after observing that the above (3) splits and the associated Galois representation $G_k \to \mathrm{GL}_{n-1}(\hat{\mathbb{Z}})$ is a direct sum of the 1-dimensional scalar representation realized as multiplication by the cyclotomic character.

So, in the case of genus 0, let us take C to be the class of meta-abelian groups, i.e., finite solvable groups of derived length ≤ 2. Then, J_C becomes the double commutator subgroup $\pi_1'' = \overline{[\pi_1', \pi_1']}$.

Notation   For any finite subset $\Lambda = \{0, 1, \infty, \lambda_1, \ldots, \lambda_m\}$ ($\lambda_i \in k$), define the multiplicative subgroup $\Gamma(\Lambda)$ of $k^\times$ to be that generated by

distribution of points on $\mathbf{P}^1$

$$\{-1, \lambda_i, 1 - \lambda_i, \lambda_i - \lambda_j \mid 1 \leq i \neq j \leq m\}.$$

Remark   The group $\Gamma(\Lambda) \subset k^\times$ is independent of the choice of (the coordinate of $\mathbf{P}^1$ such that) $\{0, 1, \infty\} \subset \Lambda$.

Then we have the following theorem.

Theorem 1 ([6])   Let Λ, Λ' be finite subsets of $\mathbf{P}^1(k)$ containing $\{0, 1, \infty\}$, and set $U = \mathbf{P}_k^1 - \Lambda$, $U' = \mathbf{P}_k^1 - \Lambda'$. Then,

$$\pi_1(U)/\pi_1'' \cong_{G_k} \pi_1(U')/\pi_1'' \implies \Gamma(\Lambda) = \Gamma(\Lambda').$$

$$(\text{not only } \cong)$$

For a finitely generated multiplicative subgroup $\Gamma \subset k^\times$, it is known as Siegel's theorem (cf. [5]) that the set of solutions (x, y) to the equation $x + y = 1$ ($x, y \in \Gamma$) is (effectively) finite. From this follows that, for any fixed U, there are only a finite number of U' satisfying the assumption of Theorem 1.

Note 4

§3. Rigidity of π_1 with no modding out   When a nonsingular algebraic curve U/k is not complete, the fundamental group $\pi_1(U \otimes \bar{k})$ is a free profinite group of finite rank. The famous paper of Belyi [1] considers the group extension

$$1 \longrightarrow \pi_1 \longrightarrow \pi_1(U) \longrightarrow G_k \longrightarrow 1$$

without taking reduction of π_1 modulo any nontrivial subgroups. (It seems that there are not many other papers treating it in this way.)

In this note, we should like to report the following nature of $\pi_1(U)$. If the complex Riemann surface associated with U is of type (g, n), then one finds a subset $\mathcal{I}$ in $\pi_1 = \Pi_{g,n}$ which is the conjugacy union of the inertia subgroups corresponding to n punctures. It can be written in the presentation of §1 (Π) as

$$\mathcal{I} = \{x\gamma_i^a x^{-1} \mid x \in \Pi_{g,n}, a \in \hat{\mathbb{Z}}, i = 1, 2, \ldots, n\}.$$

Note that this is only a subset of π_1 (not closed under the group operation) of the form of a union of certain conjugacy classes. We shall call $\mathcal{I}$ the *inertia subset* of $\pi_1(U)$. Then we have the following result.

Key lemma ([7])   The inertia subset $I \subset \pi_1$ can be characterized in terms   Note 5
of the Galois augmentation $\pi_1(U) \twoheadrightarrow G_k$ by a "non-abelian" weight filtration. Therefore, any Galois compatible isomorphism of topological groups
$f: \pi_1(U) \xrightarrow[G_k]{\sim} \pi_1(U')$ should precisely keep their inertia subsets $I \subset \pi_1(U)$ and
$I' \subset \pi_1(U')$, i.e., should induce $f(I) = I'$.

This lemma enables us to control *in a purely group-theoretical way*, say,
open immersions of algebraic curves, or residue fields of cusps on finite etale
covers of a curve. To prove this lemma, we need to use properly non-abelian
phases such as: any open subgroup of a free profinite group is again profinite
free; the ranks of those open subgroups increase in proportion to their indices
(Schreier's formula). Therefore we could not prove similar characterization
of inertia in the meta-abelian quotients of π_1. Still, we can show the similar
lemma for the case of pro-l fundamental groups, where the proof turns out
rather simpler than the profinite case as a consequence of a strong property of
the pro-l free groups. However, for the pro-nilpotent π_1 (which is the direct
product of $\pi_1^{\text{pro-}l}$ for all primes l), the corresponding lemma is not true. But if
we take larger pro-solvable or profinite π_1, then, the lemma holds true again.
By virtue of such characterization of inertia subsets in full profinite (or pro-
solvable) π_1 beyond pro-nilpotent π_1, one could "interrelate" information of
inertia subsets of $\pi_1^{\text{pro-}l}$ among different primes l. This was an important point
to enable us to scoop up a powerful essence of profinite non-abelianity, that
has led us to the following result.

Theorem 2 ([7])   Let U and U' be genus zero curves defined over a number
field k with $\pi_1(U_{\mathbb{C}}^{\text{an}})$ non-abelian. Then,

$$\pi_1(U) \cong_{G_k} \pi_1(U') \Longrightarrow U \cong_k U'.$$

Note   In the above statement, the converse implication $\Leftarrow$ trivially holds.
Note   By rigidity we mean uniqueness rather than finiteness.

§4. Case of elliptic curves   Let E be an elliptic curve with origin $O \in E(k)$.
Then, $\pi_1(E)$ is an extension of the absolute Galois group G_k by $\pi_1(E \otimes \bar{k})$, where
the kernel group can be identified with $H_1(E \otimes \bar{k}, \hat{\mathbb{Z}}) \cong \hat{\mathbb{Z}}^{\oplus 2}$:

$$1 \longrightarrow \quad \pi_1 \quad \longrightarrow \pi_1(E) \longrightarrow G_k \longrightarrow 1.$$
$$\wr\|$$
$$\hat{\mathbb{Z}}^{\oplus 2}$$

The existence of the rational point $O \in E(k)$ implies its splitness, so that considering the equivalence class of the above group extension is equivalent to

considering that of the associated Galois representation

$$G_k \longrightarrow \mathrm{GL}_2(\hat{\mathbb{Z}}) = \prod_l \mathrm{GL}(T_l E_{\bar{k}}).$$

Then, when E has no complex multiplication, Faltings' theorem (Tate conjecture) implies that the group extension $\pi_1(E)/G_k$ determines the elliptic curve E/k. However, when E does have complex multiplication, it is not necessarily the case. For example, one can construct lots of pairs of elliptic curves (E, E') over a sufficiently large number field k with $E_{\mathbb{C}}^{\mathrm{an}} \not\cong E_{\mathbb{C}}'^{\mathrm{an}}$ admitting k-isogeny maps $f: E \to E'$, $g: E' \to E$ with mutually prime degrees. For such a pair (E, E'), the Galois representations $G_k \to \mathrm{GL}(T_l E_{\bar{k}})$ and $G_k \to \mathrm{GL}(T_l E'_{\bar{k}})$ turn out to be equivalent for every l through f or g, although E and E' are not isomorphic over k.

Thus we are led to removing the origin O from E so as to consider $\pi_1(E - O)$ instead of $\pi_1(E)$. Note that $\pi_1(E - O)$ is a group extension of G_k by a free profinite group $\hat{F}_2$ of rank 2 (which is obviously non-abelian).

Theorem 3   Let (E, O), (E', O') be elliptic curves defined over k. Then,

$$\pi_1(E - O) \cong_{G_k} \pi_1(E' - O') \implies E \cong_k E'.$$

Over the curve $E - O$ one has an etale cover $E - {}_4E$ (where ${}_4E$ is a divisor on the elliptic curve E consisting of 16 geometric points of order dividing 4) which can be regarded naturally as a Kummer cover of $\mathbf{P}^1 - \{0, 1, \infty, \lambda\}$. Using this trick, the proof of Theorem 3 may be reduced to the case of genus 0.   <u>Note 6</u>

§5. On automorphisms

According to theorems by Neukirch, Ikeda, Iwasawa, Uchida (cf. [8]), for algebraic number fields k, k', we know not only

$$\pi_1(\mathrm{Spec}\, k) \cong \pi_1(\mathrm{Spec}\, k') \iff \mathrm{Spec}\, k \cong \mathrm{Spec}\, k'$$

$$(\text{i.e., } \mathrm{Gal}(\overline{\mathbb{Q}}/k) \cong \mathrm{Gal}(\overline{\mathbb{Q}}/k') \iff k \text{ and } k' \text{ are } \mathbb{Q}\text{-isomorphic}),$$

but also

$$\mathrm{Out}(\pi_1(\mathrm{Spec}\, k)) \cong \mathrm{Aut}(k).$$

Regarding this as 0-dimensional case, we may expect the following for algebraic curves.

Problem X   For U/k an algebraic curve with $\pi_1(U_{\mathbb{C}}^{\mathrm{an}})$ non-abelian, could   <u>Note 7</u>

$$\frac{\mathrm{Aut}_{G_k}\pi_1(U)}{\mathrm{Inn}\,\pi_1} \cong \mathrm{Aut}_k(U)$$

hold true? (In fact, Grothendieck conjectures such a phenomenon for a more general class of morphisms.)

Now, let us consider $U = \mathbf{P}_{\mathbf{Q}}^1 - \{0, 1, \infty\}$. The exact sequence

$$1 \longrightarrow \pi_1 \longrightarrow \pi_1(U) \longrightarrow G_{\mathbf{Q}} \longrightarrow 1$$

yields a big Galois representation

$$\varphi_{\mathbf{Q}} : G_{\mathbf{Q}} \longrightarrow \mathrm{Out}(\pi_1).$$

Presenting π_1 by use of the loops in the figure as

$$\pi_1 = \langle x, y, z \mid xyz = 1 \rangle_{\mathrm{top}},$$

one can lift the representation $\varphi_{\mathbf{Q}}$ to a unique representation $\tilde{\varphi}_{\mathbf{Q}} : G_{\mathbf{Q}} \to \mathrm{Aut}(\pi_1)$ whose Galois image lies in

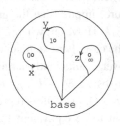

$$\mathrm{Brd}(\pi_1) := \left\{ f \in \mathrm{Aut}(\pi_1) \;\middle|\; \begin{array}{l} \exists \alpha \in \hat{\mathbb{Z}}^\times, \exists s \in \pi_1, \exists t \in \pi_1' \text{ s.t.} \\ \qquad f(x) = sx^\alpha s^{-1}; \\ \qquad f(y) = ty^\alpha t^{-1}; \\ \qquad f(z) = z^\alpha. \end{array} \right\}$$

(Belyi [1]). Recalling that Belyi showed the injectivity of $\tilde{\varphi}_{\mathbf{Q}}$, one of our next interests is to ask "how much the representation image $\mathrm{Im}\,\tilde{\varphi}_{\mathbf{Q}}$ is smaller than $\mathrm{Brd}\,\pi_1$?" This last question is indeed related to the above mentioned Problem X as follows.

Proposition 4   To verify Problem X for $U = \mathbf{P}_{\mathbf{Q}}^1 - \{0, 1, \infty\}$ affirmatively, it is necessary and sufficient to show that the centralizer of $\mathrm{Im}\,\tilde{\varphi}_{\mathbf{Q}}$ in $\mathrm{Brd}(\pi_1)$ is $\{1\}$.

Here, the center of $\mathrm{Im}\,\tilde{\varphi}_{\mathbf{Q}} \approx G_{\mathbf{Q}}$ is known to be trivial. One can also show that the centralizer Z in the above proposition has trivial image in $\mathrm{Brd}(\pi_1^{(l)}/\pi_1^{(l)''})$ $(\pi_1^{(l)} = \pi_1^{\mathrm{pro}\text{-}l})$ for every prime l. Finally, we note that the problem of seeking the image of pro-l Galois representation $\varphi_{\mathbf{Q}}^{(l)} : G_{\mathbf{Q}} \to \mathrm{Brd}(\pi_1^{(l)})$ has been approached by several other authors by the use of lower central filtration (see e.g., [4]).

<u>Note 8</u>

Complementary notes

The text above is one of the earliest publications of anabelian research, in the late 1980s in Japanese, which indicates some of the atmosphere of the dawn of investigations around Grothendieck's conjecture on fundamental groups of "anabelian" curves. As indicated in the introduction of [6], the research started under the strong influence of techniques from studies of Galois representations in $\pi_1(\mathbf{P}^1 - \{0, 1, \infty\})$ by Ihara [3] and Anderson–Ihara [9], of inverse Galois problems (especially Mike Fried's intensive use of the branch cycle argument

[17]) and analogous results in Hodge theory by Hain and Pulte [20]. In this complementary section, we shall add several notes to describe miscellaneous facts and later developments (with apologies for missing citations of many important works to be mentioned).

Throughout these notes, k denotes a number field (a finite extension of $\mathbb{Q}$).

Note 1

Here $\mathrm{Aut}_U(Y)$ should be understood to denote the opposite group of the covering transformation group of Y over U. As is well known, the theory of etale fundamental groups was established by A. Grothendieck and his collaborators in SGA1 [18]. Especially, the notion of *Galois category* in loc. cit. presents axioms unifying classical Galois theory of covers of a topological space and that of field extensions. This serves as a base for introducing our main mathematical object of study – an 'arithmetic' fundamental group equipped with a mixed structure as a group extension of the absolute Galois group of a number field by the profinite completion of a discrete fundamental group of the associated complex manifold.

Note 2

The fundamental conjecture of anabelian geometry was posed in Grothendieck's letter to Faltings [19] for hyperbolic curves (i.e., nonsingular algebraic curves of negative Euler characteristic). The references [19] and [2] had not been published for many years until the appearance of the proceedings volume [41] edited by L. Schneps and P. Lochak. As described in [19], Grothendieck proposed to study "extraordinary rigidity" of those arithmetic fundamental groups as the non-abelian analog of Faltings' theorem for abelian varieties [16]:

$$\mathrm{Hom}_k(A, A') \otimes \mathbb{Z}_l \xrightarrow{\sim} \mathrm{Hom}_{G_k}(T_lA, T_lA').$$

The fundamental conjecture (saying that the geometry of "anabelian varieties" should be reconstituted from their arithmetic fundamental groups) is only one aspect of his circle of ideas (anabelian philosophy) generalizing Belyi's theorem [1] to a vast theme extending from "Grothendieck dessin d'enfant" to "Galois–Teichmüller Lego". Grothendieck used the term "anabelian" to indicate "very far from abelian groups" ([2], p. 14). Typical candidates for anabelian varieties are hyperbolic curves, successive fiber spaces of them (Artin elementary neighborhoods), and moduli spaces of hyperbolic curves. At a very early stage, virtual center-triviality of geometric profinite fundamental groups was studied as a main feature of anabelianity (e.g., [32], [33], [24]). Belyi's injectivity result has been generalized to arbitrary hyperbolic curves by Matsumoto [28] and Hoshi–Mochizuki [22]. The status of the fundamental conjecture has been pushed forward to ideal solutions by Tamagawa [45] and Mochizuki [30]. See Note 7. The Galois–Teichmüller Lego philosophy has been taken up by Drinfeld [14] and Ihara [23] in the genus zero case by introducing what is called the Grothendieck–Teichmüller group $\widehat{GT}$. For a survey including higher genus formulations of $\widehat{GT}$, see [27].

Note 3

If U/k is π_1-equivalent over k to U'/k, then $H_1(U \otimes \bar{k}) \cong H_1(U' \otimes \bar{k})$ as G_k-modules. Then, as their weight (-1) quotients (obtained by modding out the weight (-2) submodules after tensoring with $\mathbb{Z}_l$), the l-adic Tate modules of the Jacobian varieties of the smooth compactifications of U and of U' turn out to be equivalent G_k-modules. Faltings' theorem [16], together with the good reduction criterion of Neron–Ogg–Shafarevich, implies then finiteness of those complete curves which have G_k-equivalent l-adic Tate modules. When the genus ≥ 2, Faltings' theorem also guarantees finiteness of points of bounded degree (Mordell conjecture), and so finiteness of possible punctures giving the same arithmetic fundamental group. To see finiteness of such possible punctures on genus 1 curves, however, would involve the problem of bounding heights of punctured points from a given class in $\text{Ext}_{G_k}(T_l E, \mathbb{Z}_l^r(1))$ (if we restrict ourselves to the use of only $H_1(U)$). This has apparently not been confirmed yet. We refer to a recent important paper of M. Kim [25] that, in turn, uses anabelian ideas to deduce finiteness of Diophantine problems.

Note 4

The clue to this modest result was to observe the Galois action on H_1 of the elementary abelian $(\mathbb{Z}/N\mathbb{Z})^{n-1}$-cover of $\mathbf{P}^1_{\bar{k}}-\{n \text{ points}\}$ so as to look at the kernel of its weight (-2) part, i.e., the kernel of the Galois permutation of cusps on the cover. A subtle point is that the Galois action depends on the choice of k-models of that cover, so that we need to extract a common invariant for all those possible k-models. At one other point, a technical lemma was employed "$k^\times \cap k(\mu_{l^r})^{\times l^r} = k^{\times l^r}$ for a number field $k \ni \sqrt{-1}$ and prime powers l^r" from Rubin's paper (*Invent. math.* 89 (1987), 511–526, lemma 5.7), accidentally found on the library bookshelves of the University of Tokyo. On a later day, I found the lemma traced back to Weil [47] chap. XIII, §8 lemma 9 (p. 273), while Rubin gave a much simpler proof by using Galois cohomology.

In the simplest case $\Lambda = \{0, 1, \infty, \lambda\}$, one has the multiplicative group $\Gamma(\Lambda) = \langle -1, \lambda, 1 - \lambda \rangle \subset k^\times$. When $k = \mathbb{Q}$ or many quadratic fields, this can determine the isomorphism class of $\mathbf{P}^1 - \{0, 1, \infty, \lambda\}$. But already when $k = \mathbb{Q}(\sqrt{2})$, it fails: $\lambda = -1 + \sqrt{2}$, $\lambda' = (2 - \sqrt{2})/2$ giving the same $\Gamma(\Lambda)$ but non-isomorphic

$\mathbf{P}^1 - \Lambda$ ([6], example (4.6)). Efforts to improve the results of [6] required that we treat the Galois permutations of cusps directly, not only the kernel information about them. This motivated the group-theoretical characterization of the inertia subgroups in arithmetic fundamental groups described in §3, Key lemma. See also the next note.

Note 5

In the non-abelian free profinite group $\Pi_{g,n}$ ($n \geq 1$), the inertia subgroups over n punctures form a union $\mathcal{I}$ of conjugacy classes. Any individual inertia subgroup $I \subset \mathcal{I}$ is isomorphic to $\hat{\mathbb{Z}}$, which has a big normalizer in $\pi_1(U)$ of the form of an extension of the Galois group by $\hat{\mathbb{Z}}(1)$ (branch cycle argument). The Key lemma asserts the converse of this property, i.e., $\mathcal{I}$ can be characterized as the weight (-2) subset with weight (-1) complement in $\Pi_{g,n}$. This *an-abelian weight filtration* was introduced in [7], [33] with certain techniques that pay careful attention to open neighborhoods.

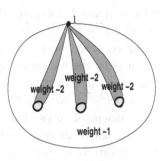

It should be noted that this, in turn, can be used for a *purely group-theoretical characterization* of the set of sectional homomorphisms *at infinity*

$$\mathrm{Sect}_\infty := \left\{ s \colon G_k \to \pi_1(U) \;\middle|\; \begin{array}{c} s(G_k) \text{ lies in a decomposition} \\ \text{subgroup at a cusp} \end{array} \right\}$$

as the set of those sections $s \colon G_k \to \pi_1(U)$ each of which has a nontrivial pro-cyclic subgroup in $\Pi_{g,n}$ stabilized and acted on by $s(G_k)$ via multiplication by the cyclotomic character. This set includes sections arising from tangential base points formulated by Deligne [13] and Anderson–Ihara [9]. Recently, Esnault–Hai [15] shed new light on the set Sect_∞ and its cardinality. See also Koenigsmann [26] and Stix [43] for related discussions.

Note 6

The following argument was behind this passage. Let (E, O) be an elliptic curve over a number field k with lambda invariant $\lambda \in \bar{k}$. The 2-isogeny and the 4-isogeny induce etale covers $E -_2 E$ and $E -_4 E$ of the punctured curve $E - \{O\}$ respectively. Their function fields over $\bar{k}$ can be written as

$$\bar{k}(E -_2 E) = \bar{k}(t, \sqrt{t(t-1)(t-\lambda)}),$$
$$\bar{k}(E -_4 E) = \bar{k}(\sqrt{t}, \sqrt{t-1}, \sqrt{t-\lambda}).$$

Let $\Delta = (\mathbb{Z}/2\mathbb{Z})^2$ be the covering group between $E -_4 E$ and $E -_2 E$ over $\bar{k}$, and regard the l-adic homology group $H_1(E - {}_4E, \mathbb{Z}_l)$ as a $(G_k \cdot \Delta)$-module. Then, (after taking a finite extension of k if necessary) the maximal Δ-coinvariant torsion-free quotient fits in the following exact sequence:

$$0 \longrightarrow \mathbb{Z}_l(1)^3 \longrightarrow H_1\left(E - {}_4E, \mathbb{Z}_l\right)_\Delta \big/ _{torsion} \longrightarrow H_1(E, \mathbb{Z}_l) \longrightarrow 0.$$

The group $H_1(E - {}_4E, \mathbb{Z}_l)$ has another interpretation as the Galois group of the maximal abelian pro-l extension of $\bar{k}(E - {}_4E)$ unramified outside the divisor ${}_4E$. It has a remarkable weight (-2) *quotient* corresponding to the Galois extension $\bar{k}(\sqrt[2l^\infty]{t}, \sqrt[2l^\infty]{t-1}, \sqrt[2l^\infty]{t-\lambda})$, which provides a canonical splitting of the above sequence. Once we know the sequence splits, we can recover the splitting uniquely by the weight argument, from which follows the group-theoretical characterization of the series of subgroups of $\pi_1(E - \{O\})$ corresponding to $\bar{k}(\sqrt[2l^n]{t}, \sqrt[2l^n]{t-1}, \sqrt[2l^n]{t-\lambda})$. Plugging this into the argument of looking at Galois permutations of cusps on abelian covers of $\mathbf{P}^1 - \{0, 1, \infty, \lambda\}$ discussed in §3, we conclude reconstitution of the cross ratio class of λ (hence $j(\lambda) \in k$) solely from information about $\pi_1(E - \{O\})$. This, together with k-isogeny $E \sim_k E'$ from Faltings' theorem, implies $E \cong_k E'$. One finds related extensions of this argument in Asada [10] §5.2 and Stix [43] §10.5.

The arithmetic fundamental group $\pi_1(E - \{O\})$ is a basic and fascinating object to study as well as $\pi_1(\mathbf{P}^1 - \{0, 1, \infty\})$. For instance, an elliptic analog of Ihara's theory [3] on Jacobi sum power series has been developed in [34], [36].

Note 7
Rigidity assertion discussed in §3:

$$(\text{Equiv})_U \qquad \pi_1(U) \cong_{G_k} \pi_1(U') \implies U \cong_k U'$$

combined with the automorphism assertion of Problem X:

$$(\text{Aut})_U \qquad \frac{\text{Aut}_{G_k}\pi_1(U)}{\text{Inn}\,\pi_1} \cong \text{Aut}_k(U) : \textbf{a finite group!}$$

implies the isomorphism version of Grothendieck's conjecture:

$$\text{Isom}_k(U, U') \cong \text{Isom}_{G_k}(\pi_1(U), \pi_1(U'))/\text{Inn}(\pi_1(U' \otimes \bar{k})).$$

This has been settled by Tamagawa [45] and Mochizuki [29]. In both works, the assertions $(\text{Equiv})_V$ for finite etale covers V of U follow all together. Here remains a little open question. Does a collection of assertions $(\text{Equiv})_V$ for sufficiently many finite etale covers V of U imply $(\text{Aut})_U$ automatically?

Grothendieck suggested in [19](6) more generally to consider the mapping

$$(*) \qquad \text{Hom}_k(U, U') \longrightarrow \text{Hom}_{G_k}(\pi_1(U), \pi_1(U'))/\text{Inn}(\pi_1(U' \otimes \bar{k})).$$

The Hom-version of Grothendieck's conjecture asserts that the above mapping gives a bijection between the set of dominant k-morphisms $U \to U'$ and the set of classes of the G_k-compatible open homomorphisms $\pi_1(U) \to \pi_1(U')$. This has been settled by Mochizuki [30].

See also [38] for a review of works by Tamagawa and Mochizuki (and of the author) till 1997, where it was found important to investigate Grothendieck's conjecture after replacing the base number field k by other arithmetic fields (finite fields, sub-p-adic fields). Here, we do not enter into any more details. For further developments, see also the articles by Tamagawa, Saidi–Tamagawa, and Mochizuki contained in the book [42].

Grothendieck's section conjecture [19](7) comes from a special case of (∗) where $U = \mathrm{Spec}(k)$ and U' is a hyperbolic curve (written U instead of U'). Then, we obtain a mapping of the set of k-rational points $U(k)$ into $\mathrm{Sect}(\pi_1(U)/G_k)/\mathrm{conj}$, where Sect denotes the set of sectional homomorphisms and '/conj' means modulo conjugacy by elements of $\pi_1(U \otimes \bar{k})$. The section conjecture asserts the bijection of $U(k)$ onto those sections outside Sect_∞ of Note 5, namely,

$$U(k) \xrightarrow{\sim} \left(\mathrm{Sect}(\pi_1(U)/G_k) - \mathrm{Sect}_\infty \right)/\mathrm{conj}.$$

Injectivity for the section conjecture and its close relationship with injectivity for the general Hom-conjecture or with the above mentioned Equiv-conjecture under anticipated anabelian situations have been known at early stages of investigation (cf. [19], [32]). The section conjecture is still an open problem, but recently important evidence has appeared, such as Stix [44] and Harari–Szamuely [21].

Finally, one may consider cases where both U and U' are spectra of function fields of varieties in (∗). The 0-dimensional case is nothing but the Neukirch–Uchida theorem. For function field cases, there have been intensive studies and results initiated by Pop [40], Bogomolov [11] and their further developments (cf. the article by F. Pop in this volume).

Note 8

The left-hand side of Problem X turns out to be naturally isomorphic to the centralizer of the Galois image in $\mathrm{Out}(\Pi_{g,n})$ called the *Galois centralizer*. By virtue of the anabelian weight filtration, the Galois centralizer as well as the Galois image lies in $\mathrm{Out}^I(\Pi_{g,n})$ defined as the group of outer automorphisms of $\Pi_{g,n}$ stabilizing the inertia subset $\mathcal{I}$. The pro-l version of Problem X mentioned here considers estimating the Galois centralizer in $\Gamma_{g,n}^{\mathrm{pro}\text{-}l} := \mathrm{Out}^I(\Pi_{g,n}^{\mathrm{pro}\text{-}l})$, where we set up a certain natural filtration by normal subgroups having graded quotients $\mathrm{gr}^0 \cong \mathrm{GSp}(2g, \mathbb{Z}_l) \times S_n$ and gr^i ($i \geq 1$) isomorphic to free $\mathbb{Z}_l$-modules of finite ranks. By conjugation, each gr^i ($i \geq 1$) turns out to get a structure of "weight $(-i)$" GSp-module. Since the Galois centralizer is a priori of weight zero, it must inject into gr^0, i.e., into $\mathrm{Aut}(H_1(U \otimes \bar{k}))$ (cf. [37], [35]).

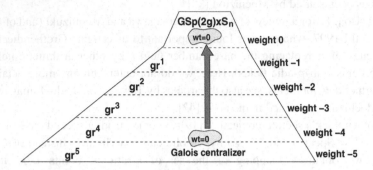

Indeed, more constraints to approximate the Galois centralizer to $\mathrm{Aut}_k(U)$ should be obtained from its commutativity with nontrivial Galois images distributed in $\Gamma_{g,n}^{\mathrm{pro}\text{-}l}$. The Galois image in $\mathrm{gr}^0 = \mathrm{GSp}(2g) \times S_n$ is within standard knowledge from the theory of l-adic Galois representations on torsion points of Jacobians (e.g., Tate conjecture proved by Faltings), which also spreads weighted carpets on gr^i ($i \geq 1$) in the above sense according to Frobenius eigen-radii (Riemann–Weil hypothesis). On the other hand, to find Galois images submerging in negative weights (called *Torelli–Galois images*) requires new knowledge about Galois representations on fundamental groups. Deligne [12] and Oda [39] suggested that one could lift Ihara's theory on $\pi_1^{\mathrm{pro}\text{-}l}(\mathbf{P}^1 - \{0, 1, \infty\})$ to any hyperbolic curve; namely, there should be a common factor for all Galois actions on pro-l fundamental groups (independent of the moduli) of hyperbolic curves. After the efforts of several authors, the last remaining case of this prediction – that of complete curves – has been settled (up to finite torsion) by Takao [46] (cf. [35], note (A4) added in English translation). Consequently, we have Torelli–Galois images in weights $-6, -10, -14, \ldots$ originated from Soule's cyclotomic characters, and find the pro-l Galois centralizer injected in $\mathrm{Sp}(2g) \times S_n$. In the original case $g = 0$, $n = 3$ considered here, it follows that the pro-l Galois centralizer for $\mathbf{P}^1 - \{0, 1, \infty\}$ coincides with S_3 as expected.

References

[Part 1] **References to [31] (1989)**

[1] G. V. Belyi, *On Galois extensions of a maximal cyclotomic field*, Izv. Akad. Nauk. SSSR **8** (1979), 267–276 (in Russian); *English transl. in* Math. USSR Izv. **14** (1980), 247–256.

[2] A. Grothendieck, *Esquisse d'un Programme*, mimeographed note 1984 (published later in [41]: Part 1, 5–48).

[3] Y. Ihara, *Profinite braid groups, Galois representations, and complex multiplications*, Ann. of Math. **123** (1986), 43–106.

[4] Y. Ihara, *Some problems on three-point ramifications and associated large Galois representations*, in "Galois representations and arithmetic algebraic geometry", Adv. Stud. Pure Math., **12**, 173–188.

[5] S. Lang, *Fundamentals of Diophantine Geometry*, Springer, 1983.

[6] H. Nakamura, *Rigidity of the arithmetic fundamental group of* $\mathbf{P}^1 - \{0, 1, \infty, \lambda\}$, Preprint 1988 (UTYO-MATH 88-21); appeared under the title: *Rigidity of the arithmetic fundamental group of a punctured projective line*, J. Reine Angew. Math. 405 (1990), 117–130.

[7] H. Nakamura, *Galois rigidity of the etale fundamental groups of punctured projective lines*, Preprint 1989 (UTYO-MATH 89-2), appeared in final form from J. Reine Angew. Math. 411 (1990), 205–216.

[8] J. Neukirch, *Über die absoluten Galoisgruppen algebraischer Zahlkölper*, Asterisque, 41/42 (1977), 67–79.

[Part 2] **References added for Complementary notes**

[9] G. Anderson, Y. Ihara, *Pro-l branched coverings of* $\mathbf{P}^1$ *and higher circular l-units*, Part 1 : Ann. of Math. **128** (1988), 271–293; Part 2: Intern. J. Math. **1** (1990), 119–148.

[10] M. Asada, *The faithfulness of the monodromy representations associated with certain families of algebraic curves*, J. Pure and Applied Algebra, **159**, 123–147.

[11] F. A. Bogomolov, *On two conjectures in birational algebraic geometry*, in "Algebraic Geometry and Analytic Geometry" (A. Fujiki et al. eds.) (1991), 26–52, Springer Tokyo.

[12] P. Deligne, *letter to Y. Ihara*, December 11, 1984.

[13] P. Deligne, *Le groupe fondamental de la droite projective moins trois points*, in "Galois group over $\mathbb{Q}$" (Y. Ihara, K. Ribet, J.-P. Serre eds.), MSRI Publ. Vol. 16 (1989), 79–297.

[14] V. G. Drinfeld, *On quasitriangular quasi-Hopf algebras and a group closely connected with* $Gal(\bar{\mathbb{Q}}/\mathbb{Q})$, Algebra i Analiz **2** (1990), 149–181 (in Russian); *English transl. in* Leningrad Math. J. **2**(4) (1991) 829–860.

[15] H. Esnault, P. H. Hai, *Packets in Grothendieck's section conjecture*, Adv. in Math., **218** (2008), 395–416.

[16] G. Faltings, *Endlichkeitssätze für abelsche Varietäten über Zahlkörpern*, Invent. Math. **73** (1983), 349–366.

[17] M. Fried, *Fields of definition of function fields and Hurwitz families — Groups as Galois groups*, Comm. in Algebra, **5** (1977), 17–82.

[18] A. Grothendieck, *Revêtement Etales et Groupe Fondamental* (SGA1), Lecture Note in Math. **224** Springer, Berlin Heidelberg New York, 1971.

[19] A. Grothendieck, *Letter to G. Faltings, June 1983*, in [41]: Part 1, 49–58.

[20] R. M. Hain, *The geometry of the mixed Hodge structures on the fundamental group*, Proc. Symp. Pure Math. **46** (1987), 247–282.

[21] D. Harari, T. Szamuely, *Galois sections for abelianized fundamental groups*, Math. Ann., **344** (2009), 779–800.

[22] Y. Hoshi, S. Mochizuki, *On the combinatorial anabelian geometry of nodally nondegenerate outer representations*, Preprint RIMS-1677, August 2009.

[23] Y. Ihara, *Braids, Galois groups and some arithmetic functions*, Proc. ICM, Kyoto, 99–120, 1990.

[24] Y. Ihara, H. Nakamura, *Some illustrative examples for anabelian geometry in high dimensions*, in [41]: Part 1, 127–138.

[25] M. Kim, *The motivic fundamental group of* $\mathbf{P}^1 \setminus \{0, 1, \infty\}$ *and theorem of Siegel*, Invent. math. **161** (2005), 629–656.

[26] J. Koenigsmann, *On the 'Section Conjecture' in anabelian geometry*, J. reine angew. Math., **588** (2005), 221–235.

[27] P. Lochak, L. Schneps, *Open problems in Grothendieck-Teichmüller theory*, Proc. Symp. Pure Math. **74** (2006), 165–186.

[28] M. Matsumoto, *On Galois representations on profinite braid groups of curves*, J. reine angew. Math. **474** (1996), 169–219.

[29] S. Mochizuki, *The profinite Grothendieck conjecture for closed hyperbolic curves over number fields*, J. Math. Sci. Univ. Tokyo, **3** (1996), 571–627.

[30] S. Mochizuki, *The local pro-p anabelian geometry of curves*, Invent. Math. **138** (1999), 319–423.

[31] H. Nakamura, *On Galois rigidity of fundamental groups of algebraic curves* (in Japanese), Report Collection of the 35th Algebra Symposium held at Hokkaido University on July 19–August 1, pp. 186–199.

[32] ——, *Galois rigidity of algebraic mappings into some hyperbolic varieties*, Internat. J. Math. **4** (1993), 421–438.

[33] ——, *Galois rigidity of pure sphere braid groups and profinite calculus*, J. Math. Sci. Univ. Tokyo **1** (1994), 71–136.

[34] ——, *On exterior Galois representations associated with open elliptic curves*, J. Math. Sci., Univ. Tokyo **2** (1995), 197–231.

[35] ——. *Galois rigidity of profinite fundamental groups* (in Japanese), Sugaku **47** (1995), 1–17; *English transl. in* Sugaku Expositions (AMS) **10** (1997), 195-215.

[36] ——, *On arithmetic monodromy representations of Eisenstein type in fundamental groups of once punctured elliptic curves*, Preprint RIMS-1691, February 2010.

[37] H. Nakamura, H. Tsunogai, *Some finiteness theorems on Galois centralizers in pro-l mapping class groups*, J. Reine Angew. Math. 441 (1993), 115–144.

[38] H. Nakamura, A. Tamagawa, S. Mochizuki, *The Grothendieck conjecture on the fundamental groups of algebraic curves*, (in Japanese), Sugaku **50** (1998), 113–129; *English transl. in* Sugaku Expositions (AMS), **14** (2001), 31–53.

[39] T. Oda, *The universal monodromy representations on the pro-nilpotent fundamental groups of algebraic curves*, Mathematische Arbeitstagung (Neue Serie) 9-15 Juin 1993, Max-Planck-Institute preprint MPI/93-57 (1993).

[40] F. Pop, *On the Galois theory of function fields of one variable over number fields*, J. reine. angew. Math., **406** (1988), 200–218.

[41] L. Schneps, P. Lochak (eds.), *Geometric Galois Actions; 1.Around Grothendieck's Esquisse d'un Programme, 2.The Inverse Galois Problem, Moduli Spaces and Mapping Class Groups*, London Math. Soc. Lect. Note Ser. **242–243**, Cambridge University Press 1997.

[42] L. Schneps (ed.), *Galois groups and fundamental groups*, MSRI Publications, **41**, 2003.

[43] J. Stix, *On cuspidal sections of algebraic fundamental groups* Preprint 2008 (arXiv:0802.4125).

[44] J. Stix, *On the period-index problem in light of the section conjecture*, Amer. J. Math., **132** (2010), 157-180.

[45] A. Tamagawa *The Grothendieck conjecture for affine curves*, Compositio Math. **109** (1997), 135–194.

[46] N. Takao, *Braid monodormies on proper curves and pro-ℓ Galois representations*, Jour Inst. Math. Jussieu, (to appear).

[47] A. Weil, *Basic Number Theory*, Grundlehren der math. Wiss. in Einzeldarstellungen, Band 144, Springer 1974.

Around the Grothendieck anabelian
section conjecture

Mohamed Saïdi

University of Exeter

Introduction

This, mostly expository, paper is built around the topic of the Grothendieck anabelian section conjecture. This conjecture predicts that splittings, or sections, of the exact sequence of the arithmetic fundamental group

$$1 \to \pi_1(\overline{X}) \to \pi_1(X) \to G_k \to 1$$

of a proper, smooth, and hyperbolic curve X, all arise from decomposition subgroups associated to rational points of X, over a base field k which is finitely generated over the prime field $\mathbb{Q}$ (cf. §2).

A birational version of this conjecture predicts that splittings of the exact sequence of absolute Galois groups

$$1 \to \mathrm{Gal}(K_X^{\mathrm{sep}}/\bar{k}.K_X) \to \mathrm{Gal}(K_X^{\mathrm{sep}}/K_X) \to G_k \to 1,$$

where K_X is the function field of the hyperbolic curve X, all arise from rational points of X in some precise way (cf. §2), under the above assumption on the base field k (cf. §2).

This conjecture is one of the main topics in anabelian geometry. It establishes a dictionary between profinite group theory, arithmetic geometry, and Diophantine geometry. This conjecture is still widely open. Only some examples are treated in the literature. The most complete achievement around this conjecture, is the proof by Koenigsmann, and by Pop, that the birational version of this conjecture holds true over p-adic local fields (cf. §5).

The main issue in investigating the section conjecture is the following. How can one produce a rational point $x \in X(k)$, from a splitting of the exact sequence of $\pi_1(X)$? This issue is completely settled in the case of birational sections over p-adic local fields by Koenigsmann, and by Pop. In this case, one produces a rational point by resorting to a local-global principle for Brauer

Non-abelian Fundamental Groups and Iwasawa Theory, eds. John Coates, Minhyong Kim, Florian Pop, Mohamed Saïdi and Peter Schneider. Published by Cambridge University Press. ©Cambridge University Press 2012.

groups of fields of transcendence degree 1 over p-adic local fields, that was proven by Lichtenbaum, in the finitely generated case, and by Pop in general. It is not clear at the time of writing, at least to the author, how to settle the above issue in the case of sections of $\pi_1(X)$.

The main idea we would like to advocate in this paper is to reduce the solution of the section conjecture to the solution of its birational version. We introduce, and investigate, the theory of cuspidalisation of sections of arithmetic fundamental groups for this purpose (cf. §4). The main aim of this theory is, starting from a section

$$s : G_k \to \pi_1(X)$$

of the exact sequence of $\pi_1(X)$, to construct a section

$$\tilde{s} : G_k \to \mathrm{Gal}(K_X^{\mathrm{sep}}/K_X)$$

of the exact sequence of $\mathrm{Gal}(K_X^{\mathrm{sep}}/K_X)$, which lifts the section s, i.e. which inserts into a commutative diagram:

$$
\begin{array}{ccc}
G_k & \xrightarrow{\ \tilde{s}\ } & \mathrm{Gal}(K_X^{\mathrm{sep}}/K_X) \\
{\scriptstyle \mathrm{id}}\downarrow & & \downarrow \\
G_k & \xrightarrow{\ s\ } & \pi_1(X)
\end{array}
$$

where the right vertical homomorphism is the natural one. If the birational version of the Grothendieck section conjecture holds true, and if starting from a section $s : G_k \to \Pi_X$ one can construct a section $\tilde{s} : G_k \to \mathrm{Gal}(K_X^{\mathrm{sep}}/K_X)$ as above which lifts s, then the Grothendieck section conjecture holds true (cf. Remarks 4.4, (ii)).

We have a natural exact sequence

$$1 \to I_X \to \mathrm{Gal}(K_X^{\mathrm{sep}}/K_X) \to \pi_1(X) \to 1,$$

where I_X is the normal subgroup of $\mathrm{Gal}(K_X^{\mathrm{sep}}/K_X)$ generated by the inertia subgroups at all geometric points of X. We exhibit the notion of (uniformly) good sections of arithmetic fundamental groups (cf. §3). Sections which arise from rational points are good sections. Our main result on the theory of cuspidalisation is that good sections $s : G_k \to \pi_1(X)$ of $\pi_1(X)$ can be lifted to sections

$$\tilde{s}^{c-ab} : G_k \to \mathrm{Gal}(K_X^{\mathrm{sep}}/K_X)^{c-ab}$$

of the natural projection $\mathrm{Gal}(K_X^{\mathrm{sep}}/K_X)^{c-ab} \twoheadrightarrow G_k$, where $\mathrm{Gal}(K_X^{\mathrm{sep}}/K_X)^{c-ab}$ is the quotient of $\mathrm{Gal}(K_X^{\mathrm{sep}}/K_X)$ defined by the following push out diagram:

$$1 \longrightarrow I_X \longrightarrow \operatorname{Gal}(K_X^{\mathrm{sep}}/K_X) \longrightarrow \pi_1(X) \longrightarrow 1$$

$$\downarrow \qquad\qquad \downarrow \qquad\qquad \mathrm{id}\downarrow$$

$$1 \longrightarrow I_X^{\mathrm{ab}} \longrightarrow \operatorname{Gal}(K_X^{\mathrm{sep}}/K_X)^{\mathrm{c-ab}} \longrightarrow \pi_1(X) \longrightarrow 1$$

and I_X^{ab} is the maximal abelian quotient of I_X, under quite general assumptions on the field k, which are satisfied by number fields, and p-adic local fields (cf. Corollary 4.7). As an application, we prove a (pro-p) version of the section conjecture over p-adic local fields, for good sections of arithmetic fundamental groups, under some additional assumptions (cf. Theorem 5.4). We state a (an unconditional) result on a semi-birational version of the p-adic section conjecture (cf. Theorem 5.6). We also prove that the existence of good sections of $\pi_1(X)$, over number fields, implies the existence of degree 1 divisor on X, under a finiteness assumption of the Tate–Shafarevich group of the jacobian of X (cf. Theorem 5.8).

Finally, in §6, we discuss a weak version of the section conjecture over p-adic local fields, which is related to the absolute anabelian geometry of hyperbolic curves over p-adic local fields.

Acknowledgement The author would like to thank the referee for his valuable comments.

1 Generalities on arithmetic fundamental groups and sections

In this section we introduce the set-up of the Grothendieck anabelian section conjecture.

1.1 Let k be a field of characteristic $p \geq 0$, and let X be a proper, smooth, geometrically connected, and hyperbolic algebraic curve over k. Let $K \overset{\mathrm{def}}{=} K_X$ be the function field of X, and η a geometric point of X above the generic point of X. Then η determines naturally an algebraic closure $\bar{k}$ of k, a separable closure K_X^{sep} of K_X, and a geometric point $\bar{\eta}$ of $\overline{X} \overset{\mathrm{def}}{=} X \times_k \bar{k}$. There exists a canonical exact sequence of profinite groups

$$(1.1) \qquad 1 \to \pi_1(\overline{X}, \bar{\eta}) \to \pi_1(X, \eta) \xrightarrow{\mathrm{pr}_X} G_k \to 1.$$

Here, $\pi_1(X, \eta)$ denotes the arithmetic étale fundamental group of X with base point η, $\pi_1(\overline{X}, \bar{\eta})$ the étale fundamental group of $\overline{X}$ with base point $\bar{\eta}$, and $G_k \overset{\mathrm{def}}{=}$

$\text{Gal}(\bar{k}/k)$ the absolute Galois group of k. We will consider the following variant of the above exact sequence (1.1). Let

$$\Sigma \subseteq \mathfrak{Primes}$$

be a non-empty subset of the set $\mathfrak{Primes}$ of all prime integers. In the case where $\text{char}(k) = p > 0$, we will assume that $p \notin \Sigma$. Write

$$\Delta_X \overset{\text{def}}{=} \pi_1(\overline{X}, \bar{\eta})^\Sigma$$

for the maximal pro-Σ quotient of $\pi_1(\overline{X}, \bar{\eta})$, and

$$\Pi_X \overset{\text{def}}{=} \pi_1(X, \eta)/\operatorname{Ker}(\pi_1(\overline{X}, \bar{\eta}) \twoheadrightarrow \pi_1(\overline{X}, \bar{\eta})^\Sigma)$$

for the quotient of $\pi_1(X, \eta)$ by the kernel of the natural surjective homomorphism $\pi_1(\overline{X}, \bar{\eta}) \twoheadrightarrow \pi_1(\overline{X}, \bar{\eta})^\Sigma$, which is a normal subgroup of $\pi_1(X, \eta)$. Thus, we have an exact sequence of profinite groups

$$(1.2) \qquad 1 \to \Delta_X \to \Pi_X \xrightarrow{\text{pr}_{X,\Sigma}} G_k \to 1.$$

We shall refer to $\pi_1(X, \eta)^{(\Sigma)} \overset{\text{def}}{=} \Pi_X$ as the geometrically pro-Σ quotient of $\pi_1(X, \eta)$, or the geometrically pro-Σ arithmetic fundamental group of X. The exact sequence (1.2) induces a natural homomorphism

$$\rho_{X,\Sigma} \colon G_k \to \operatorname{Out}(\Delta_X),$$

where $\operatorname{Out}(\Delta_X) \overset{\text{def}}{=} \operatorname{Aut}(\Delta_X)/\operatorname{Inn}(\Delta_X)$ is the group of outer automorphisms of Δ_X. For $g \in G_k$, its image $\rho_{X,\Sigma}(g)$ in $\operatorname{Out}(\Delta_X)$ is the class of the automorphism of Δ_X obtained by lifting g to an element $\tilde{g} \in \Pi_X$, and let $\tilde{g}$ act on Δ_X by inner conjugation.

A profinite group G is slim if every open subgroup of G is centre-free (cf. [7], §0). If G is a slim profinite group, we have a natural exact sequence

$$1 \to G \to \operatorname{Aut} G \to \operatorname{Out} G \to 1,$$

where the homomorphism $G \to \operatorname{Aut} G$ sends an element $g \in G$ to the corresponding inner automorphism $h \mapsto ghg^{-1}$. Moreover, if the profinite group G is finitely generated, then the groups $\operatorname{Aut}(G)$, and $\operatorname{Out}(G)$, are naturally endowed with a profinite topology, and the above sequence is an exact sequence of profinite groups. The following is an important property of the profinite groups Δ_X, and Π_X.

Lemma 1.2 *The profinite group Δ_X is slim. In particular, the exact sequence (1.2) is obtained from the following exact sequence*

$$(1.3) \qquad 1 \to \Delta_X \to \operatorname{Aut}(\Delta_X) \to \operatorname{Out}(\Delta_X) \to 1,$$

by pull back via the natural continuous homomorphism

$$\rho_{X,\Sigma} \colon G_k \to \mathrm{Out}(\Delta_X).$$

More precisely, we have a commutative diagram:

(1.4)
$$
\begin{array}{ccccccccc}
1 & \longrightarrow & \Delta_X & \longrightarrow & \mathrm{Aut}(\Delta_X) & \longrightarrow & \mathrm{Out}(\Delta_X) & \longrightarrow & 1 \\
& & \mathrm{id}\big\uparrow & & \big\uparrow & & \rho_{X,\Sigma}\big\uparrow & & \\
1 & \longrightarrow & \Delta_X & \longrightarrow & \Pi_X & \longrightarrow & G_k & \longrightarrow & 1
\end{array}
$$

where the horizontal arrows are exact, and the right square is cartesian.

Proof Well known; see for example [16], proposition 1.11. $\qquad\square$

1.3 Similarly, if we write $X \times X \overset{\mathrm{def}}{=} X \times_k X$, and $\iota \colon X \to X \times X$ for the natural diagonal embedding, then the geometric point η determines naturally (via ι) a geometric point, which we will also denote η, of $X \times X$. There exists a natural exact sequence of profinite groups

(1.5)
$$1 \to \pi_1(\overline{X \times X}, \bar\eta) \to \pi_1(X \times X, \eta) \overset{\mathrm{pr}}{\longrightarrow} G_k \to 1.$$

Here, $\pi_1(X \times X, \eta)$ denotes the arithmetic étale fundamental group of $X \times X$ with base point η, which is naturally identified with the fibre product $\pi_1(X, \eta) \times_{G_k} \pi_1(X, \eta)$, and $\pi_1(\overline{X \times X}, \bar\eta)$ is the étale fundamental group of $\overline{X \times X}$ with base point $\bar\eta$ ($\bar\eta$ is naturally induced by η), which is naturally identified with the product $\pi_1(\overline{X}, \bar\eta) \times \pi_1(\overline{X}, \bar\eta)$. Similarly, as in 1.1, we consider the maximal pro-Σ quotient

$$\Delta_{X \times X} \overset{\mathrm{def}}{=} \pi_1(\overline{X \times X}, \bar\eta)^{\Sigma}$$

of $\pi_1(\overline{X \times X}, \bar\eta)$, which is naturally identified with $\Delta_X \times \Delta_X$, and the geometrically pro-Σ quotient

$$\Pi_{X \times X} \overset{\mathrm{def}}{=} \pi_1(X \times X, \eta)^{(\Sigma)} \overset{\mathrm{def}}{=} \pi_1(X \times X, \eta)/\mathrm{Ker}(\pi_1(\overline{X \times X}, \bar\eta) \twoheadrightarrow \pi_1(\overline{X \times X}, \bar\eta)^{\Sigma})$$

of $\pi_1(X \times X, \eta)$, which is naturally identified with $\Pi_X \times_{G_k} \Pi_X$. Thus, we have a natural exact sequence

$$1 \to \Delta_{X \times X} \to \Pi_{X \times X} \to G_k \to 1.$$

1.4 Our main objects of interest are group-theoretic splittings, or sections, of the exact sequence (1.2). Let

$$s \colon G_k \to \Pi_X$$

be a continuous group-theoretic section of the natural projection

$$\mathrm{pr} \overset{\mathrm{def}}{=} \mathrm{pr}_{X,\Sigma} \colon \Pi_X \twoheadrightarrow G_k,$$

meaning that $\text{pr} \circ s \colon G_k \to G_k$ is the identity homomorphism. We will refer to $s \colon G_k \to \Pi_X$, as above, as a section of the arithmetic fundamental group Π_X. Every inner automorphism $\text{inn}^g \colon \Pi_X \to \Pi_X$ of Π_X by an element $g \in \Delta_X$, gives rise to a conjugate section $\text{inn}^g \circ s \colon G_k \to \Pi_X$. We will refer to the set

$$C[s] \overset{\text{def}}{=} \{\text{inn}^g \circ s \colon G_k \to \Pi_X\}_{g \in \Delta_X}$$

as the set of conjugacy classes of the section s.

2 Grothendieck anabelian section conjecture

We follow the same notations as in §1.

2.1 Sections of arithmetic fundamental groups arise naturally from rational points. More precisely, let $x \in X$ be a closed point with residue field $k(x)$. Let $\overline{X}$ be the universal geometrically pro-Σ étale cover of X. We have a natural morphism $\pi \colon \overline{X} \to X$ which is Galois with Galois group Π_X. Let $\tilde{x} \in \overline{X}$ be a closed point with $\pi(\tilde{x}) = x$, and $D_{\tilde{x}} \subset \Pi_X$ the decomposition group at $\tilde{x}$. Then $D_{\tilde{x}}$ is a closed subgroup of Π_X which maps isomorphically onto the open subgroup $G_{k(x)}$ of G_k via the natural projection $\text{pr} \overset{\text{def}}{=} \text{pr}_{X,\Sigma} \colon \Pi_X \twoheadrightarrow G_k$. If $\tilde{x}' \in \overline{X}$ is another closed point with $\pi(\tilde{x}') = x$ then the decomposition groups $D_{\tilde{x}}$ and $D_{\tilde{x}'}$ are conjugate by an element of Δ_X (the pre-image of x in $\overline{X}$ forms an orbit under the action of Δ_X on $\overline{X}$). We will refer to a decomposition group $D_x \overset{\text{def}}{=} D_{\tilde{x}} \subset \Pi_X$ as above as a decomposition group associated to x. Thus, D_x is only defined up to conjugation by elements of Δ_X.

In particular, let $x \in X(k)$ be a rational point. Then x determines a decomposition subgroup

$$D_x \subset \Pi_X,$$

which is defined only up to conjugation by the elements of Δ_X, and which maps isomorphically to G_k via the projection $\text{pr} \colon \Pi_X \twoheadrightarrow G_k$. Hence, the subgroup $D_x \subset \Pi_X$ determines a group-theoretic section

$$s_x \colon G_k \to \Pi_X$$

of the natural projection $\text{pr} \colon \Pi_X \twoheadrightarrow G_k$, which is defined only up to conjugation by the elements of Δ_X. Let $\overline{\text{Sec}}_{\Pi_X}$ be the set of conjugacy classes of all continuous group-theoretic sections $G_k \to \Pi_X$, of the natural projection $\text{pr} \colon \Pi_X \twoheadrightarrow G_k$, modulo inner conjugation by the elements of Δ_X (cf. 1.4). We have a natural set-theoretic map

$$\varphi_{X,\Sigma} \colon X(k) \to \overline{\text{Sec}}_{\Pi_X},$$

$$x \mapsto \varphi_{X,\Sigma}(x) \overset{\text{def}}{=} [s_x],$$

where $[s_x]$ denotes the class of a section $s_x \colon G_k \to \Pi_X$, associated to the rational point x, in $\overline{\text{Sec}}_{\Pi_X}$.

Definition 2.2 Let $s \colon G_k \to \Pi_X$ be a continuous group-theoretic section of the natural projection $\text{pr} \colon \Pi_X \twoheadrightarrow G_k$. We say that the section s is point-theoretic, or geometric, if the class $[s]$ of s in $\overline{\text{Sec}}_{\Pi_X}$ belongs to the image of the map $\varphi_{X,\Sigma}$.

The following is the main Grothendieck anabelian conjecture regarding sections of arithmetic fundamental groups. This conjecture establishes a dictionary between purely group-theoretic sections of arithmetic fundamental groups, and rational points of hyperbolic curves over finitely generated fields, in characteristic 0.

Grothendieck anabelian section conjecture (GASC, cf. [4])   Assume that k is finitely generated over the prime field $\mathbb{Q}$, and $\Sigma = \mathfrak{Primes}$. Then the map $\varphi_X \overset{\text{def}}{=} \varphi_{X,\Sigma} \colon X(k) \to \overline{\text{Sec}}_{\Pi_X}$ is bijective. In particular, every group-theoretic section of Π_X is point-theoretic in this case.

Under the assumptions of the GASC, the injectivity of the map φ_X is well known (cf. for example [8], theorem 19.1). So the statement of the GASC is equivalent to the surjectivity of the set-theoretic map φ_X, i.e. that every group-theoretic section of Π_X is point-theoretic, under the above assumptions. Note that a similar conjecture can be formulated over any field, and for any non-empty set Σ of prime integers, but one can not expect its validity in general. For example, the analog of this conjecture doesn't hold over finite fields (even if $\Sigma = \mathfrak{Primes}$). Indeed, over a finite field the natural projection $\Pi_X \twoheadrightarrow G_k$ admits group-theoretic sections, or splittings, since the profinite group G_k is free in this case. On the other hand, there are proper, smooth, geometrically connected, and hyperbolic curves over finite fields with no rational points.

2.3   Assume that k is a number field, i.e. k is a finite extension of the field of rational numbers $\mathbb{Q}$. Let v be a place of k, and denote by k_v the completion of k at v. Write

$$X_v \overset{\text{def}}{=} X \times_k k_v.$$

Let $D_v \subset G_k$ be a decomposition group at v (D_v is only defined up to conjugation), which is naturally isomorphic to the absolute Galois group G_{k_v} of k_v. By pulling back the exact sequence

$$1 \to \Delta_X \to \Pi_X \to G_k \to 1,$$

by the natural injective homomorphism $D_v \hookrightarrow G_k$, we obtain the exact sequence

$$1 \to \Delta_{X_v} \to \Pi_{X_v} \to G_{k_v} \to 1.$$

Note that there exists an isomorphism $\Delta_X \overset{\sim}{\to} \Delta_{X_v}$. In particular, a group-theoretic section $s \colon G_k \to \Pi_X$ of the arithmetic fundamental group Π_X, induces naturally a group-theoretic section

$$s_v \colon G_{k_v} \to \Pi_{X_v}$$

of Π_{X_v}, for each place v of k. Moreover, if the section s is point-theoretic, then the section s_v is point-theoretic, for each place v of k, as is easily seen. It seems quite natural to formulate an analog of the GASC over p-adic local fields.

p-adic version of Grothendieck anabelian section conjecture (p-adic GASC) Assume that k is a p-adic local field, i.e. k is a finite extension of the field $\mathbb{Q}_p$, for some prime integer p, and $\Sigma = \mathfrak{Primes}$. Then the map $\varphi_X \overset{\mathrm{def}}{=} \varphi_{X,\Sigma} \colon X(k) \to \overline{\mathrm{Sec}}_{\Pi_X}$ is bijective.

The map $\varphi_X \overset{\mathrm{def}}{=} \varphi_{X,\Sigma}$ is known to be injective in the case where k is a p-adic local field, and $p \in \Sigma$ (cf. [8], theorem 19.1). Thus, the statement of the p-adic GASC is equivalent to the surjectivity of the map φ_X in this case.

2.4 Next, we recall the definition of a system of neighbourhoods of a group-theoretic section of the arithmetic fundamental group Π_X. The profinite group Δ_X being topologically finitely generated, there exists a sequence of characteristic open subgroups

$$\cdots \subseteq \Delta_X[i+1] \subseteq \Delta_X[i] \subseteq \cdots \subseteq \Delta_X[1] \overset{\mathrm{def}}{=} \Delta_X$$

(where i ranges over all positive integers) of Δ_X, such that

$$\bigcap_{i \geq 1} \Delta_X[i] = \{1\}.$$

In particular, given a group-theoretic section $s \colon G_k \to \Pi_X$ of Π_X, we obtain open subgroups

$$\Pi_X[i, s] \overset{\mathrm{def}}{=} s(G_k).\Delta_X[i] \subseteq \Pi_X$$

(where $s(G_k)$ denotes the image of G_k in Π_X via the section s) of Π_X, whose intersection coincide with $s(G_k)$, and which correspond to a tower of finite étale (not necessarily Galois) covers

$$\cdots \to X_{i+1}[s] \to X_i[s] \to \cdots \to X_1[s] \overset{\mathrm{def}}{=} X$$

defined over k. We will refer to the set $\{X_i[s]\}_{i \geq 1}$ as a system of neighbourhoods of the section s. Note that for each positive integer i, the open subgroup $\Pi_X[i, s]$ of Π_X is naturally identified with the geometrically pro-Σ arithmetic étale fundamental group $\pi_1(X_i[s], \eta_i)^{(\Sigma)}$ (the geometric point η_i of $X_i[s]$ is naturally induced by the geometric point η of X), and sits naturally in the following exact sequence

$$1 \to \Delta_X[i] \to \Pi_X[i, s] \to G_k \to 1,$$

which inserts in the following commutative diagram:

$$
\begin{array}{ccccccccc}
1 & \longrightarrow & \Delta_X[i] & \longrightarrow & \Pi_X[i, s] & \longrightarrow & G_k & \longrightarrow & 1 \\
& & \downarrow & & \downarrow & & \text{id} \downarrow & & \\
1 & \longrightarrow & \Delta_X & \longrightarrow & \Pi_X & \longrightarrow & G_k & \longrightarrow & 1
\end{array}
$$

where the two left vertical homomorphisms are the natural inclusions. In particular, by the very definition of $\Pi_X[i, s]$, the section s restricts naturally to a group-theoretic section

$$s_i \colon G_k \to \Pi_X[i, s]$$

of the natural projection $\Pi_X[i, s] \twoheadrightarrow G_k$, which fits into the following commutative diagram:

$$
\begin{array}{ccc}
G_k & \xrightarrow{\; s_i \;} & \Pi_X[i, s] \\
\text{id} \downarrow & & \downarrow \\
G_k & \xrightarrow{\; s \;} & \Pi_X
\end{array}
$$

where the right vertical homomorphism is the natural inclusion.

Thus, a section $s \colon G_k \to \Pi_X$ gives rise to a system of neighbourhoods $\{X_i[s]\}_{i \geq 1}$, and the corresponding open subgroups $\{\Pi_X[i, s]\}_{i \geq 1}$ of Π_X inherit naturally, from the section s, sections $s_i \colon G_k \to \Pi_X[i, s]$. In investigating the point-theorecity of the section s (cf. Definition 2.2), it is important to observe not only the section s, but rather the family of sections $\{s_i\}_{i \geq 1}$. In fact, a number of important properties of the section s can be proven, using a limit argument, by observing the family of sections $\{s_i\}_{i \geq 1}$, rather than only the section s. The best illustration of this phenomenon is the following crucial observation, which is extremely important in investigating the Grothendieck anabelian section conjecture, and which is due to Tamagawa.

Lemma 2.5 *Assume that k is finitely generated over the prime field $\mathbb{Q}$, or that k is a p-adic local field. Let $s \colon G_k \to \Pi_X$ be a group-theoretic section of the natural projection* $\mathrm{pr} \colon \Pi_X \twoheadrightarrow G_k$, *and $\{X_i[s]\}_{i \geq 1}$ a system of neighbourhoods* *of the section* s *(cf. 2.4). Then the section* s *is point-theoretic if and only if* $X_i[s](k) \neq \varnothing$, *for each* $i \geq 1$.

Proof   See [16], proposition 2.8 (iv).   □

Lemma 2.5 reduces the proof of the Grothendieck anabelian section conjecture, in the case where k is finitely generated over the prime field $\mathbb{Q}$, or that k is a p-adic local field, and $\Sigma = \mathfrak{Primes}$, to proving the following implication

$$\{\overline{\mathrm{Sec}}_{\Pi_X} \neq \emptyset\} \Longrightarrow \{X(k) \neq \emptyset\}.$$

Thus, at the heart of the Grothendieck anabelian section conjecture, is the following fundamental problem.

Problem   How can one produce, under the assumptions of GASC, or the p-adic GASC, a rational point $x \in X(k)$ starting from a group-theoretic section $s : G_k \to \Pi_X$ of Π_X?

No systematic approach has yet been developed so far for attacking this problem. The only situation where a method, or a technique, is available to solve this problem positively is the method, developed by Koenigsmann, and by Pop, in the framework of the birational version of the p-adic GASC (cf. [6], [13]), and which resorts to a local-global principle for Brauer groups of fields of transcendence degree 1 over p-adic local fields (see the discussion after Theorem 5.2).

Remarks 2.6   **(i)** One of the difficulties in investigating the GASC is that, for the time being, one doesn't know how to construct sections of arithmetic fundamental groups, and hence test the validity of the conjecture on concrete examples. In fact, the GASC itself can be viewed as a "rigidity" statement. Namely, the only way one knows (so far) how to construct sections of arithmetic fundamental groups of hyperbolic curves, over finitely generated fields of characteristic 0, is via decomposition groups associated to rational points, and these should be the only sections that exist! One way, however, to construct such sections is as follows. Let X be a proper, smooth, geometrically connected, hyperbolic algebraic curve over the field k, and $\Sigma \subseteq \mathfrak{Primes}$ a non-empty set of prime integers. Consider the exact sequence

$$(1.2) \qquad\qquad 1 \to \Delta_X \to \Pi_X \xrightarrow{\ \mathrm{pr}\ } G_k \to 1,$$

where Π_X is the geometrically pro-Σ arithmetic fundamental group of X. Recall the commutative cartesian diagram:

$$
\begin{array}{ccccccccc}
1 & \longrightarrow & \Delta_X & \longrightarrow & \mathrm{Aut}(\Delta_X) & \longrightarrow & \mathrm{Out}(\Delta_X) & \longrightarrow & 1 \\
& & \mathrm{id}\big\uparrow & & \big\uparrow & & \rho_{X,\Sigma}\big\uparrow & & \\
1 & \longrightarrow & \Delta_X & \longrightarrow & \Pi_X & \xrightarrow{\ \mathrm{pr}\ } & G_k & \longrightarrow & 1
\end{array}
$$

(1.4)

In order to construct a continuous group-theoretic section $s\colon G_k \to \Pi_X$ of the natural projection $\mathrm{pr}\colon \Pi_X \twoheadrightarrow G_k$, it is equivalent to construct a continuous homomorphism $\tilde{\rho}_{X,\Sigma}\colon G_k \to \mathrm{Aut}(\Delta_X)$, which lifts the homomorphism $\rho_{X,\Sigma}\colon G_k \to \mathrm{Out}(\Delta_X)$ above, i.e. such that the following diagram commutes

$$
\begin{array}{ccc}
G_k & \xrightarrow{\tilde{\rho}_{X,\Sigma}} & \mathrm{Aut}(\Delta_X) \\
\mathrm{id}\big\downarrow & & \big\downarrow \\
G_k & \xrightarrow{\rho_{X,\Sigma}} & \mathrm{Out}(\Delta_X)
\end{array}
$$

as follows directly from the fact that the right square in the diagram (1.4) is cartesian.

(ii) In light of the Remark (i), it is possible to construct sections $s\colon G_k \to \Pi_X$ of Π_X if the image of G_k in $\mathrm{Out}(\Delta_X)$ via the natural homomorphism $\rho_{X,\Sigma}\colon G_k \to \mathrm{Out}(\Delta_X)$ is a free profinite group. Under this condition one may hope to construct non-geometric sections s, i.e. sections that do not arise from rational points. This is the method used in [5] to construct non-geometric sections in the case where k is a number field, or a p-adic local field, and $\Sigma = \{p\}$. However, this method is unlikely to produce examples of non-geometric sections in the case where $\Sigma = \mathfrak{Primes}$. Indeed, in this case the homomorphism $\rho_{X,\Sigma}\colon G_k \to \mathrm{Out}(\Delta_X)$ is injective if X is hyperbolic, and k is a number field, or a p-adic local field, as is well known.

2.7 One can formulate a birational version of the Grothendieck anabelian section conjecture as follows (see also [13]). There exists a natural exact sequence of absolute Galois groups

$$
1 \to \mathrm{Gal}(K_X^{\mathrm{sep}}/K_X.\bar{k}) \to \mathrm{Gal}(K_X^{\mathrm{sep}}/K_X) \to G_k \to 1.
$$

Let $\Sigma \subseteq \mathfrak{Primes}$ be a non-empty set of prime integers, and let

$$
\overline{G}_X \stackrel{\mathrm{def}}{=} \mathrm{Gal}(K_X^{\mathrm{sep}}/K_X.\bar{k})^{\Sigma}
$$

the maximal pro-Σ quotient of the absolute Galois group $\mathrm{Gal}(K_X^{\mathrm{sep}}/K_X.\bar{k})$. Let

$$
G_X \stackrel{\mathrm{def}}{=} \mathrm{Gal}(K_X^{\mathrm{sep}}/K_X)/\,\mathrm{Ker}(\mathrm{Gal}(K_X^{\mathrm{sep}}/K_X.\bar{k}) \twoheadrightarrow \mathrm{Gal}(K_X^{\mathrm{sep}}/K_X.\bar{k})^{\Sigma})
$$

be the maximal geometrically pro-Σ Galois group of the function field K_X. Thus, G_X sits naturally in the following exact sequence

$$1 \to \overline{G}_X \to G_X \to G_k \to 1.$$

Let $x \in X(k)$ be a rational point. Then x determines a decomposition subgroup $D_x \subset G_X$, which is defined only up to conjugation by the elements of $\overline{G}_X$, and which maps onto G_k via the natural projection $G_X \twoheadrightarrow G_k$ (cf. 2.1 for similar arguments). More precisely, D_x sits naturally in the following exact sequence

$$1 \to M_X \to D_x \to G_k \to 1,$$

where $M_X \xrightarrow{\sim} \hat{\mathbb{Z}}(1)^\Sigma$ (cf. 3.1, for the definition of the module of roots of unity M_X). The above sequence is known to be split. The set of all possible splittings, i.e. sections $G_k \to D_x$ of the above exact sequence, is a torsor under $H^1(G_k, M_X)$. The latter can be naturally identified, via Kummer theory, with the Σ-adic completion $\hat{k}^{\times,\Sigma}$ of the multiplicative group $k^\times$. Each section $G_k \to D_x$ of the natural projection $D_x \twoheadrightarrow G_k$ determines naturally a section $G_k \to G_X$ of the natural projection $G_X \twoheadrightarrow G_k$, whose image is contained in D_x.

Birational Grothendieck anabelian section conjecture (BGASC)   Assume that k is finitely generated over the prime field $\mathbb{Q}$, and that $\Sigma = \mathfrak{Primes}$. Let $s\colon G_k \to G_X$ be a group-theoretic section of the natural projection $G_X \twoheadrightarrow G_k$. Then the image $s(G_k)$ is contained in a decomposition subgroup $D_x \subset G_X$ associated to a unique rational point $x \in X(k)$. In particular, the existence of the section s implies that $X(k) \neq \varnothing$.

One can, in a similar way, formulate a p-adic version of this conjecture.

p-adic version of birational Grothendieck anabelian section conjecture (p-adic BGASC)   Assume that k is a p-adic local field, i.e. k is a finite extension of $\mathbb{Q}_p$, and $\Sigma = \mathfrak{Primes}$. Let $s\colon G_k \to G_X$ be a group-theoretic section of the natural projection $G_X \twoheadrightarrow G_k$. Then the image $s(G_k)$ is contained in a decomposition subgroup $D_x \subset G_X$ associated to a unique rational point $x \in X(k)$. In particular, the existence of the section s implies that $X(k) \neq \varnothing$.

3 Good sections of arithmetic fundamental groups

We use the same notation as in §1 and §2. We will introduce the notion of (uniformly) good sections of arithmetic fundamental groups. Point-theoretic sections of arithmetic fundamental groups (cf. Definition 2.2) are (uniformly) good sections.

3.1 Next, we recall the definition of the arithmetic Chern class associated to a group-theoretic section s of the arithmetic fundamental group Π_X (cf. [14], 1.2, and [1], for more details). In what follows all scheme cohomology groups are étale cohomology groups. First, let $\hat{\mathbb{Z}}^\Sigma$ be the maximal pro-Σ quotient of $\hat{\mathbb{Z}}$, and

$$M_X \overset{\text{def}}{=} \mathrm{Hom}(\mathbb{Q}/\mathbb{Z}, (K_X^{\text{sep}})^\times) \otimes_{\hat{\mathbb{Z}}} \hat{\mathbb{Z}}^\Sigma.$$

Note that M_X is a free $\hat{\mathbb{Z}}^\Sigma$-module of rank one, and has a natural structure of G_k-module, which is isomorphic to the G_k-module $\hat{\mathbb{Z}}(1)^\Sigma$, where the "(1)" denotes a Tate twist, i.e. G_k acts on $\hat{\mathbb{Z}}(1)^\Sigma$ via the Σ-part of the cyclotomic character. We will refer to M_X as the module of roots of unity attached to X, relative to the set of primes Σ. Let

$$\eta_X^{\text{diag}} \in H^2(X \times X, M_X)$$

be the étale Chern class, which is associated to the diagonal embedding $\iota \colon X \to X \times_k X$, or alternatively the first Chern class of the line bundle $O_{X \times X}(\iota(X))$. There exists a natural identification (cf. [7], proposition 1.1)

$$H^2(X \times X, M_X) \overset{\sim}{\to} H^2(\Pi_{X \times X}, M_X).$$

The Chern class η_X^{diag} corresponds via the above identification to an extension class

$$\eta_X^{\text{diag}} \in H^2(\Pi_{X \times X}, M_X).$$

We shall refer to the extension class η_X^{diag} as the extension class of the diagonal.

Let $s \colon G_k \to \Pi_X$ be a group-theoretic section of the natural projection $\Pi_X \twoheadrightarrow G_k$. Let

$$(3.1) \qquad\qquad 1 \to M_X \to \mathcal{D} \to \Pi_{X \times X} \to 1$$

be a group extension, whose class in $H^2(\Pi_{X \times X}, M_X)$ coincides with the extension class η_X^{diag} of the diagonal. By pulling back the group extension (3.1) by the continuous injective homomorphism

$$(s, \mathrm{id}) \colon G_k \times_{G_k} \Pi_X \to \Pi_{X \times X},$$

we obtain a natural commutative diagram:

$$
\begin{array}{ccccccccc}
1 & \longrightarrow & M_X & \longrightarrow & \mathcal{D}_s & \longrightarrow & G_k \times_{G_k} \Pi_X & \longrightarrow & 1 \\
 & & \downarrow{\scriptstyle \mathrm{id}} & & \downarrow & & \downarrow{\scriptstyle (s,\mathrm{id})} & & \\
1 & \longrightarrow & M_X & \longrightarrow & \mathcal{D} & \longrightarrow & \Pi_{X \times X} & \longrightarrow & 1
\end{array}
$$

where the right hand square is cartesian. Further, via the natural identification

$G_k \times_{G_k} \Pi_X \xrightarrow{\sim} \Pi_X$, the upper group extension $\mathcal{D}_s$ in the above diagram corresponds to a group extension (which we denote also $\mathcal{D}_s$)

$$1 \longrightarrow M_X \longrightarrow \mathcal{D}_s \longrightarrow \Pi_X \to 1.$$

We will refer to the class $[\mathcal{D}_s]$ of the extension $\mathcal{D}_s$ in $H^2(\Pi_X, M_X)$ as the extension class associated to the section s.

Definition 3.2 ((Σ)-Étale Chern class associated to a section) We define the (Σ)-étale Chern class $c(s) \in H^2(X, M_X)$ associated to the section s as the element of $H^2(X, M_X)$ corresponding to the above extension class $[\mathcal{D}_s]$, which is associated to the section s, via the natural identification $H^2(\Pi_X, M_X) \xrightarrow{\sim} H^2(X, M_X)$.

Let $\{X_i[s]\}_{i \geq 1}$ be a system of neighbourhoods of the section s, and $\{\Pi_X[i, s]\}_{i \geq 1}$ the corresponding open subgroups of Π_X (cf. 2.4). Recall that the section s restricts naturally to a group-theoretic section

$$s_i \colon G_k \to \Pi_X[i, s]$$

of the natural projection $\Pi_X[i, s] \twoheadrightarrow G_k$, which fits into the following commutative diagram:

$$
\begin{array}{ccc}
G_k & \xrightarrow{\ s_i\ } & \Pi_X[i, s] \\
{\scriptstyle \mathrm{id}}\downarrow & & \downarrow \\
G_k & \xrightarrow{\ s\ } & \Pi_X
\end{array}
$$

where the right vertical homomorphism is the natural inclusion, for each positive integer i (cf. loc. cit). One can easily observe the following lemma (see [14], lemma 1.3.1, for more details).

Lemma 3.3 *For each positive integer i, the image of the Chern class $c(s_{i+1}) \in H^2(X_{i+1}[s], M_X)$ associated to the section $s_{i+1} \colon G_k \to \Pi_X[i + 1, s]$ in $H^2(X[i, s], M_X)$, via the corestriction homomorphism $\mathrm{cor} \colon H^2(X_{i+1}[s], M_X) \to H^2(X_i[s], M_X)$, coincides with the Chern class $c(s_i)$ associated to the section $s_i \colon G_k \to \Pi_X[i, s]$.*

Definition 3.4 (Pro-(Σ)-étale Chern class associated to a section) Let

$$\varprojlim_{i \geq 1} H^2(X_i[s], M_X)$$

be the projective limit of the $H^2(X_i[s], M_X)$, where the transition homomorphisms are the corestriction homomorphisms. We define the pro-(Σ)-étale Chern class associated to the section s, relative to the system of neighbourhoods

$\{X_i[s]\}_{i\geq 1}$, as the element $\hat{c}(s) \overset{\text{def}}{=} (c(s_i))_{i\geq 1} \in \varprojlim_{i\geq 1} H^2(X_i[s], M_X)$ (cf. Lemma 3.3).

3.5 Next, we will introduce the notion of (uniformly) good sections of arithmetic fundamental groups. For each positive Σ-integer n, meaning that n is an integer which is divisible only by primes in Σ, the Kummer exact sequence in étale topology

$$1 \to \mu_n \to \mathbb{G}_m \xrightarrow{\ n\ } \mathbb{G}_m \to 1$$

induces naturally, for each positive integer i, an exact sequence of abelian groups

$$(3.2) \qquad 0 \to \text{Pic}(X_i[s])/n\,\text{Pic}(X_i[s]) \to H^2(X_i[s], \mu_n) \to_n \text{Br}(X_i[s]) \to 0,$$

which for positive integers m and n, with n divides m, fits naturally into a commutative diagram:

$$
\begin{array}{ccccccccc}
0 & \longrightarrow & \text{Pic}(X_i[s])/m\,\text{Pic}(X_i[s]) & \longrightarrow & H^2(X_i[s], \mu_m) & \longrightarrow & {}_m\text{Br}(X_i[s]) & \longrightarrow & 0 \\
 & & \downarrow & & \downarrow & & \downarrow & & \\
0 & \longrightarrow & \text{Pic}(X_i[s])/n\,\text{Pic}(X_i[s]) & \longrightarrow & H^2(X_i[s], \mu_n) & \longrightarrow & {}_n\text{Br}(X_i[s]) & \longrightarrow & 0
\end{array}
$$

where the lower and upper horizontal sequences are the above exact sequence (3.2), and the vertical homomorphisms are the natural homomorphisms. Here, $\text{Pic} \overset{\text{def}}{=} H^1(\ , \mathbb{G}_m)$ denotes the Picard group, $\text{Br} \overset{\text{def}}{=} H^2(\ , \mathbb{G}_m)$ is the Brauer–Grothendieck cohomological group, and for a positive integer n: ${}_n\text{Br} \subseteq \text{Br}$ is the subgroup of Br which is annihilated by n. By taking projective limits, the above diagram induces naturally, for each positive integer i, the following exact sequence

$$0 \to \varprojlim_{n\ \Sigma\text{-integer}} \text{Pic}(X_i[s])/n\,\text{Pic}(X_i[s]) \to$$

$$H^2(X_i[s], M_X) \to \varprojlim_{n\ \Sigma\text{-integer}} {}_n\text{Br}(X_i[s]) \to 0.$$

We will denote by

$$\text{Pic}(X_i[s])^{\wedge, \Sigma} \overset{\text{def}}{=} \varprojlim_{n\ \Sigma\text{-integer}} \text{Pic}(X_i[s])/n\,\text{Pic}(X_i[s])$$

the Σ-adic completion of the Picard group $\text{Pic}(X_i[s])$, and

$$T\,\text{Br}(X_i[s])^{\Sigma} \overset{\text{def}}{=} \varprojlim_{n\ \Sigma\text{-integer}} {}_n\text{Br}(X_i[s])$$

the Σ-Tate module of the Brauer group $\mathrm{Br}(X_i[s])$. Thus, we have a natural exact sequence

$$(3.3) \qquad 0 \to \mathrm{Pic}(X_i[s])^{\wedge,\Sigma} \to H^2(X_i[s], M_X) \to T\,\mathrm{Br}(X_i[s])^{\Sigma} \to 0.$$

In what follows we will identify $\mathrm{Pic}(X_i[s])^{\wedge,\Sigma}$ with its image in $H^2(X_i[s], M_X)$ (cf. the exact sequence (3.3)), and call it the Picard part of $H^2(X_i[s], M_X)$.

Let

$$s \colon G_k \to \Pi_X[1, s] \overset{\mathrm{def}}{=} \Pi_X$$

be a group-theoretic section of Π_X as above. For each positive integer i, let

$$s_i \colon G_k \to \Pi_X[i, s]$$

be the induced group-theoretic section of $\Pi_X[i, s]$. By pulling back cohomology classes via the section s_i, and bearing in mind the natural identifications $H^2(\Pi_X[i, s], M_X) \overset{\sim}{\to} H^2(X_i[s], M_X)$ (cf. [7], proposition 1.1), we obtain a natural (restriction) homomorphism

$$s_i^\star \colon H^2(X_i[s], M_X) \to H^2(G_k, M_X).$$

Finally, observe that if k' is a finite extension of k, and $X_{k'} \overset{\mathrm{def}}{=} X \times_k k'$, then we have a natural commutative diagram:

$$
\begin{array}{ccccccccc}
1 & \longrightarrow & \Delta_X & \longrightarrow & \Pi_{X_{k'}} \overset{\mathrm{def}}{=} \pi_1(X_{k'}, \eta)^{(\Sigma)} & \longrightarrow & G_{k'} & \longrightarrow & 1 \\
 & & {\scriptstyle \mathrm{id}}\big\downarrow & & \big\downarrow & & \big\downarrow & & \\
1 & \longrightarrow & \Delta_X & \longrightarrow & \Pi_X & \longrightarrow & G_k & \longrightarrow & 1
\end{array}
$$

where the right, and middle, vertical arrows are the natural inclusions, and the far right square is cartesian. In particular, the section $s_k \overset{\mathrm{def}}{=} s \colon G_k \to \Pi_X$ induces naturally a group-theoretic section $s_{k'} \colon G_{k'} \to \Pi_{X_{k'}}$ of $\Pi_{X_{k'}}$.

Definition 3.6 (Good and uniformly good sections of arithmetic fundamental groups) We say that the section s is a good group-theoretic section, relative to the system of neighbourhoods $\{X_i[s]\}_{i \geq 1}$, if the above homomorphisms $s_i^\star \colon H^2(X_i[s], M_X) \to H^2(G_k, M_X)$, for each positive integer $i \geq 1$, annihilate the Picard part $\mathrm{Pic}(X_i[s])^{\wedge,\Sigma}$ of $H^2(X_i[s], M_X)$. In other words, the section s is good if $\mathrm{Pic}(X_i[s])^{\wedge,\Sigma} \subseteq \mathrm{Ker}\, s_i^\star$, for each positive integer i. We say that the section s is uniformly good, relative to the system of neighbourhoods $\{X_i[s]\}_{i \geq 1}$, if the induced section $s_{k'} \colon G_{k'} \to \Pi_{X_{k'}}$ is good, relative to the system of neighbourhoods of $s_{k'}$ which is naturally induced by the $\{X_i[s]\}_{i \geq 1}$, for every finite extension k'/k.

It is easy to see that the above definition is independent of the given system of neighbourhoods $\{X_i[s]\}_{i\geq 1}$ of the section s. Indeed, let $\{Y_i[s]\}_{i\geq 1}$ be another system of neighbourhoods of the section s, and assume that the section s is good relative to the system of neighbourhoods $\{X_i[s]\}_{i\geq 1}$. Then for each $i \geq 1$, there exists $j_i \geq 1$, an integer depending on i, and a natural morphism $X_{j_i}[s] \to Y_i[s]$ (since the open subgroups $\{\Pi_X[i,s]\}_{i\geq 1}$ form a fundamental system of neighbourhoods of $\{1\}$ in Π_X). In particular, we obtain a natural commutative diagram

$$
\begin{array}{ccccc}
\mathrm{Pic}(X_{j_i}[s])^{\wedge,\Sigma} & \longrightarrow & H^2(X_{j_i}[s], M_X) & \xrightarrow{\ s_{j_i}{}^\star\ } & H^2(G_k, M_X) \\
\uparrow & & \uparrow & & \mathrm{id}\uparrow \\
\mathrm{Pic}(Y_i[s])^{\wedge,\Sigma} & \longrightarrow & H^2(Y_i[s], M_X) & \xrightarrow{\ s_i^\star\ } & H^2(G_k, M_X)
\end{array}
$$

where the horizontal maps have been already introduced, and the left and middle vertical maps are induced by the natural morphism $X_{j_i}[s] \to Y_i[s]$. From this it follows that if the image of $\mathrm{Pic}(X_{j_i}[s])^{\wedge,\Sigma}$ in $H^2(G_k, M_X)$ is zero, then the image of $\mathrm{Pic}(Y_i[s])^{\wedge,\Sigma}$ in $H^2(G_k, M_X)$ is also zero.

We will refer to a section satisfying the conditions in Definition 3.6 as good, or uniformly good, without necessarily specifying a system of neighbourhoods of the section. The notion of uniformly good sections is motivated by the fact that a necessary condition for a group-theoretic section $s\colon G_k \to \Pi_X$ to be point-theoretic, i.e. arises from a k-rational point $x \in X(k)$ (cf. Definition 2.2), is that the section s is uniformly good in the sense of Definition 3.6 (cf. [14], proposition 1.5.2).

Remarks 3.7   **(i)** If k is a p-adic local field, the conditions of goodness and uniform goodness for the section s are equivalent. Moreover, if $p \in \Sigma$, the section s is good in this case if and only if $X(k^{\mathrm{tame}}) \neq \emptyset$, where k^{tame} is the maximal tame extension of k (cf. [14], propositions 1.6.6 and 1.6.8).

(ii) If k is a number field, one has a local-global principle for (uniform) goodness. Namely, the section s is good in this case if and only if the section s_v is good, for each place v of k (cf. loc. cit., proposition 1.8.1).

Remark 3.8   Another necessary condition for the section s to be point theoretic is that the image of the pro-Chern class $\hat{c}(s) \in \varprojlim_{i\geq 1} H^2(X_i[s], M_X)$ in $\varprojlim_{i\geq 1} T\,\mathrm{Br}(X_i[s])^\Sigma$, via the natural homomorphism

$$
\varprojlim_{i\geq 1} H^2(X_i[s], M_X) \to \varprojlim_{i\geq 1} T\,\mathrm{Br}(X_i[s])^\Sigma,
$$

equals 0. In other words, if s is point theoretic, then the associated pro-Chern class $\hat{c}(s)$ (cf. Definition 3.4) lies in the Picard part $\varprojlim_{i\geq 1} \mathrm{Pic}(X_i[s])^{\wedge,\Sigma}$ of $\varprojlim_{i\geq 1} H^2(X_i[s], M_X)$. Indeed, assume that the section s is point theoretic, meaning that $s \overset{\mathrm{def}}{=} s_x \colon G_k \to \Pi_X$ arises from a k-rational point $x \in X(k)$ (cf. Definition 2.2). Then there exists a compatible system of rational points $\{x_i \in X_i[s](k)\}_{i\geq 1}$, i.e. x_{i+1} maps to x_i via the natural morphism $X_{i+1}[s] \to X_i[s]$. For every positive integer i, let $O(x_i) \in \mathrm{Pic}(X_i[s])$ be the degree 1 line bundle associated to x_i. The Chern class $c(s_i) \in H^2(X_i[s], M_X)$, which is associated to the section s_i (cf. Definition 3.2), coincides with the étale Chern class $c(x_i) \in H^2(X_i[s], M_X)$ associated to the line bundle $O(x_i)$ (cf. [7], proposition 1.6, and proposition 1.8 (iii)). Thus, the pro-Chern class $\hat{c}(s)$ is a Picard element. More precisely, it is the "pro-Picard element" induced by the $(O(x_i))_{i\geq 1}$. If the above condition is satisfied one says that the Chern class of the section s is algebraic, or that the section s is well behaved (cf. [14], proposition 1.5.1). One can easily prove that if the section s is well behaved, then the section s is good in the sense of Definition 3.6 (cf. [14], proposition 1.5.2).

4 Cuspidalisation of sections of arithmetic fundamental groups

In this section we introduce the problem of cuspidalisation of group-theoretic sections of arithmetic fundamental groups. We follow the notation of §1. In particular, $\Sigma \subseteq \mathfrak{Primes}$ is a non-empty set of prime integers, and we have the natural exact sequence

$$(1.2) \qquad 1 \to \Delta_X \to \Pi_X \xrightarrow{\mathrm{pr}_X} G_k \to 1,$$

where Π_X is the geometrically pro-Σ arithmetic fundamental group of X.

4.1 In this subsection we recall the definition of (geometrically) cuspidally central, and cuspidally abelian, arithmetic fundamental groups of affine hyperbolic curves, and the definition of cupidally abelian absolute Galois groups of function fields of curves.

4.1.1 Let $U \subseteq X$ be a non-empty open subscheme of X. The geometric point η of X determines naturally a geometric point η of U, and a geometric point $\bar{\eta}$ of $\overline{U} \overset{\mathrm{def}}{=} U \times_k \bar{k}$. Write

$$\Delta_U \overset{\mathrm{def}}{=} \pi_1(\overline{U}, \bar{\eta})^{\Sigma}$$

for the maximal pro-Σ quotient of the fundamental group $\pi_1(\overline{U}, \bar{\eta})$ of $\overline{U}$ with base point η, and

$$\Pi_U \overset{\text{def}}{=} \pi_1(U, \eta) / \operatorname{Ker}(\pi_1(\overline{U}, \bar{\eta}) \twoheadrightarrow \pi_1(\overline{U}, \bar{\eta})^\Sigma)$$

for the quotient of the arithmetic fundamental group $\pi_1(U, \eta)$ by the kernel of the natural surjective homomorphism $\pi_1(\overline{U}, \bar{\eta}) \twoheadrightarrow \pi_1(\overline{U}, \bar{\eta})^\Sigma$, which is a normal subgroup of $\pi_1(U, \eta)$. Thus, we have a natural exact sequence

$$1 \to \Delta_U \to \Pi_U \xrightarrow{\operatorname{pr}_{U,\Sigma}} G_k \to 1,$$

which fits into the following commutative diagram:

$$
\begin{array}{ccccccccc}
1 & \longrightarrow & \Delta_U & \longrightarrow & \Pi_U & \xrightarrow{\operatorname{pr}_{U,\Sigma}} & G_k & \longrightarrow & 1 \\
& & \downarrow & & \downarrow & & \text{id}\downarrow & & \\
1 & \longrightarrow & \Delta_X & \longrightarrow & \Pi_X & \xrightarrow{\operatorname{pr}_{X,\Sigma}} & G_k & \longrightarrow & 1
\end{array}
$$

where the left, and middle, vertical homomorphisms are surjective, and are naturally induced by the natural surjective homomorphisms $\pi_1(\overline{U}, \bar{\eta}) \twoheadrightarrow \pi_1(\overline{X}, \bar{\eta})$, and $\pi_1(U, \eta) \twoheadrightarrow \pi_1(X, \eta)$. Let

$$I_U \overset{\text{def}}{=} \operatorname{Ker}(\Pi_U \twoheadrightarrow \Pi_X) = \operatorname{Ker}(\Delta_U \twoheadrightarrow \Delta_X).$$

We shall refer to I_U as the cuspidal subgroup of Π_U (cf. [7], definition 1.5). It is the normal subgroup of Π_U generated by the (pro-Σ) inertia subgroups at the geometric points of $S \overset{\text{def}}{=} X \setminus U$. We have the following natural exact sequence

(4.1) $$1 \to I_U \to \Pi_U \to \Pi_X \to 1.$$

Let I_U^{ab} be the maximal abelian quotient of I_U. By pushing out the exact sequence (4.1) by the natural surjective homomorphism $I_U \twoheadrightarrow I_U^{\text{ab}}$, we obtain a natural commutative diagram:

$$
\begin{array}{ccccccccc}
1 & \longrightarrow & I_U & \longrightarrow & \Pi_U & \longrightarrow & \Pi_X & \longrightarrow & 1 \\
& & \downarrow & & \downarrow & & \text{id}\downarrow & & \\
1 & \longrightarrow & I_U^{\text{ab}} & \longrightarrow & \Pi_U^{\text{c-ab}} & \longrightarrow & \Pi_X & \longrightarrow & 1
\end{array}
$$

We will refer to the quotient $\Pi_U^{\text{c-ab}}$ of Π_U as the maximal cuspidally abelian quotient of Π_U, with respect to the natural homomorphism $\Pi_U \twoheadrightarrow \Pi_X$ (cf. [7], definition 1.5). Similarly, we can define the maximal cuspidally abelian quotient $\Delta_U^{\text{c-ab}}$ of Δ_U, with respect to the natural homomorphism $\Delta_U \twoheadrightarrow \Delta_X$, which sits in a natural exact sequence

(4.2) $$1 \to I_U^{\text{ab}} \to \Delta_U^{\text{c-ab}} \to \Delta_X \to 1.$$

Write I_U^{cn} for the maximal quotient of I_U^{ab} on which the action of Δ_X, which is naturally deduced from the exact sequence (4.2), is trivial. By pushing out the sequence (4.2) by the natural surjective homomorphism $I_U^{\mathrm{ab}} \twoheadrightarrow I_U^{\mathrm{cn}}$, we obtain a natural exact sequence

$$(4.3) \qquad 1 \to I_U^{\mathrm{cn}} \to \Delta_U^{\mathrm{c-cn}} \to \Delta_X \to 1.$$

Define

$$\Pi_U^{\mathrm{c-cn}} \overset{\mathrm{def}}{=} \Pi_U^{\mathrm{c-ab}} / \mathrm{Ker}(I_U^{\mathrm{ab}} \twoheadrightarrow I_U^{\mathrm{cn}}),$$

which sits naturally in the following exact sequence

$$(4.4) \qquad 1 \to I_U^{\mathrm{cn}} \to \Pi_U^{\mathrm{c-cn}} \to \Pi_X \to 1.$$

We shall refer to the quotient $\Pi_U^{\mathrm{c-cn}}$ of Π_U as the maximal (geometrically) cuspidally central quotient of Π_U, with respect to the natural homomorphism $\Pi_U \twoheadrightarrow \Pi_X$ (cf. loc. cit.). We have a natural commutative diagram of exact sequences:

$$
\begin{array}{ccccccccc}
1 & \longrightarrow & I_U & \longrightarrow & \Pi_U & \longrightarrow & \Pi_X & \longrightarrow & 1 \\
& & \downarrow & & \downarrow & & \mathrm{id}\downarrow & & \\
1 & \longrightarrow & I_U^{\mathrm{ab}} & \longrightarrow & \Pi_U^{\mathrm{c-ab}} & \longrightarrow & \Pi_X & \longrightarrow & 1 \\
& & \downarrow & & \downarrow & & \mathrm{id}\downarrow & & \\
1 & \longrightarrow & I_U^{\mathrm{cn}} & \longrightarrow & \Pi_U^{\mathrm{c-cn}} & \longrightarrow & \Pi_X & \longrightarrow & 1
\end{array}
$$

4.1.2 Similarly, we have a natural exact sequence of absolute Galois groups

$$1 \to G_{\bar{k}.K_X} \to G_{K_X} \to G_k \to 1,$$

where $G_{\bar{k}.K_X} \overset{\mathrm{def}}{=} \mathrm{Gal}(K_X^{\mathrm{sep}}/\bar{k}.K_X)$, and $G_{K_X} \overset{\mathrm{def}}{=} \mathrm{Gal}(K_X^{\mathrm{sep}}/K_X)$. Let

$$\overline{G}_X \overset{\mathrm{def}}{=} G_{\bar{k}.K_X}^{\Sigma}$$

be the maximal pro-Σ quotient of $G_{\bar{k}.K_X}$, and

$$G_X \overset{\mathrm{def}}{=} G_{K_X} / \mathrm{Ker}(G_{\bar{k}.K_X} \twoheadrightarrow G_{\bar{k}.K_X}^{\Sigma}),$$

which insert into the following commutative diagram of exact sequences:

$$
\begin{array}{ccccccccc}
1 & \longrightarrow & \overline{G}_X & \longrightarrow & G_X & \overset{\tilde{\mathrm{pr}}_{X,\Sigma}}{\longrightarrow} & G_k & \longrightarrow & 1 \\
& & \downarrow & & \downarrow & & \mathrm{id}\downarrow & & \\
1 & \longrightarrow & \Delta_X & \longrightarrow & \Pi_X & \overset{\mathrm{pr}_{X,\Sigma}}{\longrightarrow} & G_k & \longrightarrow & 1
\end{array}
$$

where the left vertical maps are the natural surjective homomorphisms (G_X and $\overline{G}_X$ are already defined in 2.7). Let

$$I_X \overset{\text{def}}{=} \text{Ker}(G_X \twoheadrightarrow \Pi_X) = \text{Ker}(\overline{G}_X \twoheadrightarrow \Delta_X).$$

We will refer to I_X as the cuspidal subgroup of G_X. It is the normal subgroup of G_X generated by the (pro-Σ) inertia subgroups at all geometric closed points of X. We have the following natural exact sequence

$$1 \to I_X \to G_X \to \Pi_X \to 1.$$

Let I_X^{ab} be the maximal abelian quotient of I_X. By pushing out the above sequence by the natural surjective homomorphism $I \twoheadrightarrow I^{\text{ab}}$, we obtain a natural exact sequence

(4.5) $$1 \to I^{\text{ab}} \to G_X^{\text{c-ab}} \to \Pi_X \to 1.$$

We will refer to the quotient $G_X^{\text{c-ab}}$ as the maximal cuspidally abelian quotient of G_X, with respect to the natural homomorphism $G_X \twoheadrightarrow \Pi_X$. Note that $G_X^{\text{c-ab}}$ is naturally identified with the projective limit

$$\varprojlim_U \Pi_U^{\text{c-ab}},$$

where the limit runs over all open subschemes U of X.

4.2 Next, we consider a continuous group-theoretic section $s\colon G_k \to \Pi_X$ of the natural projection $\text{pr}_X\colon \Pi_X \twoheadrightarrow G_k$.

Definition 4.3 (Lifting of group-theoretic sections)   Let $U \subseteq X$ be a non-empty open subscheme. We say a continuous group-theoretic section $s_U\colon G_k \to \Pi_U$, of the natural projection $\text{pr}_U \overset{\text{def}}{=} \text{pr}_{U,\Sigma}\colon \Pi_U \twoheadrightarrow G_k$ (meaning that $\text{pr}_U \circ s_U = \text{id}_{G_k}$), is a lifting of the section $s\colon G_k \to \Pi_X$, if s_U fits into a commutative diagram:

$$\begin{array}{ccc} G_k & \overset{s_U}{\longrightarrow} & \Pi_U \\ \text{id}\downarrow & & \downarrow \\ G_k & \overset{s}{\longrightarrow} & \Pi_X \end{array}$$

where the right vertical homomorphism is the natural one. More generally, we say that a group-theoretic section $\tilde{s}\colon G_k \to G_X$ of the natural projection $\tilde{\text{pr}} \overset{\text{def}}{=} \tilde{\text{pr}}_{X,\Sigma}\colon G_X \twoheadrightarrow G_k$ (meaning that $\tilde{\text{pr}} \circ \tilde{s} = \text{id}_{G_k}$) is a lifting of the section s, if $\tilde{s}$ fits into a commutative diagram:

$$
\begin{array}{ccc}
G_k & \xrightarrow{\tilde{s}} & G_X \\
\text{id} \downarrow & & \downarrow \\
G_k & \xrightarrow{s} & \Pi_X
\end{array}
$$

In connection with the Grothendieck anabelian section conjecture, it is natural to consider the following problem.

Cuspidalisation problem for sections of arithmetic fundamental groups
Given a group-theoretic section $s \colon G_k \to \Pi_X$ as above, and a non-empty open subscheme $U \subseteq X$, is it possible to construct a lifting $s_U \colon G_k \to \Pi_U$ of s? Similarly, given a group-theoretic section $s \colon G_k \to \Pi_X$ as above, is it possible to construct a lifting $\tilde{s} \colon G_k \to G_X$ of s?

Remarks 4.4 **(i)** One can easily verify that if the section s is point-theoretic, then the section s can be lifted to a section $s_U \colon G_k \to \Pi_U$ of the natural projection $\Pi_U \twoheadrightarrow G_k$, for every open subscheme $U \subseteq X$, and can also be lifted to a section $\tilde{s} \colon G_k \to G_X$ of the natural projection $G_X \twoheadrightarrow G_k$.

(ii) Note that a positive solution to the cuspidalisation problem, in the case where k is finitely generated over $\mathbb{Q}$ (resp. p-adic local field), and $\Sigma = \mathfrak{Primes}$, plus a positive solution to the BGASC (resp. p-adic BGASC), gives a positive solution to the GASC (resp. p-adic GASC), as follows easily from Lemma 2.5.

Our main result concerning the cuspidalisation problem is the following Theorem 4.6, which shows that good sections of arithmetic fundamental groups behave well with respect to this problem. Before stating our result, we recall some definitions.

Definition 4.5 (i) We say that the field k is slim if its absolute Galois group G_k is slim in the sense of [7], §0, meaning that every open subgroup of G_k is centre free. Examples of slim fields include number fields, and p-adic local fields (cf. [9], theorem 1.1.1). One defines in a similar way the notion of a slim profinite group G, meaning that every open subgroup of G is centre free.

(ii) We say that the field k is Σ-regular, if for every prime integer $l \in \Sigma$, and every finite extension k'/k, the l-part of the cyclotomic character $\chi_l \colon G_{k'} \to \mathbb{Z}_l^\times$ is not trivial; or equivalently, if for every prime integer $l \in \Sigma$ the image of the l-part of the cyclotomic character $\chi_l \colon G_k \to \mathbb{Z}_l^\times$ is infinite. Examples of Σ-regular fields (for every non-empty set Σ of prime integers) include number fields, p-adic local fields, and finite fields. The field k is Σ-regular if and only if, for every finite extension k'/k, the $G_{k'}$-module M_X has no non-trivial fixed elements.

The following is our main result on the cuspidalisation problem (cf. [14], theorem 2.6).

Theorem 4.6 (Lifting of uniformly good sections to cuspidally abelian arithmetic fundamental groups over slim fields) *Assume that the section* $s\colon G_k \to \Pi_X$ *is uniformly good (in the sense of Definition 3.6), and that the field k is slim, and Σ-regular (cf. Definition 4.5). Let $U \subseteq X$ be a non-empty open subscheme of X, and $\Pi_U^{\mathrm{c\text{-}ab}}$ the maximal cuspidally abelian quotient of Π_U, with respect to the natural surjective homomorphism $\Pi_U \twoheadrightarrow \Pi_X$. Then there exists a section $s_U^{\mathrm{c\text{-}ab}}\colon G_k \to \Pi_U^{\mathrm{c\text{-}ab}}$ of the natural projection $\Pi_U^{\mathrm{c\text{-}ab}} \twoheadrightarrow G_k$ which lifts the section s, i.e. which inserts into the following commutative diagram:*

$$
\begin{array}{ccc}
G_k & \xrightarrow{\ s_U^{\mathrm{c\text{-}ab}}\ } & \Pi_U^{\mathrm{c\text{-}ab}} \\
{\scriptstyle \mathrm{id}}\downarrow & & \downarrow \\
G_k & \xrightarrow{\ s\ } & \Pi_X
\end{array}
$$

Moreover, one can construct for every non-empty open subscheme $U \overset{\mathrm{def}}{=} X \setminus S$ of X a section $s_U^{\mathrm{c\text{-}ab}}\colon G_k \to \Pi_U^{\mathrm{c\text{-}ab}}$ as above (i.e. which lifts the section s), such that for every non-empty open subscheme $V \overset{\mathrm{def}}{=} X \setminus T$ of X, with $U \subseteq V$, we have the following commutative diagram:

$$
\begin{array}{ccc}
G_k & \xrightarrow{\ s_U^{\mathrm{c\text{-}ab}}\ } & \Pi_U^{\mathrm{c\text{-}ab}} \\
{\scriptstyle \mathrm{id}}\downarrow & & \downarrow \\
G_k & \xrightarrow{\ s_V^{\mathrm{c\text{-}ab}}\ } & \Pi_V^{\mathrm{c\text{-}ab}}
\end{array}
$$

where the right vertical homomorphism is the natural one.

As a corollary of the above result one obtains the following.

Corollary 4.7 (Lifting of uniformly good sections to cuspidally abelian Galois groups over slim fields) *Assume that the field k is slim, and Σ-regular (cf. Definition 4.5). Let $s\colon G_k \to \Pi_X$ be a uniformly good group-theoretic section of the natural projection $\Pi_X \twoheadrightarrow G_k$ (in the sense of Definition 3.6). Then there exists a section $s^{\mathrm{c\text{-}ab}}\colon G_k \to G_X^{\mathrm{c\text{-}ab}}$ of the natural projection $G_X^{\mathrm{c\text{-}ab}} \twoheadrightarrow G_k$, which lifts the section s, i.e. which inserts into the following commutative diagram:*

$$
\begin{array}{ccc}
G_k & \xrightarrow{\ s^{\mathrm{c\text{-}ab}}\ } & G_X^{\mathrm{c\text{-}ab}} \\
{\scriptstyle \mathrm{id}}\downarrow & & \downarrow \\
G_k & \xrightarrow{\ s\ } & \Pi_X
\end{array}
$$

4.8 Next, we would like to explain the basic idea behind the proof of Theorem 4.6. See loc. cit. for more details. Assume that the field k is slim, Σ-regular, and the section $s\colon G_k \to \Pi_X$ is uniformly good (in the sense of Definition 3.6). Let $U \overset{\text{def}}{=} X \setminus S$ be a non-empty open subscheme of X, and let $\Pi_U^{\text{c-cn}}$ be the maximal (geometrically) cuspidally central quotient of Π_U, with respect to the natural homomorphism $\Pi_U \twoheadrightarrow \Pi_X$. One would like to show, in a first step, that there exists a section $s_U^{\text{c-cn}}\colon G_k \to \Pi_U^{\text{c-cn}}$ of the natural projection $\Pi_U^{\text{c-cn}} \twoheadrightarrow G_k$ which lifts the section s, i.e. which inserts into the following commutative diagram:

$$
\begin{array}{ccc}
G_k & \xrightarrow{\; s_U^{\text{c-cn}} \;} & \Pi_U^{\text{c-cn}} \\
{\scriptstyle \text{id}}\downarrow & & \downarrow \\
G_k & \xrightarrow{\quad s \quad} & \Pi_X
\end{array}
$$

First, one treats the case where the set $S = \{x_i\}_{i=1}^n \subseteq X(k)$ consists of finitely many k-rational points. We will assume, without loss of generality, that $S = \{x\}$ consists of a single rational point $x \in X(k)$. Write $U_x \overset{\text{def}}{=} X \setminus \{x\}$. The maximal (geometrically) cuspidally central quotient $\Pi_{U_x}^{\text{c-cn}}$ of Π_{U_x}, with respect to the natural projection $\Pi_{U_x} \twoheadrightarrow \Pi_X$, sits naturally in the following exact sequence

$$(4.6) \qquad 1 \to M_X \to \Pi_{U_x}^{\text{c-cn}} \to \Pi_X \to 1$$

(cf. [7], proposition 1.8). By pulling back the group extension (4.6) by the section $s\colon G_k \to \Pi_X$, we obtain a group extension

$$(4.7) \qquad 1 \to M_X \to s^\star(\Pi_{U_x}^{\text{c-cn}}) \to G_k \to 1,$$

which inserts naturally in the following commutative diagram:

$$
\begin{array}{ccccccccc}
1 & \longrightarrow & M_X & \longrightarrow & s^\star(\Pi_{U_x}^{\text{c-cn}}) & \longrightarrow & G_k & \longrightarrow & 1 \\
& & {\scriptstyle \text{id}}\downarrow & & \downarrow & & {\scriptstyle s}\downarrow & & \\
1 & \longrightarrow & M_X & \longrightarrow & \Pi_{U_x}^{\text{c-cn}} & \longrightarrow & \Pi_X & \longrightarrow & 1
\end{array}
$$

where the right square is cartesian. The class in $H^2(\Pi_X, M_X)$ of the group extension (4.6) coincides, via the natural identification $H^2(\Pi_X, M_X) \overset{\sim}{\to} H^2(X, M_X)$, with the étale Chern class $c(x) \in H^2(\Pi_X, M_X)$ associated to the degree 1 line bundle $O(x)$ (cf. [10], lemma 4.2). The class in $H^2(G_k, M_X)$ of the group extension (4.7) coincides then with the image $s^\star(c(x))$ of the Chern class $c(x)$ via the (restriction) homomorphism $s^\star\colon H^2(X, M_X) \to H^2(G_k, M_X)$, which is naturally induced by s. This image equals 0, since the section s is assumed to be good. This follows from the very definition of goodness (cf. Definition 3.6). Thus, the group extension (4.7) admits group-theoretic splittings. A splitting of the group extension (4.7) determines a section $s_U^{c-cn}: G_k \to \Pi_U^{c-cn}$, which lifts the section s. The case where the set S consists of finitely many, not necessarily rational, points is treated in a similar way by using a descent argument, which resorts to the slimness of k, and the fact that k is Σ-regular.

In general, and in order to lift the section s to a section $s_U^{c-ab}: G_k \to \Pi_U^{c-ab}$, one uses the following description of Π_U^{c-ab}. For a finite étale Galois cover $X' \to X$, with Galois group $\mathrm{Gal}(X'/X)$, let $U' \overset{\mathrm{def}}{=} U \times_X X'$, and $\Pi_{U'}^{c-cn}$ the maximal (geometrically) cuspidally central quotient of $\Pi_{U'}$, with respect to the natural surjective homomorphism $\Pi_{U'} \twoheadrightarrow \Pi_{X'}$, which is slim. Denote by $\Pi_{U'}^{c-cn} \rtimes^{\mathrm{out}} \mathrm{Gal}(X'/X)$ the profinite group that is obtained by pulling back the exact sequence

$$1 \to \Pi_{U'}^{c-cn} \to \mathrm{Aut}(\Pi_{U'}^{c-cn}) \to \mathrm{Out}(\Pi_{U'}^{c-cn}) \to 1,$$

by the natural homomorphism:

$$\mathrm{Gal}(X'/X) \to \mathrm{Out}(\Pi_{U'}^{c-cn}).$$

Thus, we have a natural exact sequence:

$$1 \to \Pi_{U'}^{c-cn} \to \Pi_{U'}^{c-cn} \rtimes^{\mathrm{out}} \mathrm{Gal}(X'/X) \to \mathrm{Gal}(X'/X) \to 1.$$

which inserts into the following commutative diagram:

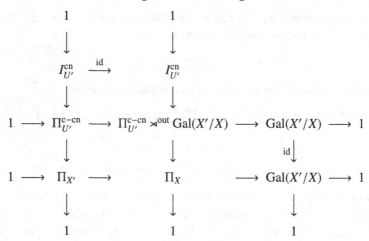

Then we have a natural isomorphism:

$$\Pi_U^{c-ab} \overset{\sim}{\to} \varprojlim_{X' \to X} \Pi_{U'}^{c-cn} \rtimes^{\mathrm{out}} \mathrm{Gal}(X'/X),$$

where the projective limit is taken over all finite étale Galois cover $X' \to X$ (cf. [14], proposition 2.5).

5 Applications to the Grothendieck anabelian section conjecture

In this section we state our main applications of the results concerning the cuspidalisation problem for sections of arithmetic fundamental groups in §4 to the Grothendieck anabelian section conjecture. As we have already mentioned (cf. Remarks 4.4(ii)), a positive answer to the cuspidalisation problem, in the case where k is finitely generated over $\mathbb{Q}$ (resp. p-adic local field), and $\Sigma = \mathfrak{Primes}$, plus a positive answer to the BGASC (resp. p-adic BGASC), implies a positive answer to the GASC (resp. p-adic GASC). The following result of Koenigsmann concerning the p-adic BGASC is fundamental (cf. [6]).

Theorem 5.1 ([6]) *The p-adic BGASC holds true. More precisely, assume that k is a p-adic local field, and $\Sigma = \mathfrak{Primes}$. Let $s\colon G_k \to G_X$ be a group-theoretic section of the natural projection $G_X \twoheadrightarrow G_k$. Then the image $s(G_k)$ is contained in a decomposition subgroup D_x associated to a unique rational point $x \in X(k)$. In particular, $X(k) \neq \varnothing$.*

This result has been strengthened by Pop, who proved the following (see [13]). For a profinite group H, and a prime integer p, we denote by H'' the maximal $\mathbb{Z}/p\mathbb{Z}$-metabelian quotient of H. Thus, H'' is the second quotient of the $\mathbb{Z}/p\mathbb{Z}$-derived series of H.

Theorem 5.2 ([13]) *Assume that k is a p-adic local field, which contains a primitive pth root of 1, and assume $p \in \Sigma$. Let $s\colon G_k'' \to G_{K_X}''$ be a group-theoretic section of the natural projection $G_{K_X}'' \twoheadrightarrow G_k''$. Then the image $s(G_k'')$ is contained in a decomposition subgroup $D_x \subset G_{K_X}''$ associated to a unique rational point $x \in X(k)$. In particular, $X(k) \neq \varnothing$. Here the $(\ )''$ of the various profinite groups are with respect to the prime p, i.e. the second quotients of the $\mathbb{Z}/p\mathbb{Z}$-derived series.*

The above theorem of Pop can be viewed as a very "minimalistic" version of the birational Grothendieck anabelian section conjecture over p-adic local fields. Note that the quotient G_k'' of G_k is finite in this case.

Let us mention few words on the proof of the above result of Pop. The technical tool in the proof, which produces a rational point $x \in X(k)$ starting from a section $s\colon G_k'' \to G_{K_X}''$, is the following. For a profinite group H, denote by H' the maximal quotient of H which is abelian and annihilated by p. Thus, H' is the first quotient of the $\mathbb{Z}/p\mathbb{Z}$-derived series of H. The existence of the section $s\colon G_k'' \to G_{K_X}''$ implies the existence of a section $s'\colon G_k' \to G_{K_X}'$ of the natural projection $G_{K_X}' \twoheadrightarrow G_k'$. Let $\tilde{L}/K_X$ be the subextension of K_X^{sep}/K_X with Galois group G_{K_X}', and L/K_X the subextension of $\tilde{L}/K_X$ corresponding to the subgroup $s'(G'_k)$ of G'_{K_X}. Then L is a field of transcendence degree 1 over k. For such a field, Pop proved a local-global principle for Brauer groups, which generalises a similar principle for function fields of curves over p-adic local fields due to Lichtenbaum, and which reads as follows. The natural diagonal homomorphism

$$\mathrm{Br}(L) \to \prod_v \mathrm{Br}(L_v),$$

where the product runs over all rank 1 valuations v of L, is injective. Now the existence of the section $s\colon G''_k \to G''_{K_X}$ implies that the natural homomorphism

$$\mathrm{Br}(k) \to \mathrm{Br}(L)$$

is injective (see [13] for more details). Let $\alpha \in \mathrm{Br}(k)$ be an element of order p. Then α survives in $\mathrm{Br}(L_v)$, i.e. its image is non-zero, for some rank 1 valuation v of L, by the above local-global principle. Pop then proves (the proof is rather technical) that v is the valuation associated to a unique k-rational point $x \in X(k)$.

5.3 Using the technique of cuspidalisation of sections of arithmetic fundamental groups, introduced in §4, one can hope to prove the p-adic GASC by reducing it to the p-adic version of the BGASC which was proved by Pop. We are able to prove that this is indeed the case, under some additional assumptions. Before stating our result we will define the notion of tame point-theoretic sections of cuspidally abelian absolute Galois groups of function fields over p-adic local fields.

Assume that k is a p-adic local field, and $p \in \Sigma$. Let G_X be the geometrically pro-Σ absolute Galois group of K_X, and $G_X^{\mathrm{c-ab}}$ the maximal cuspidally abelian quotient of G_X, with respect to the natural surjective homomorphism $G_X \twoheadrightarrow \Pi_X$ (cf. 4.1.2). Let

$$\tilde{s}\colon G_k \to G_X^{\mathrm{c-ab}}$$

be a continuous group-theoretic section of the natural projection $G_X^{\mathrm{c-ab}} \twoheadrightarrow G_k$. Let $\tilde{L}/K_X$ be the subextension of K_X^{sep}/K_X with Galois group $G_X^{\mathrm{c-ab}}$, and L/K_X the subextension of $\tilde{L}/K_X$ corresponding to the closed subgroup $\tilde{s}(G_k)$ of $G_X^{\mathrm{c-ab}}$. We say that the section $\tilde{s}$ is tame point-theoretic if the natural homomorphism

$$\mathrm{Br}(k) \to \mathrm{Br}(L)$$

is injective (cf. [14], 1.7, for more details). This condition is equivalent to the fact that the function field L admits k^{tame} rational points, where k^{tame} is the maximal tamely ramified extension of k (cf. loc. cit.). Our main result concerning the Grothendieck anabelian section conjecture over p-adic local fields is the following.

Theorem 5.4 *Assume that k is a p-adic local field which contains a primitive p-th root of 1, and $p \in \Sigma$. Let $s\colon G_k \to \Pi_X$ be a group-theoretic section of the natural projection $\Pi_X \twoheadrightarrow G_k$. Assume that s is a good section (in the sense of Definition 3.6). Then there exists a section $s^{c\text{-}ab}\colon G_k \to G_X^{c\text{-}ab}$ of the natural projection $G_X^{c\text{-}ab} \twoheadrightarrow G_k$, which lifts the section s. Furthermore, if the section $s^{c\text{-}ab}\colon G_k \to G_X^{c\text{-}ab}$ is a tame point-theoretic section (in the sense of 5.3), then $X(k) \neq \emptyset$.*

Proof The first assertion is Corollary 4.7. For a profinite group H, and a prime integer p, denote by H' the maximal quotient of H which is abelian, and annihilated by p. Thus, H' is the first quotient of the $\mathbb{Z}/p\mathbb{Z}$-derived series of H. The existence of the section $s^{c\text{-}ab}\colon G_k \to G_X^{c\text{-}ab}$ implies the existence of a section $s'\colon G_k' \to G_X'$ of the natural projection $G_X' \twoheadrightarrow G_k'$. Let $\tilde{L}/K_X$ be the sub-extension of K_X^{sep}/K_X which corresponds to the closed subgroup $\text{Ker}(G_X \twoheadrightarrow G_X')$ of G_X, and L/K_X the sub-extension of $\tilde{L}/K_X$ which corresponds to the closed subgroup $s'(G_k')$ of G_X'.

Assume that the section $s^{c\text{-}ab}\colon G_k \to G_X^{c\text{-}ab}$ is tame point-theoretic (in the sense of 5.3). Then the natural homomorphism $\text{Br}\, k \to \text{Br}\, L$ is injective. Under this assumption (which is implied, in the framework of the proof by Pop of Theorem 5.2, by the lifting property of the section $s'\colon G_k' \to G_X'$ to a section $s''\colon G_k'' \to G_X''$ which is imposed in [13]) Pop proves that the image $s'(G_k')$ is contained in a decomposition subgroup $D_x \subset G_{K_X}'$ associated to a unique rational point $x \in X(k)$ (cf. loc. cit). In particular, $X(k) \neq \emptyset$ in this case. □

With the same notations as in Theorem 5.4, the author expects that if s is a good section, then any lifting $s^{c\text{-}ab}\colon G_k \to G_X^{c\text{-}ab}$ of s (which exists by Corollary 4.7) is automatically tame point-theoretic in the sense of 5.3. (Indeed, this is the case if the section s is point-theoretic). In particular, following Theorem 5.4, every good section s should be point-theoretic, if $p \in \Sigma$. The author is unable to prove this at the time of writing this paper.

5.5 One can prove, by using cuspidalisation techniques of sections of arithmetic fundamental groups, the following (unconditional) version of the p-adic GASC.

Theorem 5.6 *Assume that k is a p-adic local field and $\Sigma = \{p\}$. Let $S \subset X$ be a set of closed points of X that is uniformly dense in X for the p-adic topology, meaning that for each finite extension k'/k, $S(k')$ is dense in $X(k')$*

for the p-adic topology. Write $\Pi_{X \setminus S}$ for the geometrically pro-Σ quotient of the arithmetic fundamental group $\pi_1(X \setminus S, \eta)$ (which is defined in a similar way as Π_X, cf. 1.1). Let $s: G_k \to \Pi_{X \setminus S}$ be a continuous group-theoretic section of the natural projection $\Pi_{X \setminus S} \twoheadrightarrow G_k$. Then s is point-theoretic, i.e. the image $s(G_k)$ in $\Pi_{X \setminus S}$ is a decomposition group $D_x \subset \Pi_{X \setminus S}$, associated to unique rational point $x \in X(k)$.

An example of a set S satisfying the assumptions of Theorem 5.6 is the set of algebraic points, in the case where X is defined over a number field

A proof of Theorem 5.6 was communicated orally to the author by A. Tamagawa. The proof relies on the idea of cuspidalisation, and consists in showing that the section $s: G_k \to \Pi_{X \setminus S}$ can be lifted to a section $\tilde{s}: G_k \to G_X$ of the natural projection $G_X \twoheadrightarrow G_k$, where G_X is the geometrically pro-Σ absolute Galois group of K_X. One then reduces the proof to the p-adic version of the BGASC that was proven by Pop. The proof uses the fact that the Galois group G_k of k is topologically finitely generated, and consists of showing that, given a (not necessarily geometrically connected) ramified Galois cover $Y \to X$ with Galois group G, one can "approximate" it by a Galois cover $Y' \to X$ with Galois group G, which is ramified only above points that are contained in S. The latter relies on an approximation argument, à la Artin, on Hurwitz spaces of covers.

5.7 In [2] sections of geometrically abelian absolute Galois groups of function fields of curves over number fields were investigated. It is shown in loc. cit. that the existence of such sections implies (in fact is equivalent to) the existence of degree 1 divisors on the curve, under a finiteness condition of the Tate–Shafarevich group of the jacobian of the curve. As an application of our results on the cuspidalisation of sections of arithmetic fundamental groups, we can prove an analogous result for good sections of arithmetic fundamental groups. Our main result concerning the Grothendieck anabelian section conjecture over number fields is the following.

Theorem 5.8 *Assume that k is a number field, and $\Sigma = \mathfrak{Primes}$. Let $s: G_k \to \Pi_X$ be a group-theoretic section of the natural projection $\Pi_X \twoheadrightarrow G_k$. Assume that s is a uniformly good section in the sense of Definition 3.6, and that the jacobian variety of X has a finite Tate–Shafarevich group. Then there exists a divisor of degree 1 on X.*

Proof Follows formally from Corollary 4.7, and [2], theorem 2.1.   □

6 On a weak form of the p-adic Grothendieck anabelian section conjecture

In this section we discuss a weak form of the p-adic GASC. We will use the following notations. Let $p > 0$ be a fixed prime integer. Let k be a p-adic local field, i.e. k is a finite extension of $\mathbb{Q}_p$, O_k its ring of integers, and F its residue field. Let X be a proper, smooth, geometrically connected, and hyperbolic curve over k. For a non-empty set of prime integers $\Sigma \subseteq \mathfrak{Primes}$, write Π_X for the geometrically pro-Σ fundamental group of X, which sits in the exact sequence

$$1 \to \Delta_X \to \Pi_X \to G_k \to 1,$$

where Δ_X is the maximal pro-Σ quotient of the fundamental group of $\overline{X} \stackrel{\text{def}}{=} X \times_k \bar{k}$ (cf. 1.1). Recall the natural map (cf. 2.1)

$$\varphi_X \stackrel{\text{def}}{=} \varphi_{X,\Sigma} \colon X(k) \to \overline{\mathrm{Sec}}_{\Pi_X}.$$

6.1 Assume that the hyperbolic k-curve X has good reduction over O_k, i.e. X extends to a smooth, proper, and relative curve $\mathcal{X}$ over O_k, and $p \notin \Sigma$. Let $\mathcal{X}_s \stackrel{\text{def}}{=} \mathcal{X} \times_{O_k} F$ be the special fibre of $\mathcal{X}$. Let ξ be a geometric point of $\mathcal{X}_s$ above the generic point of $\mathcal{X}_s$. Then ξ determines naturally an algebraic closure $\overline{F}$ of F, and a geometric point $\bar{\xi}$ of $\overline{\mathcal{X}_s} \stackrel{\text{def}}{=} \mathcal{X}_s \times_F \overline{F}$. There exists a natural exact sequence of profinite groups

$$1 \to \pi_1(\overline{\mathcal{X}_s}, \bar{\xi}) \to \pi_1(\mathcal{X}_s, \xi) \xrightarrow{\mathrm{pr}} G_F \to 1.$$

Here $\pi_1(\mathcal{X}_s, \xi)$ denotes the arithmetic étale fundamental group of $\mathcal{X}_s$ with base point ξ, $\pi_1(\overline{\mathcal{X}_s}, \bar{\xi})$ denotes the étale fundamental group of $\overline{\mathcal{X}_s} \stackrel{\text{def}}{=} \mathcal{X}_s \times_F \overline{F}$ with base point $\bar{\xi}$, and $G_F \stackrel{\text{def}}{=} \mathrm{Gal}(\overline{F}/F)$ denotes the absolute Galois group of F. Write

$$\Delta_{\mathcal{X}_s} \stackrel{\text{def}}{=} \pi_1(\overline{\mathcal{X}_s}, \bar{\xi})^\Sigma$$

for the maximal pro-Σ quotient of $\pi_1(\overline{X}, \bar{\xi})$, and

$$\Pi_{\mathcal{X}_s} \stackrel{\text{def}}{=} \pi_1(\mathcal{X}_s, \xi) / \mathrm{Ker}(\pi_1(\overline{\mathcal{X}_s}, \bar{\xi}) \twoheadrightarrow \pi_1(\overline{\mathcal{X}_s}, \bar{\xi})^\Sigma)$$

for the quotient of $\pi_1(\mathcal{X}_s, \xi)$ by the kernel of the natural surjective homomorphism $\pi_1(\overline{\mathcal{X}_s}, \bar{\xi}) \twoheadrightarrow \pi_1(\overline{X}, \bar{\xi})^\Sigma$, which is a normal subgroup of $\pi_1(\mathcal{X}, \xi))$. Thus, we have an exact sequence of profinite groups

$$(6.1) \qquad 1 \to \Delta_{\mathcal{X}_s} \to \Pi_{\mathcal{X}_s} \xrightarrow{\mathrm{pr}} G_F \to 1.$$

Moreover, after a suitable choice of the base points ξ, and η, there exists a natural commutative specialisation diagram:

$$1 \longrightarrow \Delta_X \longrightarrow \Pi_X \longrightarrow G_k \longrightarrow 1$$

(6.2)
$$\downarrow \qquad \mathrm{Sp}_X \downarrow \qquad \downarrow$$

$$1 \longrightarrow \Delta_{X_s} \longrightarrow \Pi_{X_s} \longrightarrow G_F \longrightarrow 1$$

where the left vertical homomorphism $\mathrm{Sp}\colon \Delta_X \to \Delta_{X_s}$ is an isomorphism (since we assumed $p \notin \Sigma$), and the right vertical homomorphism is the natural projection $G_k \twoheadrightarrow G_F$, as follows easily from the specialisation theory for fundamental groups of Grothendieck (cf. [3]). In fact, in the commutative diagram above (6.2), the right square is cartesian, as follows easily from the slimness of Δ_X, and the well-known criterion for good reduction of curves (cf. [14], lemma 4.2.2).

6.2 Let k_X and k_Y be two p-adic local fields, i.e. both k_X and k_Y are finite extensions of $\mathbb{Q}_p$. Let X (resp. Y) be a proper, smooth, geometrically connected and hyperbolic curve over k_X (resp. k_Y). Let $\Sigma \subseteq \mathfrak{Primes}$ be a non-empty set of prime integers, and let Π_X (resp. Π_Y) be the geometrically pro-Σ arithmetic fundamental group of X (resp. Y), which sits in the exact sequence $1 \to \Delta_X \to \Pi_X \to G_{k_X} \to 1$ (resp. $1 \to \Delta_Y \to \Pi_Y \to G_{k_Y} \to 1$). Let

$$\sigma\colon \Pi_X \xrightarrow{\sim} \Pi_Y$$

be an isomorphism between profinite groups.

Lemma 6.3 *The isomorphism σ fits into a commutative diagram:*

$$\Delta_X \longrightarrow \Delta_Y$$

$$\downarrow \qquad \qquad \downarrow$$

(6.3)
$$\Pi_X \xrightarrow{\ \sigma\ } \Pi_Y$$

$$\downarrow \qquad \qquad \downarrow$$

$$G_{k_X} \longrightarrow G_{k_Y}$$

where the horizontal maps are isomorphisms, which are naturally induced by σ. In particular, the isomorphism σ induces naturally a bijection

$$\sigma^{\mathrm{sec}}\colon \overline{\mathrm{Sec}}_{\Pi_X} \xrightarrow{\sim} \overline{\mathrm{Sec}}_{\Pi_Y}.$$

Moreover, the natural isomorphism $G_{k_X} \xrightarrow{\sim} G_{k_Y}$ which is induced by σ preserves the inertia subgroups, i.e. maps the inertia subgroup of G_{k_X} isomorphically to the inertia subgroup of G_{k_Y}.

Proof Cf. [11], proposition 1.2.1, and lemma 1.3.8.   □

Definition 6.4 We say that the isomorphism σ is point-theoretic if the image of $X(k)$ in $\overline{\mathrm{Sec}}_{\Pi_Y}$, via the map $\sigma^{\mathrm{sec}} \circ \varphi_{X,\Sigma} \colon X(k) \to \overline{\mathrm{Sec}}_{\Pi_Y}$, coincides with $\varphi_{Y,\Sigma}(Y(k_Y))$. In other words σ is point-theoretic if it induces naturally a bijection $\varphi_{X,\Sigma}(X(k_X)) \overset{\sim}{\to} \varphi_{Y,\Sigma}(Y(k_Y))$.

It is natural, in the framework of the p-adic GASC, to consider the following question.

Question 6.5 (Weak form of the Grothendieck anabelian section conjecture over p-adic local fields) Let k_X and k_Y be two p-adic local fields. Let X (resp. Y) be a proper, smooth, geometrically connected and hyperbolic curve over k_X (resp. k_Y). Assume that $\Sigma = \mathfrak{Primes}$. Let Π_X (resp. Π_Y) be the geometrically pro-Σ fundamental group of X (resp. Y), which sits in the exact sequence $1 \to \Delta_X \to \Pi_X \to G_{k_X} \to 1$ (resp. $1 \to \Delta_Y \to \Pi_Y \to G_{k_Y} \to 1$). Let

$$\sigma \colon \Pi_X \overset{\sim}{\to} \Pi_Y$$

be an isomorphism between profinite groups. Is σ point-theoretic, in the sense of Definition 6.4?

Note that the validity of the p-adic GASC implies a positive answer to Question 6.5.

Although the pro-Σ version of the Grothendieck anabelian section conjecture may not hold over p-adic local fields, in the case where $p \notin \Sigma$ (cf. [14], proposition 4.2.1), one may ask whether the following weak form of the pro-Σ Grothendieck anabelian section conjecture still holds if $p \notin \Sigma$.

Question 6.6 Let k_X and k_Y be two p-adic local fields. Let X (resp. Y) be a proper, smooth, geometrically connected and hyperbolic curve over k_X (resp. k_Y). Let $\Sigma \subset \mathfrak{Primes}$ be a non-empty set of prime integers, with $p \notin \Sigma$. Let Π_X (resp. Π_Y) be the geometrically pro-Σ fundamental group of X (resp. Y). Let

$$\sigma \colon \Pi_X \overset{\sim}{\to} \Pi_Y$$

be an isomorphism between profinite groups. Is σ point-theoretic?

Remarks 6.7 A positive answer to Question 6.6 will imply an absolute version of the Grothendieck anabelian conjecture for smooth, proper and hyperbolic curves over p-adic local fields (cf. [12], corollary 2.9).

In connection with Question 6.6, we can prove the following.

Proposition 6.8 *Let k_X and k_Y be two p-adic local fields. Let X (resp. Y) be a proper, smooth, geometrically connected and hyperbolic curve over k_X (resp.*

k_Y). *Let* $\Sigma \subset \mathfrak{Primes}$ *be a non-empty set of prime integers, with* $p \notin \Sigma$. *Let* Π_X *(resp.* Π_Y*) be the geometrically pro-Σ fundamental group of X (resp. Y). Let*

$$\sigma: \Pi_X \stackrel{\sim}{\to} \Pi_Y$$

be an isomorphism between profinite groups. Assume that X (or Y) has good reduction over k_X (or k_Y), i.e. X (or Y) extends to a proper and smooth relative curve over the valuation ring O_{k_X} of k_X (or over the valuation ring O_{k_Y} of k_Y). Then σ is point-theoretic.

Proof First, X has good reduction over k_X if and only if Y has good reduction over k_Y, as follows easily from the well-known criterion for good reduction, and the last assertion in Lemma 6.3. Assume that X (resp. Y) extends to a proper and smooth relative curve $\mathcal{X}$ over the valuation ring O_{k_X} of k_X (resp. $\mathcal{Y}$ over the valuation ring O_{k_Y} of k_Y). Let $\mathcal{X}_s$ (resp. $\mathcal{Y}_s$) be the special fiber of $\mathcal{X}$ (resp. $\mathcal{Y}$). Recall the commutative diagrams (6.2):

$$
\begin{array}{ccccccccc}
1 & \longrightarrow & \Delta_X & \longrightarrow & \Pi_X & \longrightarrow & G_{k_X} & \longrightarrow & 1 \\
& & \downarrow & & {\scriptstyle \mathrm{Sp}_X}\downarrow & & \downarrow & & \\
1 & \longrightarrow & \Delta_{X_s} & \longrightarrow & \Pi_{X_s} & \longrightarrow & G_{F_X} & \longrightarrow & 1
\end{array}
$$

and

$$
\begin{array}{ccccccccc}
1 & \longrightarrow & \Delta_Y & \longrightarrow & \Pi_Y & \longrightarrow & G_{k_Y} & \longrightarrow & 1 \\
& & \downarrow & & {\scriptstyle \mathrm{Sp}_Y}\downarrow & & \downarrow & & \\
1 & \longrightarrow & \Delta_{Y_s} & \longrightarrow & \Pi_{Y_s} & \longrightarrow & G_{F_Y} & \longrightarrow & 1
\end{array}
$$

where F_X (resp. F_Y) is the residue field of k_X (resp. k_Y).

Let $s = s_X: G_{k_X} \to \Pi_X$ be a group-theoretic section of the natural projection $\Pi_X \twoheadrightarrow G_{k_X}$, which is point-theoretic, i.e. arises from a rational point $x \in X(k_X)$. Then s induces naturally a group-theoretic section $s': G_{k_Y} \to \Pi_Y$ of the natural projection $\Pi_Y \twoheadrightarrow G_{k_Y}$, such that the following diagram is commutative:

$$
\begin{array}{ccc}
G_{k_X} & \longrightarrow & G_{k_Y} \\
s\downarrow & & s'\downarrow \\
\Pi_X & \stackrel{\sigma}{\longrightarrow} & \Pi_Y
\end{array}
$$

where the upper arrow is the natural isomorphism which is induced by σ. We will show that s' is point-theoretic. The natural map $\mathrm{Sp}_X \circ s_X: G_{k_X} \to \Pi_{X_s}$ factorises as $G_{k_X} \twoheadrightarrow G_{F_X} \stackrel{\bar{s}}{\longrightarrow} \Pi_{X_s}$, where $\bar{s}: G_{F_X} \to \Pi_{X_s}$ is a group-theoretic section of the natural projection $\Pi_{X_s} \twoheadrightarrow G_{F_X}$. The section $\bar{s}$ is point-theoretic since s is, and corresponds to a decomposition subgroup associated to the point $\bar{x} \in X_s(F_X)$, which is the specialisation of the rational point $x \in X(k_X)$. Let

$$\cdots \subseteq \Delta_X[i+1] \subseteq \Delta_X[i] \subseteq \cdots \subseteq \Delta_X[1] \overset{\text{def}}{=} \Delta_X$$

be a family of open characteristic subgroups of Δ_X, with $\bigcap_{i \geq 1} \Delta_X[i] = \{1\}$. We also denote by $\{\Delta_X[i]\}_{i \geq 1}$ the corresponding family of open subgroups of Δ_{X_s}, via the specialisation isomorphism $\text{Sp}_X \colon \Delta_X \overset{\sim}{\to} \Delta_{X_s}$.

For a positive integer i, let

$$\Pi_X[s, i] \overset{\text{def}}{=} \Delta_X[i].s(G_{k_X}), \qquad \Pi_{X_s}[\bar{s}, i] \overset{\text{def}}{=} \Delta_X[i].\bar{s}(G_{F_X}).$$

The system of open subgroups $\{\Pi_X[s, i]\}_{i \geq 1}$ (resp. $\{\Pi_{X_s}[\bar{s}, i]\}_{i \geq 1}$) corresponds to a tower of finite étale covers

$$\cdots \to X_{i+1} \to X_i \to \cdots \to X_1 \overset{\text{def}}{=} X,$$

(resp.

$$\cdots \to X_{s,i+1} \to X_{s,i} \to \cdots \to X_{s,1} \overset{\text{def}}{=} X_s),$$

where $\{X_i\}_{i \geq 1}$ (resp. $\{X_{s,i}\}_{i \geq 1}$) form a system of neighbourhoods of the section s (resp. $\bar{s}$). Let $\Pi_Y[i] \overset{\text{def}}{=} \sigma(\Pi_X[s, i])$, and $\Pi_{Y_s}[i] \overset{\text{def}}{=} \sigma(\Pi_{X_s}[\bar{s}, i])$. The natural map $\text{Sp}_Y \circ s' \colon G_{k_Y} \to \Pi_{Y_s}$ factorises as

$$G_{k_Y} \twoheadrightarrow G_{F_Y} \overset{\bar{s}'}{\longrightarrow} \Pi_{Y_s},$$

where $\bar{s}' \colon G_{F_Y} \to \Pi_{Y_s}$ is a group-theoretic section of the natural projection $\Pi_{Y_s} \twoheadrightarrow G_{F_Y}$ (this factorisation is induced by the similar factorisation of $\text{Sp}_X \circ s$). The $\{\Pi_Y[i]\}_{i \geq 1}$ (resp. $\{\Pi_{Y_s}[i]\}_{i \geq 1}$) form a system of neighbourhoods of the section s' (resp. $\bar{s}'$), and correspond to a tower of étale covers

$$\cdots \to Y_{i+1} \to Y_i \to \cdots \to Y_1 \overset{\text{def}}{=} Y$$

(resp.

$$\cdots \to \mathcal{Y}_{s,i+1} \to \mathcal{Y}_{s,i} \to \cdots \to \mathcal{Y}_{s,1} \overset{\text{def}}{=} \mathcal{Y}_s.)$$

The section $\bar{s}'$ is naturally induced by the section $\bar{s}$ via the natural isomorphism $\Pi_{X_k} \overset{\sim}{\to} \Pi_{Y_k}$. In particular, the section $\bar{s}'$ is point-theoretic, since $\bar{s}$ is point-theoretic, as follows from the arguments of Tamagawa for characterising decomposition groups of rational points over finite fields (see, e.g., [15], §1), hence $\mathcal{Y}_{s,i}(F_Y) \neq \varnothing$. Also, by the very definition of the system of neighbourhoods $\{Y_i\}_{i \geq 1}$, each Y_i has good reduction over O_{k_Y}, and extends to a smooth and proper model over O_{k_Y} whose special fibre is $\mathcal{Y}_{s,i}$. In particular, $Y_i(k_Y) \neq \varnothing$ by the theorem of liftings of smooth points (cf. [3], exposé III, corollaire 3.3). This implies that the section s' is point-theoretic by Lemma 2.5. $\qquad \square$

In connection to Question 6.6, and under the condition that the curve X (or Y) has potentially good reduction, one may hope to use Proposition 6.8, plus a descent argument, to give a positive answer to Question 6.6 in this case. The author has no idea, at the time of writing this paper, on how to perform such a descent argument.

References

[1] Esnault, H., Wittenberg, O., Remarks on cycle classes of sections of the fundamental group. Mosc. Math. J. 9 (2009), no. 3, 451-467.

[2] Esnault, H., Wittenberg, O., On abelian birational sections. Journal of the American Mathematical Society, 23 (2010), no. 3, 713–724.

[3] Grothendieck, A., Revêtements étales et groupe fondamental. Séminaire de géométrie algébrique du Bois Marie 1960–61.

[4] Grothendieck, A., Brief an G. Faltings (German with an English translation on pp. 285–293). London Math. Soc. Lecture Note Ser., 242, Geometric Galois actions, 1, 49-58, Cambridge University Press, Cambridge, 1997.

[5] Hoshi, Y., Existence of nongeometric pro-p Galois sections of hyperbolic curves. To appear in Publications of RIMS.

[6] Koenigsmann, J., On the section conjecture in anabelian geometry. J. Reine Angew. Math. 588 (2005), 221–235.

[7] Mochizuki, S., Absolute anabelian cuspidalizations of proper hyperbolic curves. J. Math. Kyoto Univ. 47 (2007), no. 3, 451–539.

[8] Mochizuki, S., The local pro-p anabelian geometry of curves. Invent. Math. 138 (1999), no. 2, 319–423.

[9] Mochizuki, S., Topics surrounding the anabelian geometry of hyperbolic curves. Galois groups and fundamental groups, 119–165, Math. Sci. Res. Inst. Publ., 41, Cambridge University Press, Cambridge, 2003.

[10] Mochizuki, S., Galois sections in absolute anabelian geometry. Nagoya Math. J. 179 (2005), 17–45.

[11] Mochizuki, S., The absolute anabelian geometry of hyperbolic curves. Galois theory and modular forms, 77–122, Dev. Math., 11, Kluwer, Boston, 2004.

[12] Mochizuki, S., Topics in absolute anabelian geometry II: decomposition groups and endomorphisms. Preprint. Available in the home web page of Shinichi Mochizuki.

[13] Pop. F., On the birational p-adic section Conjecture, Compositio Math. 146 (2010), no. 3, 621–637.

[14] Saïdi, M., Good sections of arithmetic fundamental groups. Manuscript.

[15] Saïdi, M., Tamagawa, A., A prime-to-p version of the Grothendieck anabelian conjecture for hyperbolic curves over finite fields of characteristic $p > 0$. Publ. Res. Inst. Math. Sci. 45 (2009), no. 1, 135–186.

[16] Tamagawa, A., The Grothendieck conjecture for affine curves. Compositio Math. 109 (1997), no. 2, 135–194.

From the classical to the noncommutative Iwasawa theory (for totally real number fields)

Mahesh Kakde

University College London

1 Introduction

The conjectures of Deligne [17], Beilinson [3] and Bloch–Kato [4] are a vast generalisation of the Dirichlet–Dedekind class number formula and Birch–Swinnerton-Dyer conjecture. They predict the order of arithmetic objects (such as class groups, Tate–Shafarevich groups, etc.) in terms of special values of L-functions. On the other hand, the aim of Iwasawa theory is to understand the Galois module structure of these arithmetic objects in terms of L-values. We roughly explain what may now be called classical Iwasawa theory. Let p be a prime. Let $\mathbb{Q}^{\mathrm{cyc}}$ be the cyclotomic $\mathbb{Z}_p$-extension of $\mathbb{Q}$ (see section 2). Let M be a motive over $\mathbb{Q}$. We assume that M is critical in the sense of Deligne (this means that the Euler factor at infinity $L_\infty(M, s)$ and $L_\infty(M^*(1), -s)$ are both holomorphic at $s = 0$, where M^* is the dual motive. For details see [14]). Assume that p is a good ordinary prime for M (in the sense of Greenberg [25]. This just means that the p-adic realisation of M has a finite decreasing filtration such that the action of inertia on the ith graded piece is via the ith power of the p-adic cyclotomic character). Let $\Gamma = \mathrm{Gal}(\mathbb{Q}^{\mathrm{cyc}}/\mathbb{Q}) \cong \mathbb{Z}_p$ and let $\Lambda(\Gamma)$ be the Iwasawa algebra $\mathbb{Z}_p[[\Gamma]]$ (see end of Section 2). Fix a topological generator γ of Γ. Then the Iwasawa algebra $\Lambda(\Gamma)$ is isomorphic to the power series ring $\mathbb{Z}_p[[T]]$. To such a M and p Greenberg attaches a $\Lambda(\Gamma)$-module called the Selmer module. Let us denote it by $\mathcal{S}$. Let

$$X := \mathrm{Hom}(\mathcal{S}, \mathbb{Q}_p/\mathbb{Z}_p)$$

be the Pontryagin dual of $\mathcal{S}$. It is conjectured that X is a torsion $\Lambda(\Gamma)$-module. (For example, if E is an elliptic curve defined over $\mathbb{Q}$ and p is a prime of good ordinary reduction for E and M is the motive $h^1(E)$ this was conjectured by Mazur and proven by Kato [33]). We assume this conjecture. Then the structure theory of finitely generated torsion $\Lambda(\Gamma)$-modules (see Section 3.3) gives a

Non-abelian Fundamental Groups and Iwasawa Theory, eds. John Coates, Minhyong Kim, Florian Pop, Mohamed Saïdi and Peter Schneider. Published by Cambridge University Press. ©Cambridge University Press 2012.

characteristic element $f \in \Lambda(\Gamma)$ of X. On the analytic side (see Coates [14]) one expects existence of the p-adic L-function of M, denoted by $\mathcal{L}_p$. This is expected to be an element of $J[[T]][\frac{1}{T}]$, where J is the p-adic completion of the ring of integers in the maximal unramified extension of $\mathbb{Q}_p$. The defining property of the p-adic L-function is that it interpolates the values at 0 of the complex L-function attached to the twisted motive $M(\rho)$ for almost all finite order $\rho \in \mathrm{Hom}_{\mathrm{cont}}(\Gamma, \mathbb{Z}_p^\times)$. Let n be the smallest non-negative integer such that $T^n \mathcal{L}_p \in J[[T]]$. Then the main conjecture (for the motive M) says that the ideal $T^n \mathcal{L}_p J[[T]]$ equals $f J[[T]]$.

The above formulation of the main conjecture works even when we replace $\mathbb{Q}^{cyc}$ by an extension F_∞ containing $\mathbb{Q}^{cyc}$ with $Gal(F_\infty/\mathbb{Q}) \cong H \times \Gamma$, where H is a finite abelian group of order prime to p. The main conjecture of Iwasawa is the case when M is the Tate motive $\mathbb{Q}(1)$. When the order of H is divisible by p, there isn't a nice structure theory for finitely generated torsion modules over $\mathbb{Z}_p[[H \times \Gamma]]$. To formulate the main conjecture over such extensions Kato [31] came up with a formulation using determinants. Independently Fontaine–Perrin-Riou [22] reformulated and generalised Tamagawa number conjectures (also called Bloch–Kato conjectures) using determinants instead of Tamagawa measures. Roughly, the determinant formulation replaces the Selmer group by *Selmer complex*, say SC. Put $\mathcal{G} = H \times \Gamma$. The Selmer complex SC is a perfect complex over $\Lambda(\mathcal{G})$. Using the determinants of Knudsen–Mumford one gets the invertible $\Lambda(\mathcal{G})$-module $\det_{\Lambda(\mathcal{G})}(SC)$ (see 5.4). Using a result of Tate (see proposition 2.1.3 in [23]) one knows that this invertible module is trivialisable. The main conjecture then says that there is a nice trivialisation related to L-values. This of course does not give an honest element of the Iwasawa algebra. Coates et al. [15] defined a canonical multiplicatively closed subset S^* of $\Lambda(\mathcal{G})$ satisfying Ore condition. It is conjectured in *loc. cit.* that $\Lambda(\mathcal{G})_{S^*} \otimes_{\Lambda(\mathcal{G})}^L SC$ is acyclic. One uses this to obtain an element in $K_1(\Lambda(\mathcal{G})_{S^*})$ from a trivialisation of $\det_{\Lambda(\mathcal{G})}(SC)$. We explain this in Section 6.4.

This formulation with determinants is very well suited for generalisations of Tamagawa number conjectures to motives with noncommutative coefficients. These generalisations appear in Burns–Flach [10], [11]. They use the determinants constructed by Deligne [18]. Variations of it appear in Huber–Kings [27] and Fukaya–Kato [23]. These conjectures are called Equivariant Tamagawa Number Conjectures. Here we follow Fukaya–Kato.

The aim of these notes is to give a rapid introduction to the main conjectures in noncommutative Iwasawa theory, focusing on the Iwasawa theory of the Tate motives over totally real p-adic Lie extensions, via the language used in the formulations of the Equivariant Tamagawa Number Conjectures. These notes contain nothing original. In fact, a very large portion of them is a very small portion of the article of Fukaya–Kato [23]. Nevertheless, we hope that these notes will familiarise beginners with the algebraic machinery used in the formulations of the main conjectures and serve as an introduction for the general conjectures in [10], [11], [27], [23] and [15]. The articles [49] and [21] are excellent and detailed surveys on Equivariant Tamagawa number conjectures. This list is of course by no means complete. I have just listed some of the articles that I have found useful as a student of the subject.

After setting up our notation in Section 2 we give a short survey of the classical Iwasawa main conjecture for totally real fields in Section 3. This is proven by Mazur–Wiles, Wiles and Rubin. For the generalisation we need some algebraic machinery, which we introduce in Sections 4 and 5. Section 4 contains the definition of K_0 and K_1 of a ring and Section 5 introduces the theory of determinants over noncommutative rings. In Section 6 we formulate the generalised Iwasawa main conjecture for totally real fields. This is the version of the main conjecture without p-adic L-functions. Such versions were first discussed by Huber–Kings [27]. We introduce the canonical Ore set of Coates–Fukaya–Kato–Sujatha–Venjakob [15] in Section 6.4. This conjecturally gives us a p-adic L-function. Under a certain hypothesis ($\mu = 0$) we know the validity of the noncommutative main conjecture for totally real number fields as we outline in Section 6.1. In Section 7 we indicate the generalisations to other motives again following Fukaya–Kato.

These notes grew out of the lectures given at the Newton Institute. Most of the work in proving Theorem 19 was done during this visit to the Newton Institute for the special program on 'Nonabelian Fundamental Groups in Arithmetic Geometry'. I thank the organisers of this program for inviting me to the Institute and for providing me with a great environment in which to work. I specially thank John Coates and Minhyong Kim for constant encouragement and many discussions. These notes were written during a visit to POSTECH, Pohang. I thank POSTECH for its hospitality.

2 The set up

Throughout this paper, p is an odd prime. Let F be a totally real number field. Let μ_n be the group of nth roots of 1. Let $\mu_{p^\infty} = \cup_{n \geq 0}\mu_{p^n}$. The extension $F(\mu_{p^\infty})/F$ contains a unique $\mathbb{Z}_p$ extension of F since $\mathrm{Gal}(F(\mu_{p^\infty})/F) \hookrightarrow \mathbb{Z}_p^\times \cong \mu_{p-1} \times (1 + p\mathbb{Z}_p)$. We call it the *cyclotomic $\mathbb{Z}_p$-extension* of F and denote it by F^{cyc}.

Definition 1 An admissible p-adic Lie extension of F is any totally real extension F_∞ of F satisfying:

(1) $F^{\mathrm{cyc}} \subset F_\infty$.

(2) $\mathcal{G} = \mathrm{Gal}(F_\infty/F)$ is a p-adic Lie extension.

(3) Only finitely many primes in F ramify in F_∞.

Let F_∞ be any admissible p-adic Lie extension of F. Let $H = \mathrm{Gal}(F_\infty/F^{\mathrm{cyc}})$ and let $\Gamma = \mathrm{Gal}(F^{\mathrm{cyc}}/F) \cong \mathbb{Z}_p$. Let Σ be a finite set of finite primes of F containing all the primes which ramify in F_∞. We denote by M the maximal abelian p-extension of F_∞ unramified outside primes above Σ.

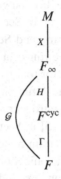

By its maximality, M is necessarily Galois over F, and thus we have a short exact sequence

$$1 \to X \to \mathrm{Gal}(M/F) \to \mathcal{G} \to 1$$

This gives an action of $\mathcal{G}$ on X: take any $g \in \mathcal{G}$ and let $\tilde{g}$ be any lift of g in $\mathrm{Gal}(M/F)$. Then g acts on X through conjugation by $\tilde{g}$. Since X is abelian this action is independent of the choice of lift. As X is pro-p we get an action of the group ring $\mathbb{Z}_p[\mathcal{G}]$ on X. Since X is compact this action extends to an action of the *Iwasawa algebra* $\Lambda(\mathcal{G}) := \mathbb{Z}_p[[\mathcal{G}]] := \varprojlim_U \mathbb{Z}_p[\mathcal{G}/U]$. Here U runs through open normal subgroups of $\mathcal{G}$. One can say that the aim of Iwasawa theory (for totally real number fields) is to understand X as a $\Lambda(\mathcal{G})$-module using special values of certain L-functions.

3 The classical main conjecture

Let us briefly describe the classical Iwasawa main conjecture as formulated by Iwasawa, Greenberg [24] and Coates [13]. In this section $\mathcal{G}$ is abelian. The short exact sequence

$$1 \to H \to \mathcal{G} \to \Gamma \to 1,$$

splits because Γ is pro-cyclic. As $\mathcal{G}$ is abelian we have an isomorphism $\mathcal{G} \cong H \times \Gamma$. We assume (for simplicity) that H is finite. Of course, if Leopoldt's conjecture holds for F and p, then H is necessarily finite. We remark that Leopoldt's conjecture is proven by Brumer [8] for all p if F is an abelian extension of $\mathbb{Q}$. We fix a topological generator γ of Γ. We have the following isomorphisms

$$\Lambda(\mathcal{G}) \cong \mathbb{Z}_p[H][[\Gamma]] \xrightarrow{\sim} \mathbb{Z}_p[H][[T]],$$

where the last ring is a power series ring in variable T with coefficients in $\mathbb{Z}_p[H]$ and the isomorphism is obtained by mapping γ to $1 + T$. We take this as an identification. For any $\chi \in \hat{H}$, a one dimensional character of H, there is a map

$$\chi : \mathbb{Z}_p[H][[T]] \to O_\chi[[T]],$$

$$\sum a_i T^i \mapsto \sum \chi(a_i) T^i.$$

Here O_χ is the ring obtained by adjoining values of χ to $\mathbb{Z}_p$. This extends to a map

$$\chi : Q(\mathbb{Z}_p[H][[T]]) \to Q(O_\chi[[T]]),$$

where for any ring R, we denote the total ring of fractions of R by $Q(R)$.

3.1 The p-adic L-function Deligne–Ribet [19], Cassou-Nogues [12] and Barsky [1] show that there exists $\zeta = \zeta(F_\infty/F) \in Q(\Lambda(\mathcal{G}))$ such that for any $\chi \in \widehat{H}$, the element $\chi(\zeta) \in Q(O_\chi[[T]])$ is the p-adic L-function, i.e. if under the p-adic cyclotomic character

$$\kappa_F : \mathrm{Gal}(F(\mu_{p^\infty})/F) \to \mathbb{Z}_p^\times, \qquad \text{if } \gamma \mapsto u,$$

then for any positive integer r divisible by $[F_\infty(\mu_p) : F_\infty]$, we have

$$\chi(\zeta)(u^r - 1) = L_\Sigma(\chi, 1 - r).$$

Here $u \equiv 1 (\mathrm{mod}\ p)$. Hence the power series $\chi(\zeta)(T)$ converges at $(u^r - 1)$. The quantity $L_\Sigma(\chi, 1 - r)$ is the value of the complex L-function associated to χ at odd negative integer (see Section 6.3). We can in fact write $\chi(\zeta)(T)$ as

$$\chi(\zeta)(T) = \frac{G_\chi(T)}{H_\chi(T)},$$

where $G_\chi(T) \in O_\chi[[T]]$ and

$$H_\chi(T) = \begin{cases} 1 & \text{if } \chi \neq 1, \\ T & \text{if } \chi = 1. \end{cases}$$

3.2 The Selmer group On the other hand $V = X \otimes \overline{\mathbb{Q}}_p$ is a finite dimensional $\overline{\mathbb{Q}}_p$ vector space (see Greenberg [24] or Iwasawa [28]). As a $\overline{\mathbb{Q}}_p \otimes \Lambda(G)$-module we have the decomposition

$$V = \oplus_{\chi \in \hat{H}} V^{(\chi)} = \oplus_{\chi \in \hat{H}} e_\chi V,$$

where

$$e_\chi = \frac{1}{|H|} \sum \chi(h) h^{-1} \in \overline{\mathbb{Q}}_p[H]$$

is the idempotent corresponding to χ. Let $f_\chi(T)$ be the characteristic polynomial of $\gamma - 1$ acting on $V^{(\chi)}$. Write $G_\chi(T)$ as

$$G_\chi(T) = \pi_\chi^{\mu_\chi} G_\chi^*(T),$$

where π_χ is a uniformiser of O_χ and $G_\chi^*(T)$ is not divisible by π_χ. The classical main conjecture for totally real fields as proven by Wiles is as follows.

Theorem 2 (Wiles)

$$G_\chi^*(T) O_\chi[[T]] = f_\chi(T) O_\chi[[T]].$$

This is theorem 1.3 in Wiles [50]. When F is abelian extension of $\mathbb{Q}$ this was proven by Mazur–Wiles [39] and by a different method by Rubin [38].

3.3 The characteristic element If the order of H is not divisible by p, then we may define μ_χ^{alg} as follows. The module $X^{(\chi)} = e_\chi(X \otimes O_\chi)$ is a finitely generated torsion $O_\chi[[T]]$-module by Greenberg [24]. There is a nice structure theory for finite generated modules over $O_\chi[[T]]$ (see [6], chapter 7). It says that if Y is a finite generated torsion $O_\chi[[T]]$-module then there is a homomorphism

$$0 \to \oplus_i O_\chi[[T]]/(f_i(T)) \xrightarrow{\alpha} Y \to D \to 0,$$

where D is a finite $O_\chi[[T]]$-module, i.e. α is a pseudo-isomorphism. The ideal of $O_\chi[[T]]$ generated by the product $\prod_i f_i(T)$ is independent of the choice of α and is called the *characteristic ideal* of Y. Any generator of this ideal is called a *characteristic element* of Y. Let $g_\chi(T)$ be a characteristic element of $X^{(\chi)}$. We write this power series as $g_\chi(T) = p^{\mu_\chi^{\text{alg}}} g_\chi^*(T)$, where the power series $g_\chi^*(T)$ is not divisible by p. It is not difficult to check that the ideal of $O_\chi[[T]]$ generated by $g_\chi^*(T)$ is same as the one generated by $f_\chi(T)$. Then Wiles also shows that

Theorem 3 (Wiles)

$$\mu_\chi^{\text{alg}} = \mu_\chi.$$

3.4 Some remarks This is theorem 1.4 in [50]. For this theorem Wiles assumes that p is an odd prime. At this point I am unaware of any good definition of μ_χ^{alg} when the order of χ is divisible by p.

In certain cases one can obtain finer information about Iwasawa modules. Kurihara [37] computes all the *Fitting ideals* of an Iwasawa module closely related to $X^{(\chi)}$ in terms of L-values (assuming the hypothesis $\mu = 0$ in Section 6.1). The 0th Fitting ideal is the characteristic ideal of the module. At present there is no analogue of higher Fitting ideals for modules over noncommutative rings.

We wish to study X as a $\Lambda(\mathcal{G})$-module whereas the theorem of Wiles describes only the isotypical parts of X. In [31] Kato proposed a generalisation of the above theorems. He called them 'generalised Iwasawa main conjecture'. We will describe these conjectures in our special situation (i.e. for totally real number fields). Fontaine–Perrin-Riou also obtained similar reformulation of the Bloch–Kato conjecture independently. Kato's philosophy is to think of the main conjectures as equivariant Bloch–Kato conjectures with big coefficients. These were later generalised by Burns–Flach and Huber–Kings and Fukaya–Kato to noncommutative situations.

4 Definition of K_0 and K_1

Throughout these notes all modules will be left modules unless otherwise is mentioned.

Definition 4 For any ring Λ, the group $K_0(\Lambda)$ is an abelian group, whose group law we denote additively, defined by the following set of generators and relations. *Generators:* $[P]$, where P is a finitely generated projective Λ-module. *Relations:* (i) $[P] = [Q]$ if P is isomorphic to Q as Λ-module, and (ii) $[P \oplus Q] = [P] + [Q]$.

It is easily seen that any element of $K_0(\Lambda)$ can be written as $[P] - [Q]$ for finitely generated projective Λ-modules P and Q, and $[P] - [Q] = [P'] - [Q']$ in $K_0(\Lambda)$ if and only if there is a finitely generated projective Λ-module R such that $P \oplus Q' \oplus R$ is isomorphic to $P' \oplus Q \oplus R$.

Definition 5 For any ring Λ, the group $K_1(\Lambda)$ is an abelian group, whose group law we denote multiplicatively, defined by the following generators and relations. *Generators:* $[P, \alpha]$, where P is a finitely generated projective Λ-module and α is an automorphism of P. *Relations:* (i) $[P, \alpha] = [Q, \beta]$ if there is an isomorphism f from P to Q such that $f \circ \alpha = \beta \circ f$, (ii) $[P, \alpha \circ \beta] = [P, \alpha][P, \beta]$, and (iii) $[P \oplus Q, \alpha \oplus \beta] = [P, \alpha][Q, \beta]$.

Here is an alternate description of $K_1(\Lambda)$. We have a canonical homomorphism $\mathrm{GL}_n(\Lambda) \to K_1(\Lambda)$ defined by mapping α in $\mathrm{GL}_n(\Lambda)$ to $[\Lambda^n, \alpha]$, where Λ^n is regarded as a set of row vectors and α acts on them from the right. Now using the inclusion maps $\mathrm{GL}_n(\Lambda) \hookrightarrow \mathrm{GL}_{n+1}(\Lambda)$ given by $g \mapsto \begin{pmatrix} g & 0 \\ 0 & 1 \end{pmatrix}$, we let

$$\mathrm{GL}(\Lambda) = \cup_{n \geq 1} \mathrm{GL}_n(\Lambda).$$

Then the canonical homomorphisms $\mathrm{GL}_n(\Lambda) \to K_1(\Lambda)$ induce an isomorphism (see, for example, [41], chapter 1)

$$\frac{\mathrm{GL}(\Lambda)}{[\mathrm{GL}(\Lambda), \mathrm{GL}(\Lambda)]} \xrightarrow{\sim} K_1(\Lambda),$$

where $[\mathrm{GL}(\Lambda), \mathrm{GL}(\Lambda)]$ is the commutator subgroup of $\mathrm{GL}(\Lambda)$. If Λ is commutative, then the determinant maps, $\mathrm{GL}_n(\Lambda) \to \Lambda^\times$, induce the determinant map

$$\det : K_1(\Lambda) \to \Lambda^\times,$$

via the above isomorphism. This gives a splitting of the canonical homomorphism $\Lambda^\times = \mathrm{GL}_1(\Lambda) \to K_1(\Lambda)$. If Λ is semilocal then Vaserstein ([47], [48]) proves that the canonical homomorphism $\Lambda^\times = \mathrm{GL}_1(\Lambda) \to K_1(\Lambda)$ is surjective. From these two facts we conclude that if Λ is a semilocal commutative ring, then the determinant map induces a group isomorphism between $K_1(\Lambda)$ and $\Lambda^\times$.

Example   If P is a compact p-adic Lie group, then $\Lambda(P)$ is a semilocal ring. Hence, in addition, if P is abelian then $K_1(\Lambda(P)) = \Lambda(P)^\times$.

5 The theory of determinants

We introduce the theory of determinants over arbitrary rings following the elementary construction of Fukaya–Kato. Determinants over arbitrary rings were first constructed by Deligne [18] using the category of virtual objects. This is used in noncommutative Equivariant Tamagawa Number Conjecture (ETNC) by Burns–Flach [10].

Let Λ be a ring with unit. Let $\mathcal{P}(\Lambda)$ be the category of finitely generated projective Λ-modules.

Definition 6   A perfect complex of Λ-modules is any object in the derived category of the category of Λ-modules which can be represented by a bounded complex of finitely generated projective Λ-modules.

Let $C(\Lambda)$ be the category of perfect complexes of Λ-modules with morphisms given by isomorphisms. Let $\mathcal{E}(\Lambda)$ be the category of short exact sequences of complexes of Λ-modules (actual complexes and not their class in the derived category) whose classes in the derived category are perfect. The morphisms in $\mathcal{E}(\Lambda)$ are quasi-isomorphisms.

5.1 Definition of the category $\mathcal{L}(\Lambda)$ We define the category $\mathcal{L}(\Lambda)$ as follows:

Objects Pairs (P, Q), where P and Q are finitely generated projective Λ-modules.

Morphisms $\mathrm{Mor}((P, Q), (P', Q')) = \emptyset$ if $[P] - [Q] \neq [P'] - [Q']$ in $K_0(\Lambda)$. If $[P] - [Q] = [P'] - [Q']$ in $K_0(\Lambda)$, then there is a finitely generated projective Λ-module R such that $P \oplus Q' \oplus R \cong P' \oplus Q \oplus R$. Let

$$G_R = \mathrm{Aut}(P' \oplus Q \oplus R) \quad \text{and} \quad I_R = \mathrm{Isom}(P \oplus Q' \oplus R, P' \oplus Q \oplus R).$$

Then the set of morphisms is defined as

$$\mathrm{Mor}((P, Q), (P', Q')) := K_1(\Lambda) \times^{G_R} I_R.$$

Here $K_1(\Lambda) \times^{G_R} I_R$ is the quotient of $K_1(\Lambda) \times I_R$ under the action of G_R given by

$$g \cdot (x, y) = (x\bar{g}, g^{-1}y),$$

where $g \in G_R$, $x \in K_1(\Lambda)$, $y \in I_R$ and $\bar{g}$ denotes the image of g in $K_1(\Lambda)$. We denote the class of (x, y) in $K_1(\Lambda) \times^{G_R} I_R$ by $[(x, y)]$. The set of morphisms between two objects is either empty or a $K_1(\Lambda)$-torsor. The set of automorphisms of any object is naturally isomorphic to $K_1(\Lambda)$.

Lemma 7 *The set* $\mathrm{Mor}((P, Q), (P', Q'))$ *is independent of the choice of R.*

Proof In this proof $i = 1, 2$. Let R_i be finitely generated projective Λ-modules such that

$$P \oplus Q' \oplus R_i \cong P' \oplus Q \oplus R_i.$$

Let $R = R_1 \oplus R_2$. Then

$$P \oplus Q' \oplus R \cong P' \oplus Q \oplus R$$

and we have

$$I_{R_i} \hookrightarrow I_R, \qquad G_{R_i} \hookrightarrow G_R.$$

This induces a maps

$$K_1(\Lambda) \times^{G_{R_i}} I_{R_i} \to K_1(\Lambda) \times^{G_R} I_R.$$

These maps are injective maps of $K_1(\Lambda)$-torsors and hence isomorphims. $\quad\square$

Composition in $\mathcal{L}(\Lambda)$ The composition of morphisms in $\mathcal{L}(\Lambda)$ is given as follows: Let (P_1, Q_1), (P_2, Q_2) and (P_3, Q_3) be objects in $\mathcal{L}(\Lambda)$ such that $[P_1] - [Q_1] = [P_2] - [Q_2] = [P_3] - [Q_3]$, hence $\mathrm{Mor}((P_1, Q_1), (P_2, Q_2))$ and $\mathrm{Mor}((P_2, Q_2), (P_3, Q_3))$ are non-empty sets. Let R_1 and R_2 be such that

$$P_1 \oplus Q_2 \oplus R_1 \cong P_2 \oplus Q_1 \oplus R_1,$$

$$P_2 \oplus Q_3 \oplus R_2 \cong P_3 \oplus Q_2 \oplus R_2.$$

Let $[(x_i, y_i)] \in K_1(\Lambda) \times^{G_{R_i}} I_{R_i} = \mathrm{Mor}((P_i, Q_i), (P_{i+1}, Q_{i+1}))$ for $i = 1, 2$. Let $R = Q_2 \oplus R_1 \oplus R_2$ and define $z \in \mathrm{Isom}(P_1 \oplus Q_3 \oplus R, P_3 \oplus Q_1 \oplus R)$ as follows

$$z \colon P_1 \oplus Q_3 \oplus Q_2 \oplus R_1 \oplus R_2 \xrightarrow{s} P_1 \oplus Q_2 \oplus R_1 \oplus Q_3 \oplus R_2 \xrightarrow{y_1 \oplus \mathrm{id}}$$

$$P_2 \oplus Q_1 \oplus R_1 \oplus Q_3 \oplus R_2 \xrightarrow{s} P_2 \oplus Q_3 \oplus R_2 \oplus Q_1 \oplus R_1 \xrightarrow{y_2 \oplus \mathrm{id}}$$

$$P_3 \oplus Q_2 \oplus R_2 \oplus Q_1 \oplus R_1 \xrightarrow{s} P_3 \oplus Q_1 \oplus Q_2 \oplus R_1 \oplus R_2.$$

Here we abuse the notation by using the same letter s for all the "switching maps" above. All other notation is hopefully self explanatory. Then

$$[(x_2, y_2)] \circ [(x_1, y_1)] = [(x_1 x_2 [(Q_2, -1)], z)].$$

We also have a bifunctor $\otimes \colon \mathcal{L}(\Lambda) \times \mathcal{L}(\Lambda) \to \mathcal{L}(\Lambda)$ defined by $(P, Q) \otimes (P', Q') = (P \oplus P', Q \oplus Q')$. This product is clearly associative. We define $\psi_{(P,Q),(P',Q')} \in \mathrm{Mor}((P, Q) \otimes (P', Q'), (P', Q') \otimes (P, Q))$ by

$$[(1, s)].$$

here $s \colon P \oplus P' \oplus Q' \oplus Q \xrightarrow{x} P' \oplus P \oplus Q \oplus Q'$ is the obvious automorphism. It is clear that $\psi_{(P',Q'),(P,Q)} \circ \psi_{(P,Q),(P',Q')} = \mathrm{id}_{(P,Q)\otimes(P',Q')}$. This clearly satisfies the pentagon and the hexagon axioms (see Knudsen [35]). This makes $\mathcal{L}(\Lambda)$ an *AC tensor category*. (For a definition of AC tensor category, see Breuning [7].) The AC tensor category $\mathcal{L}(\Lambda)$ is a *Picard category* because it is non-empty, every morphism is an isomorphism and for every object (P_0, Q_0) the functor $(P, Q) \mapsto (P_0, Q_0) \otimes (P, Q)$ is an autoequivalence of $\mathcal{L}(\Lambda)$. An inverse of (P, Q) is (Q, P). We will always denote by $(P, Q)^{-1}$ the object (Q, P).

Definition 8   There is a natural functor $\mathrm{Det}_\Lambda \colon \mathcal{P}(\Lambda)_{\mathrm{iso}} \to \mathcal{L}(\Lambda)$

$$\mathrm{Det}_\Lambda(P) := (P, 0).$$

For every short exact sequence

$$E \colon 0 \to A \to B \to C \to 0$$

in $\mathcal{P}(\Lambda)_{\mathrm{iso}}$ we define an isomorphism $\mathrm{Det}_\Lambda(E)\colon \mathrm{Det}_\Lambda(B) \to \mathrm{Det}_\Lambda(A)\otimes\mathrm{Det}_\Lambda(C)$ as $[(1,x)]$, where x is any isomorphism $B \xrightarrow{\sim} A \oplus C$ induced by the short exact sequence E.

We leave it as an elementary exercise to check that the definition of $\mathrm{Det}_\Lambda(E)$ is independent of the choice of x. For example according to our definition if $f\colon A \to A$ is an isomorphism in $\mathcal{P}(\Lambda)$ then $\mathrm{Det}_\Lambda(0 \to A \xrightarrow{f} A \to 0 \to 0)$ is $\mathrm{Det}_\Lambda(f)$ and $\mathrm{Det}_\Lambda(0 \to 0 \to A \xrightarrow{f} A \to 0)$ is $\mathrm{Det}_\Lambda(f)^{-1}$.

The functor Det_Λ is a *determinant functor* from the exact category $\mathcal{P}(\Lambda)$ to the Picard category $\mathcal{L}(\Lambda)$. One can directly show that it is a *universal determinant functor* in the sense of Deligne [18] (see also Breuning [7] and Knudsen [35]). Alternately we can use the obvious functor from $\mathcal{L}(\Lambda)$ into the category of virtual objects constructed by Deligne (such a functor exists because the category of virtual objects is universal) and then use the result of Deligne which states that if a monoidal functor between Picard categories induces an isomorphism on the π_0 and π_1 of the categories then it is an equivalence of categories (see Breuning [7]). Here π_0 of a Picard category is the set of isomorphism classes of objects in the category with the group operation given by the bifunctor $\otimes$ on the category. The π_1 of a Picard category is the group of automorphisms of any objects in the Picard category. Hence $\pi_0(\mathcal{L}(\Lambda)) = K_0(\Lambda)$ and $\pi_1(\mathcal{L}(\Lambda)) = K_1(\Lambda)$.

5.2 Extension to the derived category Let $C^\cdot$ be a bounded complex of finitely generated projective Λ-modules. Let C_{even} and C_{odd} denote the direct sum of the even and odd degree terms of $C^\cdot$ respectively. Define

$$\mathrm{Det}_\Lambda(C^\cdot) := (C_{\mathrm{even}}, C_{\mathrm{odd}}).$$

If $C^\cdot$ is acyclic then there is a canonical isomorphism

$$\mathrm{Det}_\Lambda(0) \to \mathrm{Det}_\Lambda(C^\cdot)$$

defined as follows: let I^i be the image of C^i in C^{i+1}. Then there is an exact sequence $0 \to I^{i-1} \to C^i \to I^i \to 0$. Hence there is an isomorphism

$$\mathrm{Det}_\Lambda(C^i) \cong \mathrm{Det}_\Lambda(I^{i-1}) \otimes \mathrm{Det}_\Lambda(I^i).$$

Hence we get an isomorphism $\mathrm{Det}_\Lambda(C_{\mathrm{even}}) \cong \mathrm{Det}_\Lambda(C_{\mathrm{odd}})$ which induces $\mathrm{Det}_\Lambda(0) \cong \mathrm{Det}_\Lambda(C^\cdot)$. Let $f\colon C_1^\cdot \to C_2^\cdot$ be a quasi-isomorphism. Then f induces an isomorphism $\mathrm{Det}_\Lambda(f)\colon \mathrm{Det}_\Lambda(C_1^\cdot) \to \mathrm{Det}_\Lambda(C_2^\cdot)$ as follows. Let $C_3^\cdot$ be the mapping cone of f so that we have an exact sequence

$$0 \to C_2^\cdot \to C_3^\cdot \to C_1^\cdot[1] \to 0$$

and $C_3^{\cdot}$ is acyclic. Then $\mathrm{Det}_\Lambda(0) \cong \mathrm{Det}_\Lambda(C_3^{\cdot}) \cong \mathrm{Det}_\Lambda(C_2^{\cdot}) \otimes \mathrm{Det}_\Lambda(C_1^{\cdot}[1]) = \mathrm{Det}_\Lambda(C_2^{\cdot}) \otimes \mathrm{Det}_\Lambda(C_1^{\cdot})^{-1}$. This gives an isomorphism

$$\mathrm{Det}_\Lambda(f): \ \mathrm{Det}_\Lambda(C_1^{\cdot}) \cong \mathrm{Det}_\Lambda(C_2^{\cdot}).$$

This gives an extension of the functor $\mathrm{Det}_\Lambda : \mathcal{P}(\Lambda) \to \mathcal{L}(\Lambda)$ to

$$\mathrm{Det}_\Lambda : C(\Lambda) \to \mathcal{L}(\Lambda).$$

And for every object $E: 0 \to A^{\cdot} \to B^{\cdot} \to C^{\cdot} \to 0$ in $\mathcal{E}(\Lambda)$ there is an isomorphism

$$\mathrm{Det}_\Lambda(E): \ \mathrm{Det}_\Lambda(B^{\cdot}) \to \mathrm{Det}_\Lambda(A^{\cdot}) \otimes \mathrm{Det}_\Lambda(C^{\cdot}).$$

defined as before.

Let $f, g: C_1 \to C_2$ be two quasi-isomorphisms between bounded complexes of finitely generated projective Λ-modules. If f and g are homotopic then $\mathrm{Det}_\Lambda(f) = \mathrm{Det}_\Lambda(g)$ (by corollary 2.12 in Knudsen [35]). Hence the functor Det_Λ factorises through the category $C(\Lambda)$.

5.3 Change of rings Let Λ and Λ' be rings. Let Y be a finitely generated projective Λ'-module endowed with a structure of a right Λ-module such that the actions of Λ and Λ' on Y commute. Then we have a functor

$$Y \otimes_\Lambda : \mathcal{L}(\Lambda) \to \mathcal{L}(\Lambda')$$

$$(P, Q) \mapsto (Y \otimes_\Lambda P, Y \otimes_\Lambda Q).$$

Example   For a ring homomorphism $\Lambda \to \Lambda'$, we have a functor

$$\Lambda' \otimes_\Lambda : \mathcal{L}(\Lambda) \to \mathcal{L}(\Lambda').$$

which takes $\mathrm{Det}_\Lambda(P)$ to $\mathrm{Det}_{\Lambda'}(\Lambda' \otimes_\Lambda P)$.

Example   If a homomorphism $\Lambda' \to \Lambda$ makes Λ a finitely generated Λ'-module, then

$$\Lambda \otimes_\Lambda : \mathcal{L}(\Lambda) \to \mathcal{L}(\Lambda').$$

This functor is obtained by restricting to the action of Λ' on the finitely generated projective Λ-modules. Then the induced homomorphism $K_1(\Lambda) = \mathrm{Aut}(\mathrm{Det}_\Lambda(0)) \to \mathrm{Aut}(\mathrm{Det}_{\Lambda'}(0)) = K_1(\Lambda')$ is the norm homomorphism.

5.4 Relation with Knudsen–Mumford determinants If Λ is a commutative ring then for any finitely generated projective Λ-module P we have the determinant module $\det_\Lambda(P) = \wedge^n P$ (n is the rank of P as a Λ-module) the highest exterior power. One can extend this to construct an invertible Λ-module $\det_\Lambda(C^\cdot)$ for any bounded complex of finitely generated projective Λ-modules $C^\cdot$. Roughly, the determinants of a complex are alternating tensor products of determinants of modules in the complex. However, doing this in a functorial manner leads to difficulties with signs. Knudsen–Mumford [36] show how to get around this problem. The determinant module $\det_\Lambda(C^\cdot)$ is related to $\text{Det}_\Lambda(C^\cdot)$ as follows.

If $C^\cdot$ is such that the class $[C^\cdot] := \sum_i (-1)^i [C^i]$ of $C^\cdot$ in $K_0(\Lambda)$ is 0, then

$$\det_\Lambda(C^\cdot) = \Lambda \times^{K_1(\Lambda)} \text{Isom}(\text{Det}_\Lambda(0), \text{Det}_\Lambda(C^\cdot)).$$

Here $K_1(\Lambda)$ acts on Λ via the determinant map $\det\colon K_1(\Lambda) \to \Lambda^\times$. More generally, if $C_1^\cdot$ and $C_2^\cdot$ are two bounded complexes of finitely generated projective Λ-modules such that $[C_1^\cdot] = [C_2^\cdot]$ in $K_0(\Lambda)$, then

$$\Lambda^\times \times^{K_1(\Lambda)} \text{Isom}(\text{Det}_\Lambda(C_1^\cdot), \text{Det}_\Lambda(C_2^\cdot)) \xrightarrow{\sim} \text{Isom}_\Lambda(\det_\Lambda(C_1^\cdot), \det_\Lambda(C_2^\cdot))$$

and

$$\Lambda \times^{K_1(\Lambda)} \text{Isom}(\text{Det}_\Lambda(C_1^\cdot), \text{Det}_\Lambda(C_2^\cdot)) \xrightarrow{\sim} \text{Hom}_\Lambda(\det_\Lambda(C_1^\cdot), \det_\Lambda(C_2^\cdot)).$$

5.5 Relation with characteristic elements Now let Λ be a ring of the form $O[[T]]$ as in Section 3 with O the ring of integers in a finite extension of $\mathbb{Q}_p$. Let X be a finitely generated torsion Λ-module. Then using the structure theory we may assume that $X \cong \oplus_{i=1}^n \Lambda/\Lambda f_i$. We take the following projective resolution of X

$$0 \to \Lambda^n \xrightarrow{(f_1,\ldots,f_n)} \Lambda^n \to X \to 0.$$

Let $C^\cdot$ be the complex $\Lambda^n \xrightarrow{(f_1,\ldots,f_n)} \Lambda^n$ concentrated in degree 0 and 1. Then the class of $C^\cdot$ is zero in $K_0(\Lambda)$. Hence there is an isomorphism

$$\lambda\colon \text{Det}_\Lambda(0) \to \text{Det}_\Lambda(C^\cdot).$$

After tensoring with $Q(\Lambda)$ we get

$$Q(\Lambda) \otimes_\Lambda \lambda\colon \text{Det}_{Q(\Lambda)}(0) \to \text{Det}_{Q(\Lambda)}(Q(\Lambda) \otimes_\Lambda C^\cdot) \cong \text{Det}_{Q(\Lambda)}(0).$$

Since $\text{Aut}(\text{Det}_{Q(\Lambda)}(0)) = K_1(Q(\Lambda)) = Q(\Lambda)^\times$ this isomorphism gives an element $f \in Q(\Lambda)^\times$. Since λ is unique up to the action of $K_1(\Lambda) = \Lambda^\times$, the ideal Λf is independent of any choices. It is easily seen that the principal fractional ideal Λf is the fractional ideal $\Lambda(f_1 \cdots f_n)^{-1} = \text{char}_\Lambda(C^\cdot)^{-1}$.

We use this observation to generalise the main conjecture in the next section.

6 Generalised Iwasawa main conjecture

We now return to the set up in Section 2. We take an admissible p-adic Lie extension F_∞/F with with Galois group $\mathcal{G} = \mathrm{Gal}(F_\infty/F)$. Let $\mathbb{Z}_p(1) = \varprojlim_i \mu_{p^i}$.

As a group this is just the additive group of p-adic integers but the action of $\mathrm{Gal}(\overline{F}/F)$ is through the p-adic cyclotomic character

$$g \cdot x = \kappa_F(g)x.$$

More conceptually one thinks of $\mathbb{Z}_p(1)$ as a Galois invariant lattice in the p-adic realisation of the Tate motive $\mathbb{Q}(1)$. Put $T := T(F_\infty/F) := \Lambda(\mathcal{G}) \otimes \mathbb{Z}_p(1)$ where $\Lambda(\mathcal{G})$ acts on T through the first factor. The group $\mathrm{Gal}(\overline{F}/F)$ acts on T by

$$g \cdot (x \otimes y) = (x\overline{g}^{-1} \otimes g(y)).$$

Here $\overline{g}$ denote the image in $\mathcal{G}$ of $g \in \mathrm{Gal}(\overline{F}/F)$.

Remark For more general conjectures as in Fukaya–Kato, Burns–Flach, Huber–Kings, etc., one takes the $\Lambda(\mathcal{G})$-modules obtained by tensoring $\Lambda(\mathcal{G})$ with Galois invariant lattices in p-adic realisations of motives (see Section 7).

6.1 Galois cohomology Let G be any pro-finite group and let A be a topological abelian group endowed with a continuous action of G. Let $C(G,A)$ be the complex of continuous cochains of G with values in A. We denote by $R\Gamma(G,A)$ the complex $C(G,A)$ regarded as an object of the derived category. Let $H^m(G,A)$ be the mth cohomology group of the complex $C(G,A)$.

Let F_Σ be the maximal extension of F unramified outside Σ. Let $U = \mathrm{Spec}(O_F[\frac{1}{\Sigma}])$. If $G = \mathrm{Gal}(F_\Sigma/F)$, then we denote $C(G,A)$, $R\Gamma(G,A)$ and $H^m(G,A)$ by $C(U,A)$, $R\Gamma(U,A)$ and $H^m(U,A)$ respectively. If A is a finite group of p-power order then $R\Gamma(U,A)$ is same as the etale cohomology groups $R\Gamma(U_{\mathrm{et}},A)$, where A denotes the sheaf on etale site of U corresponding to A.

For any prime v of F, we let F_v denote the completion of F at v. If $G = \mathrm{Gal}(F_v/F)$, then we denote $C(G,A)$, $R\Gamma(G,A)$ and $H^m(G,A)$ by $C(F_v,A)$, $R\Gamma(F_v,A)$ and $H^m(F_v,A)$ respectively.

We now define the compact support cohomology. Let $f \colon A^{\cdot} \to B^{\cdot}$ be a map of complexes. Then mapping fibre of f is defined as the complex $C^{\cdot}$ with $C^n = A^n \oplus B^{n-1}$. If d_A and d_B are differential of $A^{\cdot}$ and $B^{\cdot}$ respectively then the differential d_C of $C^{\cdot}$ is given by

$$d_C(a,b) = (-d_A(a), f(a) + d_B(b)).$$

The mapping fibre sits in an exact triangle

$$C^{\cdot} \to A^{\cdot} \to B^{\cdot} \to .$$

We define $C_c(U, A)$ to be the mapping fibre of $C(U, A) \rightarrow \oplus_{v \notin U} C(F_v, A)$. We denote by $R\Gamma_c(U, A)$ the complex $C_c(U, A)$ regarded as an object in the derived category, and $H_c^m(U, A)$ be the mth cohomology of $C_c(U, A)$. This is called the compact support cohomology of U with coefficients in A.

6.2 The Selmer complex In our situation we take the Selmer complex $C :=$ $C(F_\infty/F) := R\Gamma_c(U, T)$. This is a perfect complex of $\Lambda(\mathcal{G})$-modules. Using the duality we compute the cohomology groups as

$$H^i(C) = H_c^i(U, T) = \begin{cases} \mathbb{Z}_p & \text{if } i = 3, \\ X & \text{if } i = 2, \\ 0 & \text{otherwise.} \end{cases}$$

Recall that X is the Galois group $\text{Gal}(M/F_\infty)$ of the maximal abelian p-extension of F_∞ unramified outside Σ.

Proposition 9 *Let ρ be any p-adic Artin representation of $\mathcal{G}$, i.e. a homomorphism*

$$\rho: \mathcal{G} \rightarrow GL(V),$$

with open kernel and V a finite dimensional vector space over a finite extension L of $\mathbb{Q}_p$. Let r be any positive integer divisible by $[F_\infty(\mu_p) : F_\infty]$. Then V is a $\Lambda(\mathcal{G})$-module through right action of $\Lambda(\mathcal{G})$ on V using $\rho\kappa_F^r$. Then $R\Gamma_c(U, V^*(1 - r))$ is acyclic. Hence

$$V \otimes_{\Lambda(\mathcal{G})} C(F_\infty/F)$$

is acyclic.

Proof First note that $V \otimes C(F_\infty/F) = R\Gamma_c(U, V^*(1 - r))$. Hence the second assertion follows from the first one. Moreover, using the calculations above, we conclude that $H_c^i(U, V^*(1 - r)) = 0$ for $i \neq 2, 3$. Using the duality we must show that $H^0(U, V(r))$ and $H^1(U, V(r))$ are both 0. Since r is positive $H^0(U, V(r)) = 0$. Using the inflation-restriction sequence we may assume that $\text{Gal}(F_U/F)$ acts trivially on V. Hence we need to show that $H^1(U, L(r))$ is 0. But Soulé [46] shows that

$$K_{2r-1}\left(O_F[\frac{1}{\Sigma}]\right) \otimes_{\mathbb{Z}} L \rightarrow H^1(U, L(r))$$

is surjective. On the other hand, since r is a positive even integer and F is a totally real number field, Borel's calculation (see [5]) shows that the group $K_{2r-1}(O_F[\frac{1}{\Sigma}])$ is finite. $\qquad\square$

6.3 The main conjecture Let L be a finite extension of $\mathbb{Q}_p$. Let O_L be the ring of integers of L. Fix an isomorphism $\iota\colon \overline{\mathbb{Q}}_p \to \mathbb{C}$. Let r be a positive integer divisible by $[F_\infty(\mu_p) : F_\infty]$. For any p-adic Artin representation

$$\rho\colon \mathcal{G} \to GL_n(O_L),$$

we get the complex L-function $L(\iota \circ \rho, s)$ and the L-function $L_\Sigma(\iota \circ \rho, s)$ with Euler factors at primes in Σ removed. The result of Seigel [44] and Klingen [34] shows that for any even positive integer r the values $L(\iota \circ \rho, 1 - r)$ and $L_\Sigma(\iota \circ \rho, 1 - r)$ are algebraic and non-zero. Hence they give p-adic numbers (for details see Coates–Lichtenbaum [16], section 1). We denote these p-adic numbers simply by $L(\rho, 1 - r)$ and $L_\Sigma(\rho, 1 - r)$.

Conjecture 10 (Generalised main conjecture) There exists an isomorphism

$$\zeta = \zeta(F_\infty/F)\colon \operatorname{Det}_{\Lambda(\mathcal{G})}(0) \to \operatorname{Det}_{\Lambda(\mathcal{G})}(C(F_\infty/F))^{-1},$$

such that for any p-adic Artin representation $\rho\colon \mathcal{G} \to \mathrm{GL}(V)$ and any positive integer r divisible by $[F_\infty(\mu_p) : F_\infty]$, the map

$$\operatorname{Det}_L(0) \to \operatorname{Det}_L(R\Gamma_c(U, V^*(1 - r)))^{-1} = \operatorname{Det}_L(0).$$

induced by ζ is $L_\Sigma(\rho, 1 - r) \in L^\times = K_1(L) = \operatorname{Aut}(\operatorname{Det}_L(0))$.

Proposition 11 (Fukaya–Kato, Iwasawa) $[C(F_\infty/F)] = 0$ *in* $K_0(\Lambda(\mathcal{G}))$.

Proof Let H' be an open normal pro-p subgroup of H. Then proposition 9.1.3 in Bass [2] gives an injection

$$K_0(\Lambda(\mathcal{G})) \to K_0(\Lambda(\mathcal{G}/H')).$$

This map send the class $[P]$ of a finitely generated projective $\Lambda(\mathcal{G})$-module P to the class in $K_0(\Lambda(\mathcal{G}/H'))$ of the finitely generated projective $\Lambda(\mathcal{G}/H')$-module $\Lambda(\mathcal{G}/H') \otimes_{\Lambda(\mathcal{G})} P$. Note that

$$\Lambda(\mathcal{G}/H') \otimes_{\Lambda(\mathcal{G})} C(F_\infty/F) = C(F_\infty^{H'}/F).$$

Hence it is enough to show that the class of $C(F_\infty^{H'}/F)$ is 0 in $K_0(\Lambda(\mathcal{G}/H'))$. Next we use the injection

$$K_0(\Lambda(\mathcal{G}/H')) \to K_0(Q(\Lambda(\mathcal{G}/H'))).$$

Since $Q(\Lambda(\mathcal{G}/H')) \otimes_{\Lambda(\mathcal{G}/H')} C(F_\infty^{H'}/F)$ is an acyclic complex (we have computed the cohomologies and they are easily shown to be torsion by the result of Iwasawa [24]), its class in $K_0(Q(\Lambda(\mathcal{G}/H')))$ is 0. □

This is the basic evidence in favour of the conjecture and it says that there is an isomorphism

$$\mathrm{Det}_{\Lambda(\mathcal{G})}(0) \to \mathrm{Det}_{\Lambda(\mathcal{G})}(C(F_\infty/F))^{-1}$$

or in other words the "line bundle" $\mathrm{Det}_{\Lambda(\mathcal{G})}(R\Gamma_c(U, T))$ is trivial. The conjecture says that there is a canonical or special isomorphism or trivialisation normalised using L-values. Such a special isomorphism is called *zeta element* by Kato.

Note that we have said nothing about uniqueness in the conjecture. A zeta element for an admissible p-adic Lie extension may not be unique by itself. But if we consider all admissible p-adic Lie extension of a totally real number field F and demand that the zeta elements for them are compatible in a certain sense then we get uniqueness as we now show. Let F^+ denote the maximal totally real extension of F.

Proposition 12 (Fukaya–Kato) *The homomorphism*

$$\varprojlim_{F'} K_1(\mathbb{Z}_p[\mathrm{Gal}(F'/F)]) \to \varprojlim_{F'} K_1(\mathbb{Q}_p[\mathrm{Gal}(F'/F)]),$$

where F' runs through finite extensions of F contained in F^+, is injective.

Proof For any finite group G, let G_r be the subset of all elements of G whose orders are prime to p. Let $\mathbb{Z}_p[G_r]$ be the free $\mathbb{Z}_p$-module generated by G_r. Then G acts on $\mathbb{Z}_p[G_r]$ by conjugation. By theorem 12.10 in Oliver [41] there is a canonical surjection

$$H_2(G, \mathbb{Z}_p[G_r]) \to \ker(K_1(\mathbb{Z}_p[G]) \to K_1(\mathbb{Q}_p[G])).$$

Hence we must show that $\varprojlim_{F'} H_2(\mathrm{Gal}(F'/F), \mathbb{Z}_p[\mathrm{Gal}(F'/F)_r]) = 0$. By taking the Pontryagin dual $\mathrm{Hom}_{\mathrm{cont}}(\,\cdot\,, \mathbb{Q}_p/\mathbb{Z}_p)$ we need to show that

$$\varinjlim_{F'} H^2(\mathrm{Gal}(F'/F), \mathrm{Map}(\mathrm{Gal}(F'/F)_r, \mathbb{Q}_p/\mathbb{Z}_p)) = 0.$$

This limit is equal to

$$\varinjlim_{F'} H^2(\mathrm{Gal}(F^+/F), \mathrm{Map}(\mathrm{Gal}(F'/F)_r, \mathbb{Q}_p/\mathbb{Z}_p)).$$

Put $X = \mathrm{Map}(\mathrm{Gal}(F'/F)_r, \mathbb{Q}_p/\mathbb{Z}_p)$. Consider the inflation restriction exact sequence

$$H^1(\mathrm{Gal}(F^+, X)^{\mathrm{Gal}(F^+/F)} \to H^2(\mathrm{Gal}(F^+/F), X) \to H^2(F, X).$$

Since $\mathrm{Gal}(\overline{F}/F^+)$ acts trivially on X and p is odd we have

$$H^1(F^+, X) = \mathrm{Hom}(\mathrm{Gal}(\overline{F}/F^+), X) = 0.$$

Hence we must show that $H^2(F, X) = 0$. Since $\mathrm{Gal}(\overline{F}/F')$ acts trivially on X, we have $H^2(F', X) = 0$ by the result of Schneider [43] that $H^2(F, \mathbb{Q}_p/\mathbb{Z}_p) = 0$. As the p cohomological dimension of F is 2 the trace map

$$H^2(F', X) \to H^2(F, X)$$

is surjective. Hence $H^2(F, X) = 0$ as required. □

Corollary 13 *Assume that $\zeta(F_\infty/F)$ exists for all admissible p-adic Lie extensions F_∞/F. For $F \subset F_\infty \subset F'_\infty$ with F'_∞/F_∞ finite if we assume that the norm homomorphism*

$$K_1(\Lambda(\mathrm{Gal}(F'_\infty/F))) \xrightarrow{\mathrm{norm}} K_1(\Lambda(\mathrm{Gal}(F_\infty/F))),$$

maps $\zeta(F'_\infty/F)$ to $\zeta(F_\infty/F)$. Then $\zeta(F_\infty/F)$ are unique.

Proof Fukaya–Kato (proposition 1.5.1 in [23]. See also lemma 19 in [30]) show that we have a natural isomorphism

$$K_1(\Lambda(\mathcal{G})) \xrightarrow{\sim} \varprojlim_U K_1(\mathbb{Z}_p[\mathcal{G}/U]),$$

where U runs through all open normal subgroup of $\mathcal{G}$. Hence kernel of the map

$$\varprojlim_{F_\infty} K_1(\Lambda(\mathrm{Gal}(F_\infty/F))) \to \varprojlim_{F_\infty} K_1(\Lambda_{\mathbb{Q}_p}(\mathrm{Gal}(F_\infty/F))),$$

is trivial. Here F_∞ runs through all admissible p-adic Lie extensions of F and for any profinite group P, we denote by $\Lambda_{\mathbb{Q}_p}(P)$ the inverse limit

$$\varprojlim_U \mathbb{Q}_p[P/U],$$

U running through all open normal subgroups of P. The structure of the ring $\Lambda_{\mathbb{Q}_p}(\mathrm{Gal}(F_\infty/F))$ is very simple and so is the structure of its K_1 group. Put $\Lambda_{\mathbb{Q}_p} = \Lambda_{\mathbb{Q}_p}(\mathrm{Gal}(F_\infty/F))$. The map

$$\mathrm{Det}_{\Lambda_{\mathbb{Q}_p}}(0) \to \mathrm{Det}_{\Lambda_{\mathbb{Q}_p}}(\Lambda_{\mathbb{Q}_p} \otimes C(F_\infty/F)),$$

induced by $\zeta(F_\infty/F)$ is uniquely determined by the interpolation property. Hence we get the corollary. □

6.4 The canonical Ore set The above formulation is very nice but it is not clear (at least to me) how one constructs elements in a K_1 torsor. However, following Coates–Fukaya–Kato–Sujatha–Venjakob [15] we can get an honest element in the K_1 group of the total ring of fractions of $\Lambda(\mathcal{G})$. Note that if the cohomology groups of $R\Gamma_c(U, T)$ are torsion as $\Lambda(\mathcal{G})$-modules then $Q(\Lambda(\mathcal{G})) \otimes R\Gamma_c(U, T)$ is acyclic and $\zeta(F_\infty/F)$ induces an automorphism of $\mathrm{Det}_{Q(\Lambda(\mathcal{G}))}(0)$

$$\mathrm{Det}_{Q(\Lambda(\mathcal{G}))}(0) \to \mathrm{Det}_{Q(\Lambda(\mathcal{G}))}(Q(\Lambda(\mathcal{G})) \otimes R\Gamma_c(U, T)) \cong \mathrm{Det}_{Q(\Lambda(\mathcal{G}))}(0),$$

which is given by an element in $K_1(Q(\Lambda(G)))$. We may call this element the p-adic L-function.

We remark that if G is one dimensional as a p-adic Lie group (equivalently, if the subgroup H is finite) then, as we have already mentioned, we know that $R\Gamma_c(U, T)$ is torsion by the result of Iwasawa. If the dimension of G is bigger than one then Ochi and Venjakob [40] prove that $R\Gamma_c(U, T)$ is $\Lambda(G)$-torsion.

If the dimension of G is bigger than one and if U is an open subgroup of H, then the natural map from $K_1(\Lambda(G))$ to $K_1(\Lambda(G/U))$ does not extend to the map between K_1 groups of the total ring of fractions (simply because there are non-zero divisors in $\Lambda(G)$ which map to 0 in $\Lambda(G/U)$). Hence the compatibility as in corollary 13 cannot be obtained. If we can localise at a smaller multiplicatively closed set to get the p-adic L-function then we can get around this difficulty. Following [15] we introduce such a set.

Definition 14 Let S be the subset of $\Lambda(G)$ defined by

$$S := S(G, H) := \{f \in \Lambda(G) | \Lambda(G)/\Lambda(G)f \text{ is a finitely generated } \Lambda(H)\text{-module}\}$$

The following proposition is proven in [15].

Proposition 15 *S is a multiplicatively closed subset of $\Lambda(G)$ and does not contain any zero-divisors. It also satisfies the left and right Ore condition, i.e. for any $s \in S$ and any $r \in \Lambda(G)$ there exists t_1, t_2 in S and w_1, w_2 in $\Lambda(G)$ such that*

$$sw_1 = rt_1, \qquad w_2 s = t_2 r.$$

Definition 16 Let $S^* = \cup_{n \geq 0} p^n S$ be the multiplicatively closed subset of $\Lambda(G)$ generated by S and p. From the above proposition it is clear that S^* does not contain any zero-divisors and is a left and right Ore set.

Since $\Lambda(G)$ is a noncommutative ring in general it is not always possible to localise at multiplicatively closed subset. The Ore conditions are precisely the conditions needed to localise as we do in the case of commutative rings. Hence we get injections

$$\Lambda(G) \to \Lambda(G)_S \to \Lambda(G)_{S^*}.$$

6.1 Known results

Definition 17 We say that F_∞/F satisfies the hypothesis $\mu = 0$ if there exists a pro-p open subgroup H' of H such that the Galois group over $F_\infty^{H'}$ of the maximal abelian p-extension of $F_\infty^{H'}$ unramified outside Σ is a finitely generated $\mathbb{Z}_p$-module.

Of course, this is a special case of a general conjecture of Iwasawa asserting that, for every finite extension K of $\mathbb{Q}$, the Galois group over K^{cyc} of the maximal unramified abelian p-extension of K^{cyc} is a finitely generated $\mathbb{Z}_p$-module. When K is an abelian extension of $\mathbb{Q}$, this conjecture is proven by Ferrero–Washington [20] and by a different method by Sinnott [45].

Lemma 18 $X = \text{Gal}(M/F_\infty)$ *is finitely generated over* $\Lambda(H)$ *if and only if* F_∞/F *satisfies the hypothesis* $\mu = 0$.

Proof Let H' be any pro-p open subgroup of H. Thus $\Lambda(H')$ is a local ring. It follows from Nakayama's lemma that X is finitely generated over $\Lambda(H)$ if and only if $X_{H'}$ is a finitely generated $\mathbb{Z}_p$-module. Let F_Σ denote the maximal pro-p extension of F_∞ which is unramified outside primes above Σ, and let $K_\infty = F_\infty^{H'}$. Then we have the inflation-restriction exact sequence

$$0 \to H^1(H', \mathbb{Q}_p/\mathbb{Z}_p) \to H^1(\text{Gal}(F_\Sigma/K_\infty), \mathbb{Q}_p/\mathbb{Z}_p)$$
$$\to H^1(\text{Gal}(F_\Sigma/F_\infty), \mathbb{Q}_p/\mathbb{Z}_p)^{H'} \to H^2(H', \mathbb{Q}_p/\mathbb{Z}_p).$$

As H' is a p-adic Lie group, $H^i(H', \mathbb{Q}_p/\mathbb{Z}_p)$ are cofinitely generated $\mathbb{Z}_p$-modules for all $i \geq 0$. Moreover, since $\text{Gal}(F_\Sigma/F)$ acts trivially on $\mathbb{Q}_p/\mathbb{Z}_p$, we have

$$H^1(\text{Gal}(F_\Sigma/L), \mathbb{Q}_p/\mathbb{Z}_p) = \text{Hom}(\text{Gal}(M_L/L), \mathbb{Q}_p/\mathbb{Z}_p),$$

for every intermediate field L with $F_\Sigma \supset L \supset F$; here M_L denotes the maximal abelian p-extension of L which is unramified outside Σ_L. We conclude from the above exact sequence that $X_{H'}$ is a finitely generated $\mathbb{Z}_p$-module if and only if $\text{Gal}(M_{K_\infty}/K_\infty)$ is a finitely generated $\mathbb{Z}_p$-module. The conclusion of the lemma is now plain since $\text{Gal}(M_{K_\infty}/K_\infty)$, being a finitely generated $\mathbb{Z}_p$-module, is precisely the hypothesis $\mu = 0$ for some open subgroup H'. $\square$

It is proven in [15] that a finitely generated $\Lambda(\mathcal{G})$ is S-torsion if and only if it is finitely generated over $\Lambda(H)$. Hence under the hypothesis $\mu = 0$ we know that the cohomologies of $C(F_\infty/F)$ are S torsion, i.e. $\Lambda(\mathcal{G})_S \otimes_{\Lambda(\mathcal{G})} C(F_\infty/F)$ is acyclic. Hence the generalised main conjecture would give the p-adic L-function as an element of $K_1(\Lambda(\mathcal{G})_S)$.

Theorem 19 *Let* F_∞/F *be an admissible p-adic Lie extension satisfying the hypothesis* $\mu = 0$. *Then the main conjecture is true for* F_∞/F.

A strategy for proving this theorem was given by Kato and Burns [9]. It crucially uses the theorem of Wiles mentioned in Section 3. Using this strategy partial results were obtained in [32], [29] and [26]. The full result is obtained in [30]. A special case is proven in Ritter–Weiss [42] which is extended to the general case in Burns [9].

Questions (1) Can one prove the generalised main conjecture (for totally real number fields) just assuming that the complex $C(F_\infty/F)$ is S^* torsion? (2) Can one show that the complex $C(F_\infty/F)$ is S^* torsion? This is weaker than the conjecture of Iwasawa on vanishing of μ invariants.

7 Generalisations

We briefly present the generalisation of the above conjectures for other motives. Again we follow Fukaya–Kato.

(*) Let Λ be a ring with an ideal I such that Λ/I^n is finite of p-power order and $\Lambda \cong \varprojlim_n \Lambda/I^n$.

Example Let L be a finite extension of $\mathbb{Q}_p$ and let O_L be the ring of integers in L. Let G be a profinite group with open normal topologically finitely generated pro-p subgroup P. Then the Iwasawa algebra $O_L[[G]]$ satisfies the above condition (*). The ideal I can be taken to be the kernel of the map

$$O_L[[G]] \to O_L[G/P]/(p).$$

Let T be a finitely generated projective Λ module with a continuous Λ-linear action of $\mathrm{Gal}(\overline{\mathbb{Q}}/\mathbb{Q})$ that is unramified at almost all primes (i.e. for almost all primes the inertia subgroups act trivially on T). Let Σ be a finite set of finite primes of $\mathbb{Q}$ containing p and all primes at which T is ramified and put $U = \mathrm{Spec}(\mathbb{Z}[\frac{1}{\Sigma}])$. Then $R\Gamma_c(U, T)$ is a perfect complex of Λ-modules (proposition 1.6.5 [23]). If we choose a different finite set Σ' of finite primes (still containing p and all primes at which T is ramified) and put $U' = \mathrm{Spec}(\mathbb{Z}[\frac{1}{\Sigma'}])$ then there is a canonical isomorphism between $\mathrm{Det}_\Lambda(R\Gamma_c(U, T))$ and $\mathrm{Det}_\Lambda(R\Gamma_c(U', T))$. Put

$$\Delta_\Lambda(T) = \mathrm{Det}_\Lambda(R\Gamma_c(U, T))^{-1}.$$

Then $\Delta_\Lambda(T)$ is independent of the choice of U.

Conjecture 20 (Fukaya–Kato) There is a unique way to associate an isomorphism

$$\zeta_\Lambda(T)\colon \mathrm{Det}_\Lambda(0) \to \Delta_\Lambda(T)$$

to each pair (Δ, T) as above, satisfying the following conditions.

(1) For pairs $(\Lambda, T_1), (\Lambda, T_2)$ and (Λ, T_3) and an exact sequence

$$0 \to T_1 \to T_2 \to T_3 \to 0,$$

of Λ-modules, the canonical isomorphism

$$\Delta_\Lambda(T_2) \cong \Delta_\Lambda(T_1) \otimes \Delta_\Lambda(T_3)$$

sends $\zeta_\Lambda(T_2)$ to $\zeta_\Lambda(T_1) \cdot \zeta_\Lambda(T_3)$.

(2) Let (Λ, T) be a pair as above. Let Λ' be another ring satisfying $(*)$. Let Y be a finitely generated projective Λ' module with a compatible right action of Λ. Put $T' = Y \otimes_\Lambda T$. Then the canonical isomorphism

$$Y \otimes_\Lambda \Delta_\Lambda(T) \cong \Delta_{\Lambda'}(T'),$$

maps $\zeta_\Lambda(T)$ to $\zeta_{\Lambda'}(T')$.

(3) If M is a $\mathbb{Q}$-motive over $\mathbb{Q}$ and let T_p be a Galois stable lattice in the p-adic realisation M_p of M. Let F be a p-adic Lie extension of $\mathbb{Q}$ (possible finite) unramified outside a finite set of primes of $\mathbb{Q}$. Put $G = \text{Gal}(F/\mathbb{Q})$. Let $\Lambda = \mathbb{Z}_p[[G]]$ and put $T = \Lambda \otimes_{\mathbb{Z}_p} T_p$. Then $\zeta_\Lambda(T)$ is related in a precise sense to the special values of L-functions of twists of M.

Remark We do not make the third condition precise here. For details see Fukaya–Kato [23] or Venjakob [49]. Fukaya–Kato prove the uniqueness part of the statement and our proof of Corollary 13 follows closely their proof. The above conjecture is usually called the "Equivariant Tamagawa Number Conjecture".

References

[1] Barsky, D. Fonctions zeta p-adiques d'une classe de rayon des corps de nombres totalement reels. *Groupe de travail d'analyse ultraletrique*, 5(16):1–23, 1977-1978.

[2] Bass, Hyman. *Algebraic K-theory*. W. A. Benjamin, Inc., New York-Amsterdam, 1968.

[3] Beilinson, A. Higher regulators and values of L-functions. *J. Soviet Math.*, 30:2036–2070, 1985.

[4] Bloch, S. and Kato, K. L-functions and tamagawa numbers of motives. In P. Cartier, L. Illusie, N. M. Katz, G. Laumon, Yu. Manin, and Kenneth A. Ribet, editors, *The Grothendieck Festschrift*, volume 1, pages 333–400. Birkhauser Boston, Boston, 1990.

[5] Borel, Armand. Stable real cohomology of arithmetic groups. *Ann. Sci. École Norm. Sup. (4)*, 7:235–272 (1975), 1974.

[6] Bourbaki, Nicolas. *Commutative Algebra. Chapters 1–7*. Elements of Mathematics (Berlin). Springer-Verlag, 1989.

[7] Breuning, Manuel. Determinant functors on triangulated categories. *Journal of K-Theory: K-Theory and its Applications to Algebra, Geometry and Topology*, 2010.

[8] Brumer, Armand. On the units of algebraic number fields. *Mathematika*, 14:121–124, 1967.

[9] Burns, D. On main conjectures in non-commutative Iwasawa theory and related conjectures. Preliminary version, 2010.

[10] Burns, D. and Flach, M. Tamagawa numbers for motives with (non-commutative) coefficients. *Doc. Math.*, 6:501–570, 2001.

[11] Burns, D. and Flach, M. Tamagawa numbers for motives with (noncommutative) coefficients. II. *Amer. J. Math.*, 125(3):475–512, 2003.

[12] Cassou-Noguès, P. Valeurs aux entiers négatifs des fonctions zêta et fonctions zêta p-adiques. *Invent. Math.*, 51(1):29–59, 1979.

[13] Coates, J. p-adic L-functions and Iwasawa's theory. In A. Frohlich, editor, *Algebraic Number Fields: L-functions and Galois properties*. Academic Press, London, 1977.

[14] Coates, J. Motivic p-adic L-functions. In J. Coates and M. J. Taylor, eds., *L-functions and arithmetic (Durham, 1989)*, volume 153, pages 141–172. Cambridge University Press, 1991.

[15] Coates, J., Fukaya, T., Kato, K., Sujatha, R. and Venjakob, O. The GL_2 main conjecture for elliptic curves without complex multiplication. *Publ. Math. IHES*, 2005.

[16] Coates, J. and Lichtenbaum, S. On l-adic zeta functions. *Ann. of Math.*, 98:498–550, 1973.

[17] Deligne, P.. Valeurs de fonctions L et periodes d'integrales. In Borel, Armand and Casselman, W., editors, *Automorphic forms, representations and L-functions*, number 2 in Proc. Sympos. Pure Math., Oregon State Univ., Corvallis, Ore. 1977, pages 313–346, 1979.

[18] Deligne, P.. Le déterminant de la cohomologie. In Kenneth A. Ribet, editor, *Current trends in arithmetical algebraic geometry (Arcata, Calif., 1985)*, volume 67 of *Contemp. Math.*, pages 93–177. Amer. Math. Soc., 1987.

[19] Deligne, P. and Ribet, Kenneth A. Values of abelian L-functions at negative integers over totally real fields. *Inventiones Math.*, 59:227–286, 1980.

[20] Ferrero, B. and Washington, L.C. The Iwasawa invariant μ_p vanishes for abelian number fields. *Ann. of Math.*, 109:377–395, 1979.

[21] Flach, M. The equivariant Tamagawa number conjecture: a survey. In D. Burns, J. Sands, and D. Solomon, eds., *Stark's conjecture: recent work and new directions*, volume 358 of *Contemp. Math.*, pages 79–125. Amer. Math. Soc., 2004.

[22] Fontaine, J.-M. and Perrin-Riou, B. Autour des conjectures de Bloch et Kato. III. le case général. *C. R. Acad. Sci Paris Sér. I Math.*, 313(7):421–428, 1991.

[23] Fukaya, T. and Kato, K. A formulation of conjectures on p-adic zeta functions in non-commutative Iwasawa theory. In N. N. Uraltseva, editor, *Proceedings of the St. Petersburg Mathematical Society*, volume 12, pages 1–85, March 2006.

[24] Greenberg, R. On p-adic L-functions and cyclotomic fields – II. *Nagoya Math. J.*, 67:139–158, 1977.

[25] Greenberg, R. Iwasawa theory for motives. In J. Coates and M. J. Taylor, editors, *L-functions and arithmetic (Durham, 1989)*, volume 153, pages 211–233. Cambridge University Press, 1991.

[26] Hara, T. Inductive construction of the *p*-adic zeta functions for non-commutative *p*-extensions of totally real fields with exponent *p*. http://arxiv.org/abs/ 0908.2178v2, 2010.

[27] Huber, A. and Kings, G. Equivariant Bloch-Kato conjecture and non-abelian Iwasawa main conjecture. In *Proceedings of the International Congress of Mathematicians*, volume 2. Higher Ed. Press, Beijing, 2002.

[28] Iwasawa, K. On Z_l-extensions of algebraic number fields. *Ann. of Math.*, 98(2):246–326, 1973.

[29] Kakde, M. *Proof of the Main Conjecture of Noncommutative Iwasawa Theory for Totally Real Number Fields in Certain Cases*. PhD thesis, Cambridge University, 2008. Accepted in Journal of Algebraic Geometry.

[30] Kakde, Mahesh. The main conjecture of Iwasawa theory for totally real fields. http://arxiv.org/abs/1008.0142,2010.

[31] Kato, K. Lectures on the approach to Iwasawa theory for Hasse-Weil *L*-functions via B_{dR}. I. In *Arithmetic algebraic geometry*, LNM, 1553, pages 50–163. Springer-Verlag, Berlin, 1993.

[32] Kato, K. Iwasawa theory of totally real fields for Galois extensions of Heisenberg type. Very preliminary version, 2006.

[33] Kato, Kazuya. *p*-adic Hodge theory and values of zeta functions of modular forms. *Asterisque*, (295):117–290, 2004.

[34] Klingen, H. Über die werte der Dedekindschen Zetafunktionen. *Math. Ann.*, pages 265–272, 1962.

[35] Knudsen, Finn F. Determinant functors on exact categories and their extensions to categories of bounded complexes. *Michigan Math. J.*, 50(2):407–444, 2002.

[36] Knudsen, Finn F. and Mumford, David. The projectivity of the moduli space of stable curves. I. Preliminaries on "det" and "Div". *Math. Scand.*, (1):19–55, 1976.

[37] Kurihara, Masato. On the structure of ideal class groups of CM-fields. *Doc. Math.*, (Extra Vol.):539–563, 2003.

[38] Lang, Serge. *Cyclotomic Fields I and II (with an appendix by Karl Rubin)*. GTM, 121. Springer-Verlag, New York, 1990.

[39] Mazur, B., and Wiles, A. Class fields of abelian extensions of $\mathbb{Q}$. *Invent. Math.*, 76(2):179–330, 1984.

[40] Ochi, Yoshihiro and Venjakob, Otmar. On the ranks of Iwasawa modules over *p*-adic Lie extensions. *Mathematical Proceedings of the Cambridge Philosophical Society*, 135:25–43, 2003.

[41] Oliver, R. *Whitehead Groups of Finite Groups*. Number 132 in London Mathematical Society Lecture Note Series. Cambridge University Press, 1988.

[42] Ritter, J. and Weiss, A. On the 'main conjecture' of equivariant Iwasawa theory. http://arxiv.org/abs/1004.2578, April 2010.

[43] Schneider, Peter. Über gewisse Galoiscohomologiegruppen. *Math. Z.*, 168(2), 1979.

[44] Seigel, C. Über die Fourierschen Koeffizienten von Modulformen. *Göttingen Nachr.*, 3:15–56, 1970.

[45] Sinnott, W. On the μ-invariant of the Γ-tranform of a rational function. *Invent. Math.*, 75(2):273–282, 1984.

[46] Soulé, C. *K*-théorie des anneaux d'entiers de corps de nombres et cohomologie étale. *Invent. Math.*, 55(3):251–295, 1979.

[47] Vaserstein, L. N. On stabilization for general linear groups over a ring. *Math. USSR Sbornik*, 8:383–400, 1969.

[48] Vaserstein, L. N. On the Whitehead determinant for semi-local rings. *J. Algebra*, 283:690–699, 2005.

[49] Venjakob, O. From the Birch and Swinnerton-Dyer conjecture to non-commutative Iwasawa theory via equivariant Tamagawa number conjecture – a survey. In D. Burns, K. Buzzard and J. Nekovávr, eds., *L-functions and Galois representations*, volume 320 of *London Mathematical Society Lecture Note Series*, pages 333–380. Cambridge University Press, 2007.

[50] Wiles, A. The Iwasawa conjecture for totally real fields. *Ann. of Math.*, 131(3):493–540, 1990.

On the $\mathfrak{M}_H(G)$-conjecture

J. Coates

University of Cambridge

R. Sujatha

Tata Institute of Fundamental Research

1 Introduction

Let p be any prime number, and let G be a compact p-adic Lie group with a closed normal subgroup H such that G/H is isomorphic to the additive subgroup of p-adic integers $\mathbb{Z}_p$. Write $\Lambda(G)$ (respectively, $\Lambda(H)$) for the Iwasawa algebra of G (respectively, H) with coefficients in $\mathbb{Z}_p$. As was shown in [5], there exists an Ore set in $\Lambda(G)$ which enables one to define a characteristic element, with all the desirable properties, for a special class of torsion $\Lambda(G)$-modules, namely those finitely generated left $\Lambda(G)$-modules W such that $W/W(p)$ is finitely generated over $\Lambda(H)$; here $W(p)$ denotes the p-primary submodule of W. This simple piece of pure algebra leads to a class of deep arithmetic problems, which will be the main concern of this paper. We shall loosely call these problems *the $\mathfrak{M}_H(G)$- conjectures*, and it should be stressed that their validity is essential even for the formulation of the main conjectures of non-commutative Iwasawa theory.

Let F be a finite extension of $\mathbb{Q}$, and F_∞ a Galois extension of F satisying (i) $G = \mathrm{Gal}(F_\infty/F)$ is a p-adic Lie group, (ii) F_∞/F is unramified outside a finite set of primes of F, and (iii) F_∞ contains the cyclotomic $\mathbb{Z}_p$-extension of F, which we denote by F^{cyc}. Take $H = \mathrm{Gal}(F_\infty/F^{\mathrm{cyc}})$, so that G/H is isomorphic to $\mathbb{Z}_p$. Let A be an abelian variety defined over F with good ordinary reduction at all places v of F dividing p. The original $\mathfrak{M}_H(G)$-conjecture, made in [5], asserts that the dual of the Selmer group of A over F_∞, which we denote by $X(A/F_\infty)$, has the property that $X(A/F_\infty)/X(A/F_\infty)(p)$ is finitely generated as a module over $\Lambda(H)$. When H is finite, this is precisely the classical conjecture of Mazur [17]. However, for infinite H, it is much stronger than what is probably the simplest generalization of Mazur's conjecture, namely the assertion that $X(A/F_\infty)$ is $\Lambda(G)$-torsion. We also stress that the $\mathfrak{M}_H(G)$-conjecture requires no hypothesis at all about $X(A/F^{\mathrm{cyc}})(p)$ being finite. Nevertheless, we

Non-abelian Fundamental Groups and Iwasawa Theory, eds. John Coates, Minhyong Kim, Florian Pop, Mohamed Saïdi and Peter Schneider. Published by Cambridge University Press. ©Cambridge University Press 2012.

have to confess that all results we can prove to date about the conjecture require us to assume that A is isogenous over F to an abelian variety A' over F with $X(A'/F^{\mathrm{cyc}})(p)$ finite. In §3, we give several equivalent formulations of the $\mathfrak{M}_H(G)$-conjecture in the first case beyond Mazur's conjecture, namely when $H = \mathbb{Z}_p$. In particular, even in the special case when E is an elliptic curve over $\mathbb{Q}$ with complex multiplication by an order in the ring of integers of an imaginary quadratic field K, and p is a prime of good ordinary reduction for E, we still do not know how to prove the $\mathfrak{M}_H(G)$-conjecture for E over the abelian extension $F_\infty = \mathbb{Q}(E_{p^\infty})$ of K, where E_{p^∞} denotes the group of all p-power division points on E.

In the final part of the paper, we propose a version of the $\mathfrak{M}_H(G)$-conjecture for the dual of the Selmer group of an ordinary Hida family over a p-adic Lie extension F_∞ of $\mathbb{Q}$, which, as above, is assumed to be unramified outside a finite set of primes, and to contain the cyclotomic $\mathbb{Z}_p$-extension of $\mathbb{Q}$. In formulating this conjecture, we have taken what we feel is the simplest and most natural definition of the Selmer group of the Hida family over a p-adic Lie extension, although, to our knowledge, it does not seem to appear in the existing literature on the Iwasawa theory of Hida families. Again, the interest of such an $\mathfrak{M}_H(G)$-conjecture in this setting is that it enables one to define a characteristic element for the dual Selmer group, and thus opens the way to formulating a non-commutative main conjecture for such Hida families. We do not attempt to give a precise formulation of such a main conjecture here, but hope to return to this question in a subsequent paper.

In conclusion, we wish to warmly thank Ralph Greenberg, both for pointing out to us a blunder in an earlier version of this paper, and for many helpful comments on the present text.

2 Statement of the conjecture

In this section, we recall the classical $\mathfrak{M}_H(G)$-conjecture and discuss some related material. Throughout, p will be an odd prime, and $\mathbb{Q}_p$, $\mathbb{Z}_p$ the field of p-adic numbers, and the ring of p-adic integers. We write K for a fixed finite extension of $\mathbb{Q}_p$, and O for the ring of integers of K. Let F be a finite extension of $\mathbb{Q}$. We define a Galois extension F_∞ of F to be an *admissible p-adic Lie extension of F* if (i) $G = \mathrm{Gal}(F_\infty/F)$ is a p-adic Lie group, (ii) F_∞/F is unramified outside a finite set of primes of F, and (iii) F_∞ contains the cyclotomic $\mathbb{Z}_p$- extension of F, which we shall always denote by F^{cyc}. We shall assume throughout this paper that F_∞ is an admissible p-adic Lie extension of F, and put

$$G = \mathrm{Gal}(F_\infty/F), \quad H = \mathrm{Gal}(F_\infty/F^{\mathrm{cyc}}), \quad \Gamma = G/H = \mathrm{Gal}(F^{\mathrm{cyc}}/F). \quad (1)$$

Let Σ be any finite set of primes of F, which contains all non-archimedean primes of F which ramify in F_∞. Note that Σ always contains all primes of F dividing p, since F_∞ contains F^{cyc}. Write

$$\Lambda_O(G) = \varprojlim O[G/U], \quad (2)$$

where U runs over all open normal subgroups of G, for the Iwasawa algebra of G with coefficients in O. When $O = \mathbb{Z}_p$, we simply put $\Lambda(G)$. Let

$$\chi_F : \mathrm{Gal}(\bar{F}/F) \to \mathbb{Z}_p^\times$$

be the character giving the action of the absolute Galois group of F on the group μ_{p^∞} of all p-power roots of unity. We assume that we are given a continuous Galois representation

$$\rho : \mathrm{Gal}(\bar{F}/F) \to \mathrm{Aut}_K(V), \quad (3)$$

where V is a two dimensional vector space over K. Moreover, we shall always assume that ρ satisfies:

Condition R1 ρ is unramified outside a finite set of primes of F.

Condition R2   For each place v of F dividing p, there exists a subspace $F_v^+ V$ dimension one, which is stable under the action of the decomposition group at v, and which is such that the inertial subgroup at v acts trivially on $V/F_v^+ V$, and on $F_v^+ V$ by χ_F^n for some integer $n > 0$.

The two basic examples of such a Galois representation (and we shall only consider one or the other of these after this section of the paper) are given by:

(i) $V = V_p(E)$, where E is an elliptic curve defined over an arbitrary finite extension F of $\mathbb{Q}$, and having good ordinary reduction at all places v of F dividing p;

(ii) V is the Galois representation attached to a primitive Hecke eigenform for $\mathrm{GL}_2/\mathbb{Q}$, which is ordinary at p, relative to some fixed embedding of the algebraic closure of $\mathbb{Q}$ into $\bar{\mathbb{Q}}_p$.

By compactness, V will always contain a free O-submodule T, which is stable under the action of $\mathrm{Gal}(\bar{F}/F)$. We fix, once and for all, such an O-*lattice* T, and define

$$A = V/T. \quad (4)$$

For each place v of F dividing p, define F_v^+A to be the image of F_v^+V in A. We now recall Greenberg's definition of the Selmer group attached to A. Let $\mathcal{L}$ denote any Galois extension of F containing F^{cyc} (we shall only need to define the Selmer group of A for such extensions). For each non-archimedean prime w of $\mathcal{L}$, define $\mathcal{L}_w$ to be the union of the completions at w of the finite extensions of F contained in $\mathcal{L}$. Write I_w for the inertial subgroup of $\text{Gal}(\bar{\mathcal{L}}_w/\mathcal{L}_w)$. Also, when w lies above p, put $F_w^+A = F_v^+A$, where v is the restriction of w to F. We then define the Selmer group of T over $\mathcal{L}$ by

$$\text{Sel}(T/\mathcal{L}) = \text{Ker}\Big(H^1(\mathcal{L}, A) \to \prod_{w \nmid p} H^1(\mathcal{L}_w, A) \times \prod_{w \mid p} H^1(\mathcal{L}_w, A/F_w^+A)\Big). \quad (5)$$

Since $\mathcal{L} \supset F^{\text{cyc}}$, it is well known (see [4]) that, when $V = V_p(E)$ for an elliptic curve over F with

$$T = T_p(E) = \varprojlim E_{p^n},$$

the above definition does indeed give the classical Selmer group of E over $\mathcal{L}$. As usual, we write

$$X(T/\mathcal{L}) = \text{Hom}\,(\text{Sel}(T/\mathcal{L}), \mathbb{Q}_p/\mathbb{Z}_p) \quad (6)$$

for the compact Pontryagin dual of $\text{Sel}(T/\mathcal{L})$. Of course, both $\text{Sel}(T/\mathcal{L})$ and $X(T/\mathcal{L})$ are endowed with canonical left actions of $\text{Gal}(\mathcal{L}/F)$, and, since A is also an O-module, these extend by continuity to left module structures over the ring $\Lambda_O(\text{Gal}(\mathcal{L}/F))$. It is not difficult to show, using condition R1, that $X(T/\mathcal{L})$ is finitely generated as a left module over $\Lambda_O(\text{Gal}(\mathcal{L}/F))$.

The starting point of the deeper study of these Iwasawa modules is a natural generalization of an old conjecture of Mazur [17] (Mazur considered only the special case when $V = V_p(E)$ with E an elliptic curve defined over F). However, to state a general conjecture, we must impose two further conditions on the representation ρ, namely the following.

Condition R3 The representation ρ is odd, in the sense that, for each real place v of F, the determinant of $\rho(c_v)$ is equal to -1, where c_v denotes a complex conjugation at v.

If v is a non-archimedean place of F, write Nv for the cardinality of the residue field of v, and Frob_v for the arithmetic Frobenius in $\text{Gal}(\bar{F}_v/F_v)/I_v$.

Condition R4 ρ is pure, in the sense that, for each finite place v of F where ρ is unramified, the eigenvalues of $\rho(\text{Frob}_v)$ are algebraic numbers having absolute value $(Nv)^{n/2}$ in all complex embeddings, where n is the positive integer occurring in Condition R2.

The following is then a special case of a conjecture of Greenberg [9] extending Mazur's conjecture.

Conjecture M   Assume ρ satisfies Conditions R1, R2, R3 and R4. Then $X(T/F^{\text{cyc}})$ is $\Lambda_O(\Gamma)$-torsion.

The principal result to date in support of the conjecture is due to Kato [12], who has proven it when F is abelian over $\mathbb{Q}$, and ρ is modular in the sense that it is the Galois representation attached to a primitive Hecke eigenform for GL_2 over $\mathbb{Q}$.

There is an evident generalization of Conjecture M to an admissible p-adic Lie extension. We say that a left $\Lambda_O(G)$-module W is $\Lambda_O(G)$-torsion if every element of W has an annihilator in $\Lambda_O(G)$ which is not a zero divisor. It is then natural to conjecture that, assuming ρ satisfies Conditions R1, R2, R3, and R4, $X(T/F_\infty)$ should be $\Lambda_O(G)$-torsion for every admissible p-adic Lie extension F_∞/F (see [13] for a proof of this conjecture in certain interesting cases when G has dimension 2). However, such a generalization of Conjecture M does not allow us, at least in our present state of knowledge, to define a suitable characteristic element for $X(T/F_\infty)$. Indeed, it seems to be impossible in this generality to define such a characteristic element, which is related to the G-Euler characteristic of W in the way that one would expect (see the examples discussed at the end of [6]). To overcome this difficulty, the following stronger version of Conjecture M, when G has dimension > 1, was introduced in [5]. Fix a local parameter π of O. Let $W(\pi)$ denote the $\Lambda_O(G)$-submodule consisting of all elements of W, which are annihilated by a power of π. It has long been known (see [17]) that examples of $X(T/F^{\text{cyc}})$ occur with $X(T/F^{\text{cyc}})(\pi)$ being large, in the sense that it is not pseudo-null as a $\Lambda_O(\Gamma)$-module.

Conjecture $\mathfrak{M}_H(G)$ Assume ρ satisfies Conditions R1, R2, R3 and R4. Then, for every admissible p-adic Lie extension F_∞ of F, we have that $X(M/F_\infty)/X(M/F_\infty)(\pi)$ is finitely generated over $\Lambda_O(H)$.

We spend the remainder of this section briefly discussing general evidence for this conjecture, in the spirit of §5 of [5]. We begin by establishing the best known result in support of it.

Theorem 2.1   *Assume ρ satisfies Conditions R1, R2, R3 and R4. Suppose further that there exists a finite extension L of F contained in F_∞ such that $X(T/L^{\text{cyc}})$ is a finitely generated O-module and $\text{Gal}(F_\infty/L)$ is pro-p. Then $X(T/F_\infty)$ is finitely generated over $\Lambda_O(H)$. In particular, Conjecture $\mathfrak{M}_H(G)$ is valid for $X(T/F_\infty)$.*

It seems reasonable to expect that, for fixed T and fixed base field F, $X(T/F^{\text{cyc}})$

should be finitely generated over O for all sufficiently large good ordinary primes p, but a proof of this assertion, even for $F = \mathbb{Q}$, still seems far away. Nevertheless, it can often be verified numerically for small primes p, as is illustrated by the following.

Example Let E be the elliptic curve $X_1(11)$, given by

$$y^2 + y = x^3 - x^2, \tag{7}$$

and take $F = \mathbb{Q}$ and ρ the p-adic representation of $\mathrm{Gal}(\bar{\mathbb{Q}}/\mathbb{Q})$ in $\mathrm{Aut}(V_p(E))$, where p is any prime of good ordinary reduction for E ($p = 3, 5, 7, \ldots$). Numerical calculations due to T. Dokchitser, V. Dokchitser and C. Wuthrich show that $X(T_3(E)/\mathbb{Q}(\mu_{3^\infty})) = 0$, $X(T_7(E)/\mathbb{Q}(\mu_{7^\infty})) = \mathbb{Z}_7$, and, as E has a rational point of order 5, it is easy to prove theoretically that $X(T_5(E)/\mathbb{Q}(\mu_{5^\infty})) = 0$. Thus Theorem 2.1 shows that, for the primes $p = 3, 5, 7$, Conjecture $\mathfrak{M}_H(G)$ is valid for any admissible p-adic Lie extension F_∞ of $\mathbb{Q}$, which contains $\mathbb{Q}(\mu_{p^\infty})$, and which is such that $\mathrm{Gal}(F_\infty/\mathbb{Q}(\mu_p))$ is pro-p. For example, for all three primes, we can take F_∞ to be the field obtained by adjoining to $\mathbb{Q}$ all p-power roots of unity and all p-power roots of some fixed integer $m > 1$. As another example, we can take $F_\infty = \mathbb{Q}(E_{5^\infty})$ for the prime $p = 5$.

To prove Theorem 2.1, we need a preliminary lemma. As before, let Σ be a finite set of primes of F which contain the infinite primes and those that ramify in F_∞. Enlarging Σ if necessary, we assume from now on that Σ also contains all non-archimedean primes of F which are ramified for ρ. Thus the extension $F(A)/F$ is unramified outside Σ and the archimedean primes of F. Let F_Σ denote the maximal extension of F unramified outside Σ and the archimedean primes of F, and, for each subfield $\mathcal{L}$ of F_Σ, write

$$\mathcal{G}_\Sigma(\mathcal{L}) = \mathrm{Gal}(F_\Sigma/\mathcal{L}). \tag{8}$$

Lemma 2.2 Assume that $\mathcal{L} \supset F^{\mathrm{cyc}}$. Then

$$H^1(\mathcal{G}_\Sigma(\mathcal{L}), A) = \mathrm{Ker}\left(H^1(\mathcal{L}, A) \to \prod_{w \nmid \Sigma} H^1(\mathcal{L}_w, A)\right). \tag{9}$$

Proof Let ϕ be any cocycle representative of a class in the right hand side of (9). Take any place w of $\mathcal{L}$ which does not lie above a place in Σ. Since the restriction of ϕ to the decomposition group of w is a coboundary, and w is unramified in the extension $\mathcal{L}(A)/\mathcal{L}$, we conclude that the restriction of ϕ to the inertia subgroup of w must be zero. Hence ϕ belongs to the left hand side of (9). Conversely, suppose that ψ is any element of the left hand side of (9). For each place w of $\mathcal{L}$, not above Σ, write ψ_w for the restriction of ψ to $H^1(\mathcal{L}_w, A)$. Then ψ_w must belong to $H^1(\mathrm{Gal}(\mathcal{L}_w^{\mathrm{nr}}/\mathcal{L}_w), A)$, where $\mathcal{L}_w^{\mathrm{nr}}$ denotes the maximal unramified extension of w. But, as $\mathcal{L} \supset F^{\text{cyc}}$, $\text{Gal}(\mathcal{L}_w^{\text{nr}}/\mathcal{L}_w)$ has profinite order prime to p, whence $H^1(\text{Gal}(\mathcal{L}_w^{\text{nr}}/\mathcal{L}_w), A) = 0$. Thus $\psi_w = 0$, and the proof is complete. $\qquad\square$

It is convenient to introduce the following notation. Note that, if $\mathcal{L}$ is a finite extension of F^{cyc}, there are only finitely many places of $\mathcal{L}$ above each non-archimedean place of F. For such an extension and a place v of F, define $J_v(T/\mathcal{L})$ to be either

$$J_v(T/\mathcal{L}) = \bigoplus_{w|v} H^1(\mathcal{L}_w, A) \quad \text{or} \quad \bigoplus_{w|v} H^1(\mathcal{L}_w, A/F_w^+A), \qquad (10)$$

according as v does not or does divide p. If $\mathcal{L}$ is an infinite extension of F^{cyc}, one defines $J_v(T/\mathcal{L})$ to be the inductive limit under restriction of the $J_v(T/\mathcal{L}')$ where $\mathcal{L}'$ runs over all finite extensions of F^{cyc} contained in $\mathcal{L}$. We then have the following immediate corollary of Lemma 2.2.

Corollary 2.3 *For each extension $\mathcal{L}$ of F^{cyc} contained in F_Σ, we have the exact sequence*

$$0 \to \text{Sel}(T/\mathcal{L}) \to H^1(\mathcal{G}_\Sigma(\mathcal{L}), A) \to \bigoplus_{v \in \Sigma} J_v(T/\mathcal{L}). \qquad (11)$$

Assume for the rest of this section that ρ satisfies Conditions R1, R2, R3 and R4. Let L be any finite extension of F contained in F_∞, and put

$$H_L = \text{Gal}(F_\infty/L^{\text{cyc}}), \qquad \Gamma_L = \text{Gal}(L^{\text{cyc}}/L). \qquad (12)$$

The dual of the restriction map

$$\beta_L: H^1(\mathcal{G}_\Sigma(L^{\text{cyc}}), A) \to H^1(\mathcal{G}_\Sigma(F_\infty), A)^{H_L}$$

gives rise to a canonical $\mathbb{Z}_p$-homomorphism

$$r_L: X(T/F_\infty)_{H_L} \to X(T/L^{\text{cyc}}). \qquad (13)$$

Lemma 2.4 Ker r_L and Coker r_L *are finitely generated O-modules.*

Proof Applying (11) for $\mathcal{L} = L^{\text{cyc}}$ and $\mathcal{L} = F_\infty$, we obtain a commutative diagram with exact rows

$$
\begin{array}{ccccccc}
0 & \longrightarrow & \text{Sel}(T/F_\infty)^{H_L} & \longrightarrow & H^1(\mathcal{G}_\Sigma(F_\infty), A)^{H_L} & \longrightarrow & \bigoplus_{v\in\Sigma} J_v(T/F_\infty)^{H_L} \\
& & \alpha_L \uparrow & & \beta_L \uparrow & & \uparrow \delta_L = \underset{v\in\Sigma}{\oplus} \delta_{L,v} \\
0 & \longrightarrow & \text{Sel}(T/L^{\text{cyc}}) & \longrightarrow & H^1(\mathcal{G}_\Sigma(L^{\text{cyc}}), A) & \longrightarrow & \bigoplus_{v\in\Sigma} J_v(T/L^{\text{cyc}}).
\end{array}
$$

Thus, as Σ is a finite set, it suffices via the snake lemma to show that both Ker β_L and Ker $\delta_{L,v}$ are cofinitely generated $\mathbb{Z}_p$-modules. But

$$\text{Ker } \beta_L = H^1(H_L, A(F_\infty)), \quad \text{where } A(F_\infty) = A^{\mathcal{G}_\Sigma(F_\infty)}$$

and, when v does not divide p

$$\operatorname{Ker} \delta_L = \bigoplus_{w|v} H^1(\operatorname{Gal}(F_{\infty,w}/L_w^{\mathrm{cyc}}), A(F_{\infty,w})),$$

where w runs over the finite set of places of L^{cyc} above v, and we have also written w for some fixed place of F_∞ above w; when v divides p, a similar assertion holds with A replaced by A/F_v^+A. The conclusion of the lemma follows on noting that, for any p-adic Lie group $\mathcal{H}$ and any $\mathbb{Z}_p$-cofinitely generated $\mathcal{H}$-module W, all of the cohomology groups $H^i(\mathcal{H}, W)$ are cofinitely generated $\mathbb{Z}_p$-modules.

We can now prove Theorem 2.1. Thus we assume that there exist a finite extension L of F contained in F_∞ such that $X(T/L^{\mathrm{cyc}})$ is a finitely generated O-module, and H_L is pro-p. We conclude from Lemma 2.4 that $X(T/F_\infty)_{H_L}$ is a finitely generated O-module. As H_L is pro-p, $\Lambda_O(H_L)$ is a local ring. Hence the compact version of Nakayama's lemma shows that $X(T/F_\infty)$ is a finitely generated $\Lambda_O(H_L)$-module, and thus certainly a finitely generated $\Lambda(H)$-module, completing the proof. $\qquad\square$

For every algebraic extension $\mathcal{L}$ of F^{cyc}, define

$$Y(T/\mathcal{L}) = X(T/\mathcal{L})/X(T/\mathcal{L})(\pi). \tag{14}$$

Thus the $\mathfrak{M}_H(G)$-conjecture asserts that $Y(T/F_\infty)$ is finitely generated over $\Lambda_O(H)$.

Proposition 2.5 *Assume the* $\mathfrak{M}_H(G)$*-conjecture holds for* T/F_∞*. Then, for every finite extension L of F, with $L \subset F_\infty$, $X(T/L^{\mathrm{cyc}})$ is $\Lambda_O(\Gamma_L)$-torsion, where* $\Gamma_L = \operatorname{Gal}(L^{\mathrm{cyc}}/L)$*.*

Proof We have the commutative diagram with exact rows

$$\begin{array}{ccccc}
X(T/F_\infty)_{H_L} & \longrightarrow & Y(T/F_\infty)_{H_L} & \longrightarrow & 0 \\
r_L \downarrow & & s_L \downarrow & & \\
X(T/L^{\mathrm{cyc}}) & \longrightarrow & Y(T/L^{\mathrm{cyc}}) & \longrightarrow & 0,
\end{array} \tag{15}$$

where s_L is induced by r_L. By Lemma 2.4, Coker r_L is a finitely generated O-module, and so (15) shows that Coker s_L is finitely generated over O. Our hypothesis that $Y(T/F_\infty)$ is finitely generated over $\Lambda_O(H)$, and the fact that H_L is of finite index in H, imply that $Y(T/F_\infty)_{H_L}$ is also finitely generated over O, and thus $Y(T/L^{\mathrm{cyc}})$ must be finitely generated over O. This is equivalent to the assertion that $X(T/L^{\mathrm{cyc}})$ is $\Lambda_O(\Gamma_L)$-torsion, completing the proof. $\qquad\square$

In view of Proposition 2.5, it is natural to ask whether one can prove the $\mathfrak{M}_H(G)$-conjecture for T/F_∞ if we know that $X(T/L^{\mathrm{cyc}})$ is $\Lambda_O(\Gamma_L)$-torsion for every finite extension L of F contained in F_∞. Unfortunately, this is unknown at present. We shall give some rather weak partial results in this direction in the next section.

3 Additional evidence for the $\mathfrak{M}_H(G)$-conjecture

When the Galois group of our admissible p-adic Lie extension F_∞/F has dimension 1, the $\mathfrak{M}_H(G)$-conjecture is, of course, equivalent to the generalized Mazur conjecture (Conjecture M). In this section, we will give some additional evidence for the $\mathfrak{M}_H(G)$-conjecture in the first interesting case, namely when G is pro-p of dimension 2, and has no elements of order p. For simplicity, we shall assume that our Galois representation ρ comes from an elliptic curve E defined over F, having good ordinary reduction at all primes v of F above p. Thus the Conditions R1, R2, R3 and R4 are very well known to hold in this case. We take the $\mathbb{Z}_p$-lattice $T = T_p(E) = \varprojlim E_{p^n}$ inside $V_p(E)$, and write $X(E/F_\infty)$, $X(E/F^{\mathrm{cyc}})$ for the corresponding Selmer groups, which coincide with the duals of the classical Selmer groups over these fields. We recall that

$$Y(E/\mathcal{L}) = X(E/\mathcal{L})/X(E/\mathcal{L})(p)$$

for any algebraic extension $\mathcal{L}$ of F^{cyc}.

We assume for the rest of the section that F_∞/F is such that G is pro-p of dimension 2, and has no element of order p. In such a case, $H = \mathbb{Z}_p$, and G is a semi-direct or direct product of two copies of $\mathbb{Z}_p$. We recall that a finitely generated torsion left $\Lambda(G)$-module W is said to be *pseudonull* if $\mathrm{Ext}^1_{\Lambda(G)}(W, \Lambda(G)) = 0$. If W is an arbitrary finitely generated torsion $\Lambda(G)$-module, it is shown in [21] that there exists a $\Lambda(G)$-homomorphism

$$f \colon W(p) \to \bigoplus_{i=1}^{t} \Lambda(G)/p^{r_i}\Lambda(G),$$

where all the r_i are integers ≥ 1, and $\mathrm{Ker}\, f$ and $\mathrm{Coker}\, f$ are pseudonull. We then define the μ_G-*invariant* of W by

$$\mu_G(W) = \sum_{i=1}^{t} r_i. \tag{16}$$

Theorem 3.1 *Let F_∞/F be an admissible p-adic Lie extension with G pro-p of dimension 2, having no elements of order p. Assume that $X(E/F^{\mathrm{cyc}})$ is $\Lambda(\Gamma)$-torsion. Then $X(E/F_\infty)$ is $\Lambda(G)$-torsion, and*

$$\mu_G(X(E/F_\infty)) = \mu_\Gamma(X(E/F^{\mathrm{cyc}})) - \mu_\Gamma(Y(E/F_\infty)_H). \tag{17}$$

Corollary 3.2 *Under the same hypotheses as Theorem 3.1, $X(E/F_\infty) \in \mathfrak{M}_H(G)$ if and only if*

$$\mu_G(X(E/F_\infty)) = \mu_\Gamma(X(E/F^{\mathrm{cyc}})). \tag{18}$$

We first show that the theorem implies the corollary. Assume that $X(E/F_\infty)$ belongs to $\mathfrak{M}_H(G)$, whence $Y(E/F_\infty)_H$ is a finitely generated $\mathbb{Z}_p$-module. Thus $\mu_\Gamma(Y(E/F_\infty)_H) = 0$, and (18) follows from (17). Conversely, assume that (18) holds, whence $\mu_\Gamma(Y(E/F_\infty)_H) = 0$ by (17). Applying the diagram (15) and Lemma 2.4 for $L = F$, we see that our hypothesis that $X(T/F^{\mathrm{cyc}})$ is $\Lambda(\Gamma)$-torsion implies that $Y(E/F_\infty)_H$ is a finitely generated $\mathbb{Z}_p$-module. Hence, as $H = \mathbb{Z}_p$, Nakayama's lemma shows that $Y(E/F_\infty)$ is finitely generated over $\Lambda(H)$, completing the proof of the corollary. □

The next well-known result is needed for the proof of Theorem 3.1.

Proposition 3.3 *Under the hypotheses of Theorem 3.1, we have*

$$H_i(H, W) = 0 \quad (i \geq 1)$$

for the three H-modules $W = X(E/F_\infty)$, $X(E/F_\infty)(p)$, and $Y(E/F_\infty)$.

We only sketch the proof as the arguments in it are well known (see [6, remark 2.6 and proof of proposition 2.13], [13]). Since $H = \mathbb{Z}_p$, we have $H_i(H, A) = 0$ for any compact $\Lambda(H)$-module A and all $i \geq 2$. Hence it suffices to show that

$$H_1(H, X(E/F_\infty)) = 0, \tag{19}$$

because both $X(E/F_\infty)(p)$ and $Y(E/F_\infty)$ $(= p^r X(E/F_\infty)$ for some integer $r \geq 0)$ are submodules of $X(E/F_\infty)$. Let $\lambda_{E/L}$ denote the right hand map in the exact sequence (11). It is well known that our hypothesis of $X(E/F^{\mathrm{cyc}})$ being $\Lambda(\Gamma)$-torsion implies that $\lambda_{E/F^{\mathrm{cyc}}}$ is surjective, (see [13, theorem 7.2], for instance), and that $H^2(\mathcal{G}_\Sigma(F^{\mathrm{cyc}}), E_{p^\infty}) = 0$. Now the Hochschild–Serre spectral sequence gives an exact sequence

$$H^2(\mathcal{G}_\Sigma(F^{\mathrm{cyc}}), E_{p^\infty}) \to H^2(\mathcal{G}_\Sigma(F_\infty), E_{p^\infty})^H \to H^2(H, H^1(\mathcal{G}_\Sigma(F_\infty), E_{p^\infty}), \tag{20}$$

and as $H \simeq \mathbb{Z}_p$, it follows that the group on the right is trivial, whence the group in the middle of this sequence must be zero. It then follows from Nakayama's lemma that

$$H^2(\mathcal{G}_\Sigma(F_\infty), E_{p^\infty}) = 0. \tag{21}$$

We next use the commutative diagram in the proof of Lemma 2.4 with $\mathcal{L} =$

F^{cyc}. As $H = \mathbb{Z}_p$, one sees easily that the right vertical map δ_F must be surjective. Since $\lambda_{E/F^{\mathrm{cyc}}}$ is surjective, it follows that we have the exact sequence

$$0 \to \mathrm{Sel}(E/F_\infty)^H \to H^1(\mathcal{G}_\Sigma(F_\infty), E_{p^\infty})^H \overset{\gamma_{F_\infty}}{\to} \underset{v \in \Sigma}{\oplus} J_v(F_\infty)^H \to 0. \qquad (22)$$

Also $H^1(H, H^1(\mathcal{G}_\Sigma(F_\infty), E_{p^\infty})) = 0$ because of the exact sequence

$$H^2(\mathcal{G}_\Sigma(F^{\mathrm{cyc}}), E_{p^\infty}) \to H^1(H, H^1(\mathcal{G}_\Sigma(F_\infty), E_{p^\infty})) \to H^3(H, E_{p^\infty})$$

coming from the Hochschild–Serre spectral sequence. It now follows easily from equation (22) (see [6, proof of lemma 2.5]) that λ_{E/F_∞} is surjective and $H^1(H, \mathrm{Sel}(E/F_\infty)) = 0$, completing the proof.

We can now prove Theorem 3.1. Since it was shown in the proof of the previous proposition that λ_{E/F_∞} is surjective and (6) is valid, a well known argument (see [3, theorem 4.12] shows that $X(E/F_\infty)$ is $\Lambda(G)$-torsion. For any G-module A and Γ-module B, define

$$\chi(G, A) = \prod_{i=0}^{2} (\#H_i(G, A))^{(-1)^i}, \quad \chi(\Gamma, B) = \prod_{i=0}^{1} (\#H_i(\Gamma, B))^{(-1)^i}.$$

As $H_1(H, X(E/F_\infty)(p)) = 0$, we conclude from the Hochschild–Serre spectral sequence that

$$\chi(G, X(E/F_\infty)(p)) = \chi(\Gamma, X(E/F_\infty)(p)_H). \qquad (23)$$

Using the well-known connexion between Euler characteristics and μ-invariants (see [6, proof of prop. 2.12]), (23) is equivalent to

$$\mu_G(X(E/F_\infty)) = \mu_\Gamma(X(E/F_\infty)(p)_H). \qquad (24)$$

But we have a commutative diagram with exact rows

$$\begin{array}{ccccccccc}
0 & \longrightarrow & X(E/F_\infty)(p)_H & \longrightarrow & X(E/F_\infty)_H & \longrightarrow & Y(E/F_\infty)_H & \longrightarrow & 0 \\
 & & {\scriptstyle r_{1,F}}\downarrow & & {\scriptstyle r_F}\downarrow & & {\scriptstyle r_{2,F}}\downarrow & & \\
0 & \longrightarrow & X(E/F^{\mathrm{cyc}})(p) & \longrightarrow & X(E/F^{\mathrm{cyc}}) & \longrightarrow & Y(E/F^{\mathrm{cyc}}) & \longrightarrow & 0,
\end{array}$$

and $\mathrm{Ker}\,(r_F)$ and $\mathrm{Coker}\,(r_F)$ are finitely generated $\mathbb{Z}_p$-modules by Lemma 2.4. Also $Y(E/F^{\mathrm{cyc}})$ is a finitely generated $\mathbb{Z}_p$-module because $X(E/F^{\mathrm{cyc}})$ is $\Lambda(\Gamma)$-torsion. Thus $\mu_\Gamma(Y(E/F_\infty)_H) = \mu_\Gamma(\mathrm{Coker}\,(r_{1,F}))$. As $\mu_\Gamma(\mathrm{Ker}\,(r_{1,F})) = 0$, assertion (17) follows. This completes the proof of Theorem 3.1. $\qquad \square$

For the remainder of this section, we shall assume that we are in the more restrictive case when G is abelian, and so topologically isomorphic to $\mathbb{Z}_p^2$. We pick a lifting of Γ to G, which we also denote by Γ, so that we then have $G = H \times \Gamma$. For each integer $n \geq 0$, let H_n be the unique closed subgroup of H of index p^n.

Theorem 3.4 *Assume that $G = H \times \Gamma$. Let M be a finitely generated torsion $\Lambda(G)$-module such that (i) $\mu_G(M) = 0$, and (ii) $(M)_{H_n}$ is $\Lambda(\Gamma)$-torsion for all $n \geq 0$. Then*

$$\mu_\Gamma\left((M)_{H_n}\right) \leq c.(n+1) \quad (n = 0, 1, 2 \ldots) \tag{25}$$

where $c > 0$ does not depend on n.

Proof Fix topological generators γ_1, γ_2 of Γ and H, respectively. There is a unique topological isomorphism of $\mathbb{Z}_p$-algebras from $\Lambda(G)$ onto $\mathbb{Z}_p[[T_1, T_2]]$, which sends γ_1 to $1 + T_1$, and γ_2 to $1 + T_2$, and we shall henceforth identify these two rings via this isomorphism. For each $n \geq 0$, put

$$\omega_n(T_2) = (1 + T_2)^{p^n} - 1. \tag{26}$$

We recall also that $\Lambda(G)$ is a unique factorization domain. Now let M be any torsion $\Lambda(G)$-module satisfying conditions (i) and (ii) of Theorem 3.4. By the structure theory (see [2, chapter 7], we have an exact sequence of $\Lambda(G)$-modules

$$0 \to \bigoplus_{i=1}^{r} \Lambda(G)/f_i\Lambda(G) \to M \to D \to 0, \tag{27}$$

where the f_i are non-zero elements of $\Lambda(G)$, and D is pseudonull. Moreover, it is well known and easy to see the conditions (i) and (ii) are equivalent to the respective assertions that

$$(p, f_i) = 1, \quad (f_i, \omega_n(T_2)) = 1, \quad (i = 1, \ldots, r), n \geq 0. \tag{28}$$

Taking H_n-coinvariants of the exact sequence (27), we obtain

$$\bigoplus_{i=1}^{r} \Lambda(G)/(f_i, \omega_n(T_2)) \to M_{H_n} \to D_{H_n} \to 0.$$

As the module on the left of this exact sequence is $\Lambda(\Gamma)$-torsion, we see immediately that the theorem follows from Propositions 3.5 and 3.7 below. $\quad\square$

Proposition 3.5 *Assume that $M = \Lambda(G)/f\Lambda(G)$, where f is a non-zero element of $\Lambda(G)$ such that $(f, p) = 1$ and $(f, \omega_n(T_2)) = 1$ for all $n \geq 0$. Then*

$$\mu_\Gamma\left((M)_{H_n}\right) \leq c'.(n+1) \quad (n = 0, 1, 2, \ldots), \tag{29}$$

where $c' > 0$ does not depend on n.

Before giving the proof of Proposition 3.5, we establish a preliminary lemma, which is well known, but for which we have been unable to find a suitable reference.

Lemma 3.6 *Let $M = \Lambda(G)/f\Lambda(G)$, where f is a non-zero element of $\Lambda(G)$ satisfying $(f, \omega_n(T_2)) = 1$ for some integer $n \geq 0$. Put*

$$h_n(T_1) = \prod_{\zeta \in \mu_{p^n}} f(T_1, \zeta - 1), \qquad (30)$$

where the product is taken over all ζ in μ_{p^n}. Then $h_n(T_1)\Lambda(\Gamma)$ is the characteristic ideal of the torsion $\Lambda(\Gamma)$-module $(M)_{H_n}$.

Proof Let O be the ring of integers of $\mathbb{Q}_p(\mu_{p^n})$, and put $M_O = M \otimes_{\mathbb{Z}_p} O$. Write $\Lambda_O(G)$, $\Lambda_O(\Gamma)$ for the Iwasawa algebras of G and Γ with coefficients in O, so that $\Lambda_O(G) = O[[T_1, T_2]]$, and $\Lambda_O(\Gamma) = O[[T_1]]$. Since $(f, \omega_n(T_2)) = 1$, we have an exact sequence of $\Lambda(\Gamma)$-modules

$$0 \to \Lambda_O(G)_{H_n} \xrightarrow{\alpha_n} \Lambda_O(G)_{H_n} \to (M_O)_{H_n} \to 0, \qquad (31)$$

where α_n is the map induced by multiplication by f. Put

$$B_n = \bigoplus_{\zeta \in \mu_{p^n}} \Lambda_O(\Gamma).$$

We have the $\Lambda_O(\Gamma)$-homomorphism

$$\theta_n \colon \Lambda_O(G)_{H_n} \to B_n$$

defined by $\theta_n(v(T_1, T_2) \bmod H_n) = (v(T_1, \zeta - 1))$, where ζ runs over μ_{p^n}. Clearly θ_n is injective. Since both $\Lambda_O(G)_{H_n}$ and B_n have rank p^n as $\Lambda_O(\Gamma)$-modules, it follows that A_n is $\Lambda_O(\Gamma)$-torsion, where $A_n = \mathrm{Coker}\,\theta_n$. Define

$$\beta_n \colon B_n \to B_n$$

by $\beta_n(k_\zeta(T_1)) = (f(T_1, \zeta - 1)k_\zeta(T_1))$, and let $\delta_n \colon A_n \to A_n$ be the $\Lambda_O(\Gamma)$-homomorphism induced by β_n. As A_n is $\Lambda_O(\Gamma)$-torsion, and as characteristic ideals multiply along exact sequences, we conclude from the exact sequence

$$0 \to \mathrm{Ker}\,\delta_n \to A_n \xrightarrow{\delta_n} A_n \to \mathrm{Coker}\,\delta_n \to 0, \qquad (32)$$

that $\mathrm{Ker}\,\delta_n$ and $\mathrm{Coker}\,\delta_n$ have the same characteristic ideals as $\Lambda_O(\Gamma)$-modules. On the other hand, recalling (31) and that α_n, β_n are injective, a simple application of the snake lemma shows that we have the exact sequence of $\Lambda_O(\Gamma)$-modules

$$0 \to \mathrm{Ker}\,\delta_n \to (M_O)_{H_n} \to \mathrm{Coker}\,\beta_n \to \mathrm{Coker}\,\delta_n \to 0.$$

We conclude that $(M_O)_{H_n}$ and $\mathrm{Coker}\,\beta_n$ have the same characteristic ideals as $\Lambda_O(\Gamma)$-modules. But $h_n(T_1)\Lambda_O(\Gamma)$ is obviously the characteristic ideal of $\mathrm{Coker}\,\beta_n$. Hence $(M_O)_{H_n}$ has characteristic ideal $h_n(T_1)\Lambda_O(\Gamma)$. This completes the proof of Lemma 3.6.   □

We now prove Proposition 3.5. Thus f will now denote a non-zero element of $\Lambda(G)$ such that $(f, p) = 1$ and $(f, \omega_n(T_2)) = 1$ for all $n \geq 0$. Let $h_n(T_1)$ be the element of $\mathbb{Z}_p[[T_1]]$ defined by (30), so that $h_n(T_1)$ is a characteristic element for the $\Lambda(\Gamma)$-module $(M)_{H_n}$, where $M = \Lambda(G)/f\Lambda(G)$. As usual, write ord for the order function on $\bar{\mathbb{Q}}_p^\times$, normalized so that $\text{ord}(p) = 1$. Similarly, if $g = \sum_{n=0}^\infty c_n T^n$ is any power series with coefficients c_n being p-adic integers in $\bar{\mathbb{Q}}_p$, we define $\text{ord}(g) = \inf_{n \geq 0} \text{ord}(c_n)$. Now

$$\mu_\Gamma((M)_{H_n}) = \text{ord}(h_n(T_1)). \tag{33}$$

We estimate the right hand side of this equation as follows. We can write

$$f(T_1, T_2) = \sum_{m=0}^\infty b_m(T_2) T_1^m.$$

Since $f \notin p\Lambda(G)$, we can find an integer $k \geq 0$ such that $b_0(T_2), \ldots, b_{k-1}(T_2)$ belong to $p\mathbb{Z}_p[[T_2]]$, but $b_k(T_2)$ does not belong to $p\mathbb{Z}_p[[T_2]]$. Hence we can write

$$b_k(T_2) = \sum_{r=0}^\infty e_r T_2^r \quad (e_r \in \mathbb{Z}_p),$$

and find an integer $s \geq 0$ such that $e_0, \ldots, e_{s-1}$ are in $p\mathbb{Z}_p$, but e_s is in $\mathbb{Z}_p^\times$. Now fix some integer $n_0 \geq 1$ such that

$$\phi(p^{n_0}) > s. \tag{34}$$

Assume that t is any integer $\geq n_0$. We claim that

$$\text{ord}(b_k(\zeta - 1)) = s/\phi(p^t) \quad (\zeta \in \mu_{p^t} \setminus \mu_{p^{t-1}}). \tag{35}$$

To prove this, we simply note that

$$\text{ord}(e_s(\zeta - 1)^s) = s/\phi(p^t) \quad (\zeta \in \mu_{p^t} \setminus \mu_{p^{t-1}})$$

and this order is < 1 by virtue of (34) and the fact that $t \geq n_0$. On the other hand, it is clear from the definition of s that, for all integers $r \neq s$, we have either $\text{ord}(e_r(\zeta - 1)^r) \geq 1$ for $r < s$, or $\text{ord}(e_r(\zeta - 1)^r) > s/\phi(p^t)$ for $r > s$, whence (35) is clear. It follows from (35) that

$$\text{ord}(f(T_1, \zeta - 1)) \leq s/\phi(p^t) \quad (t \geq n_0, \zeta \in \mu_{p^t} \setminus \mu_{p^{t-1}}). \tag{36}$$

For $n \geq n_0$, define

$$g_n(T_1) = h_n(T_1)/h_{n_0-1}(T_1). \tag{37}$$

We deduce immediately from (33) that, for all $n \geq n_0$, we have

$$\text{ord}(g_n(T_1)) \leq \sum_{t=n_0}^n (s/\phi(p^t)) \cdot \phi(p^t) = s(n - n_0).$$

As $\mathrm{ord}(h_{n_0-1}(T_1))$ does not depend on n, it follows that

$$\mathrm{ord}(h_n(T_1)) \leq c'(n+1) \quad (n \geq 0),$$

where c' does not depend on n. In view of (35), the proof of Proposition 3.5 is complete.   □

Proposition 3.7 *Let M be any pseudonull $\Lambda(G)$-module. Then we have*

$$\mu_\Gamma(M_{H_n}) \leq c'' \tag{38}$$

where c'' does not depend on n.

Proof Put $M' = M/M(p)$. Since there is an exact sequence

$$M(p)_{H_n} \to M_{H_n} \to M'_{H_n} \to 0,$$

it is clear that it suffices to establish (38) in the two cases $M = M'$ and $M = M(p)$. Assume first that $M(p) = 0$. As M is pseudonull, it has (Krull) dimension 0 or 1 over $\Lambda(G)$. If M has dimension 0, M is finite and so $c'' = 0$. Hence we may assume that M has dimension 1. Since multiplication by p is injective on M, a standard result in dimension theory asserts that M/pM must have dimension 0, and so is finite. Thus M is a finitely generated $\mathbb{Z}_p$-module by Nakayama's lemma, and so $\mu_\Gamma(M_{H_n}) = 0$ for all $n \geq 0$.

Suppose next that $M = M(p)$. Let Γ_n denote the unique closed subgroup of Γ of index p^n. Then clearly

$$\mu_\Gamma(M_{H_n}) = \mu_{\Gamma_n}(M_{H_n})/p^n. \tag{39}$$

On the other hand, since M_{H_n} is annihilated by a power of p, a well-known formula asserts that

$$\mu_{\Gamma_n}(M_{H_n}) = \mathrm{ord}(\#((M_{H_n})_{\Gamma_n}) - \mathrm{ord}(\#((M_{H_n})^{\Gamma_n})).$$

Note that $(M_{H_n})_{\Gamma_n} = M_{G_n}$, where G_n is the unique closed subgroup of G topologically generated by $\gamma_1^{p^n}$ and $\gamma_2^{p^n}$. Hence

$$\mu_{\Gamma_n}(M_{H_n}) \leq \mathrm{ord}(\#(M_{G_n})). \tag{40}$$

But theorem 2.3 of [7] for $d = 2$ asserts that

$$\mathrm{ord}(\#(M_{G_n})) \leq c'' p^n, \tag{41}$$

where c'' does not depend on n. Combining the estimates (40) and (41) with formula (39), the assertion (29) follows. This completes the proof of Proposition 3.7, and so also of Theorem 3.4.   □

We end this section by establishing a purely cyclotomic criterion for the

$\mathfrak{M}_H(G)$-conjecture. We continue to assume that $G \sim \mathbb{Z}_p^2$, and fix a lifting of Γ to G so that $G = H \times \Gamma$. Define L_∞ to be the fixed field of the subgroup Γ. Thus L_∞/F is a $\mathbb{Z}_p$-extension, and we define L_n to be its unique sub-extension of degree p^n over F. Since $L_\infty \cap F^{\mathrm{cyc}} = F$, $\mathrm{Gal}(L_n^{\mathrm{cyc}}/L_n)$ can be identified with Γ for all $n \geq 0$.

Theorem 3.8 *Assume that* $G = H \times \Gamma$, *and that* $X(E/L_n^{\mathrm{cyc}})$ *is* $\Lambda(\Gamma)$-*torsion for all* $n \geq 0$. *Then* $X(E/F_\infty)$ *belongs to* $\mathfrak{M}_H(G)$ *if and only if, for all* $n \geq 0$, *we have*

$$\mu_\Gamma(X(E/L_n^{\mathrm{cyc}})) \geq p^n \mu_\Gamma(X(E/F^{\mathrm{cyc}})). \tag{42}$$

Proof Put

$$\mu_G(X(E/F_\infty)) = \alpha, \quad \mu_\Gamma(X(E/F^{\mathrm{cyc}})) = \beta. \tag{43}$$

Let $G_n = \mathrm{Gal}(F_\infty/L_n)$, so that G_n is a subgroup of G of index p^n. As $\Lambda(G)$ is a free $\Lambda(G_n)$-module of rank p^n, we have

$$\mu_{G_n}(X(E/F_\infty)) = p^n \alpha. \tag{44}$$

If $X(E/F_\infty)$ belongs to $\mathfrak{M}_H(G)$, then we can apply Corollary 3.2 to all the sub-extensions $F_\infty/L_n(n = 0, 1, \cdots)$. In view of (44), it follows immediately that

$$\alpha = \beta, \quad \mu_\Gamma(X(E/L_n^{\mathrm{cyc}})) = \alpha.p^n \qquad (n \geq 0), \tag{45}$$

and so (42) always holds with equality. For the argument in the other direction, note first that it follows from (17) for the extension F_∞/L_n that

$$\mu_\Gamma(X(E/L_n^{\mathrm{cyc}})) = \mu_{G_n}(X(E/F_\infty)) + \mu_\Gamma(Y(E/F_\infty)_{H_n}) \tag{46}$$

where $H_n = \mathrm{Gal}(F_\infty/L_n^{\mathrm{cyc}})$. One sees easily that $Y(E/F_\infty)_{H_n}$ is $\Lambda(\Gamma)$-torsion, because $X(E/L_n^{\mathrm{cyc}})$ is assumed to be $\Lambda(\Gamma)$-torsion (cf. the proof of Proposition 3.3). By definition, $\mu_G(Y(E/F_\infty)) = 0$. We conclude from (46) and Theorem 3.4 that

$$\mu_\Gamma(X(E/L_n^{\mathrm{cyc}})) \leq \alpha.p^n + c.(n+1) \quad (n \geq 0), \tag{47}$$

where c is some constant ≥ 0, which does not depend on n. Now suppose that (42) holds, i.e.

$$\mu_\Gamma(X(E/L_n^{\mathrm{cyc}})) \geq \beta.p^n \quad (n \geq 0). \tag{48}$$

Combining (47) and (48), and dividing both sides by p^n, it follows that

$$\beta \leq \alpha + c.(n+1)/p^n \quad (n \geq 0) \tag{49}$$

But $c.(n+1)/p^n \to 0$ as $n \to \infty$, and so we conclude that $\beta \leq \alpha$. On the other hand (17) shows that $\alpha \leq \beta$, and so $\beta = \alpha$. Thus, by Corollary 3.2, $X(E/F_\infty) \in \mathfrak{M}_H(G)$, and the proof is complete.   □

In fact, the proof also shows that $X_(E/F_\infty)$ belongs to $\mathfrak{M}_H(G)$ if and only if (42) holds with equality for all $n \geq 0$. In some special cases, one can prove the inequality (42) for some values of n. For example, if the kernel of multiplication by p on $X(E/L_n^{cyc})$ is of finite index in $X(E/L_n^{cyc})(p)$, it can be shown that (42) holds for this n. However, we have to confess that we do not know how to prove it for all $n \geq 0$ for a single elliptic curve E/F which is not isogenous to some elliptic curve E'/F with $\mu_\Gamma(X(E'/F)) = 0$.

We are very grateful to R. Greenberg for pointing out to us the kind of difficulty which has to be overcome to prove (42). As earlier, identify $\Lambda(G)$ with $\mathbb{Z}_p[[T_1, T_2]]$, where $1 + T_1$ corresponds to a topological generator γ_1 of Γ, and $1 + T_2$ corresponds to a topological generator γ_2 of H. Define

$$W = \mathbb{Z}_p[[T_1, T_2]]/(T_2 - p).$$

One verifies immediately that $W(p) = 0$, and that W is a free $\mathbb{Z}_p[[T_1]]$-module of rank 1. Moreover, the topological generator γ_2 of H acts on W via $1 + p$, and thus

$$W_{H_n} = \mathbb{Z}_p/p^{n+1}\mathbb{Z}_p[[T_1]]. \tag{50}$$

Also $H_1(H_n, W) = W^{H_n} = 0$ for all $n \geq 0$. Thus W has many similar properties to the $\Lambda(G)$-module $X(E/F_\infty)$. However, it is clear that W does not belong to the category $\mathfrak{M}_H(G)$. Moreover,

$$\mu_\Gamma(W_{H_0}) = 1, \quad \mu_\Gamma(W_{H_n}) = n + 1, \tag{51}$$

and so the analogue of the inequality (42) is certainly false for all $n \geq 1$. Hence to establish the $\mathfrak{M}_H(G)$-conjecture for $X(E/F_\infty)$ we need, in particular, to rule out the possibility that a module like W could be pseudo-isomorphic to a direct summand of $X(E/F_\infty)$ as a $\Lambda(G)$- module.

4 Hida families over p-adic Lie extensions

The aim of this section is to propose a definition for the Selmer group attached to an ordinary primitive R-adic form in the sense of Hida [14], [15] over every admissible p-adic Lie extension of $\mathbb{Q}$. Our definition is the naive one, and differs somewhat from special cases considered in the existing literature (see [10], [19]). In order to keep technicalities from Hida theory to a minimum, we impose throughout some simplifying technical assumptions about our R-adic form, but we believe that all we do can be extended to the most general class of R-adic forms.

Fix an embedding

$$i: \bar{\mathbb{Q}} \hookrightarrow \bar{\mathbb{Q}}_p, \tag{52}$$

where $\bar{\mathbb{Q}}$ denotes the algebraic closure of $\mathbb{Q}$ in $\mathbb{C}$. Throughout, K will denote a suitably large finite extension of $\mathbb{Q}_p$, and O the ring of integers of K. Let

$$R = O[[W]], \tag{53}$$

where W is an indeterminate. For each $k \in \mathbb{Z}$, $\xi_k: R \rightarrow O$ will denote the continuous O-algebra homomorphism such that $\xi_k(W) = (1 + p)^k - 1$. Write $P_k = ((1 + p)^k - (1 + W))R$ for the kernel of ξ_k.

Let N be an integer prime to p, and ψ a Dirichlet character modulo Np. We shall take the somewhat restricted definition of a Hida family of cusp forms given by the following. An R-adic cusp form $\mathcal{F}$ of level N and character ψ will be a formal power series

$$\mathcal{F} = \sum_{n=1}^{\infty} a_n(W)q^n, \qquad \text{with } a_n(W) \in R \ (n = 0, 1, \dots),$$

such that, for all integers $k \geq 2$,

$$F_k = \sum_{n=1}^{\infty} a_n((1 + p)^k - 1)q^n \tag{54}$$

is the image under i of the q-expansion of a classical cusp form of weight k, level Np, and character $\psi\omega^{-k}$; here ω denotes the Teichmüller character modulo p. We shall always assume that $\mathcal{F}$ is *primitive* in the sense that (i) $a_1(W) = 1$, (ii) each specialisation F_k ($k \geq 2$) is an eigenform for the Hecke operators, and (iii) each F_k ($k \geq 2$) is N-new.

Assume from now on that $\mathcal{F}$ is such a primitive R-adic form of tame level N. We will always suppose that $\mathcal{F}$ is *ordinary*, meaning that $a_p(W)$ is a unit in R, or equivalently that $a_p((1 + p)^k - 1)$ is a unit in O for some integer $k \geq 2$. A fundamental theorem of Hida [14] attaches a canonical Galois representation to such a primitive, ordinary R-adic form $\mathcal{F}$. Let $Q(R)$ be the field of fractions of R. Then there exists a vector space $V_{\mathcal{F}}$ of dimension two over $Q(R)$, and a representation

$$\rho_{\mathcal{F}}: \text{Gal}(\bar{\mathbb{Q}}/\mathbb{Q}) \rightarrow \text{Aut}_{Q(R)}(V_{\mathcal{F}}) \tag{55}$$

which is absolutely irreducible and unramified outside a finite set of primes. Moreover, we assume that $\rho_{\mathcal{F}}$ is continuous in the sense that there exists a finitely generated R-submodule $\mathcal{D}$ of $V_{\mathcal{F}}$ which is stable under the action of $\mathrm{Gal}(\bar{\mathbb{Q}}/\mathbb{Q})$, and has the property that the induced map

$$\rho_{\mathcal{F}}: \mathrm{Gal}(\bar{\mathbb{Q}}/\mathbb{Q}) \to \mathrm{Aut}_R(\mathcal{D})$$

is continuous when $\mathcal{D}$ is given the $\mathfrak{m}$-adic topology, where $\mathfrak{m}$ denotes the maximal ideal of R (see [14]). This notion of continuity suffices for our purposes, but R. Greenberg has kindly pointed out to us the paper [11], where a subtler notion of continuity of such representations is discussed.

By an R-lattice in $V_{\mathcal{F}}$, we mean a finitely generated free R-submodule of $V_{\mathcal{F}}$ which has R-rank two, and which is stable under the action of $\mathrm{Gal}(\bar{\mathbb{Q}}/\mathbb{Q})$. We are grateful to R. Greenberg for the remark that, since we are assuming that $R = O[[W]]$, such a Galois invariant free R-lattice always exists. Indeed, let $\mathcal{D}$ be the R-submodule of V discussed above, and define

$$\mathcal{T} = \bigcap_{\mathfrak{p}} \mathcal{D}_{\mathfrak{p}},$$

where $\mathfrak{p}$ runs over the prime ideals of height one of F, and $\mathcal{D}_{\mathfrak{p}}$ denotes the localization of $\mathcal{D}$ at $\mathfrak{p}$. Then $\mathcal{T}$ is plainly stable under the action of $\mathrm{Gal}(\bar{\mathbb{Q}}/\mathbb{Q})$, and is reflexive, hence free, as an R-module.

In what follows, $\mathcal{T}$ will denote any R-lattice in $V_{\mathcal{F}}$, which is stable under the action of $\mathrm{Gal}(\bar{\mathbb{Q}}/\mathbb{Q})$, and we write $\rho_{\mathcal{T}}$ for the corresponding representation in $\mathrm{Aut}_R(\mathcal{T})$. One of the fundamental properties of the representation $\rho_{\mathcal{T}}$ is the following. Put

$$T_k = \mathcal{T}/P_k\mathcal{T}, \tag{56}$$

so that T_k is a free O-module of rank two. Composing $\rho_{\mathcal{T}}$ with the canonical map arising from the surjection of $\mathcal{T}$ onto T_k, we obtain, for each integer $k \geq 2$, a new representation

$$\rho_{F_k}: \mathrm{Gal}(\bar{\mathbb{Q}}/\mathbb{Q}) \to \mathrm{Aut}_O(T_k) = \mathrm{GL}_2(O). \tag{57}$$

Then, for all integers $k \geq 2$, this representation ρ_{F_k} is the Galois representation attached to the classical modular form (cf. (54)), whose lattice T_k is given by (56).

As $\mathcal{F}$ is assumed to be ordinary, Wiles [22] has shown that there exists an R-submodule $F^+\mathcal{T}$ of $\mathcal{T}$ of R-rank one, which is stable under the action of $\mathrm{Gal}(\bar{\mathbb{Q}}_p/\mathbb{Q}_p)$, and which is such that the quotient R-module $\mathcal{T}/F^+\mathcal{T}$ is unramified as a $\mathrm{Gal}(\bar{\mathbb{Q}}_p/\mathbb{Q}_p)$-module. If k is an integer ≥ 2, we can define F^+T_k to be the image of $F^+\mathcal{T}$ in $T_k = \mathcal{T}/P_k\mathcal{T}$. Then the action of $\mathrm{Gal}(\bar{\mathbb{Q}}_p/\mathbb{Q}_p)$ on T_k is clearly unramified.

In order to work with the Galois cohomology of discrete R-modules, we define

$$\hat{R} = \mathrm{Hom}\,(R, \mathbb{Q}_p/\mathbb{Z}_p), \tag{58}$$

and endow $\hat{R}$ with the R-module structure given by $(rf)(x) = f(rx)$, for $r \in R$, $f \in \hat{R}$, and $x \in R$. Define

$$\mathcal{A} = \mathcal{T} \otimes_R \hat{R}. \tag{59}$$

Moreover, defining $F^+\mathcal{A}$ to be $F^+\mathcal{T} \otimes_R \hat{R}$, let

$$\mathcal{B}_p = \mathcal{A}/F^+\mathcal{A}. \tag{60}$$

Now take F_∞ to be any admissible p-adic Lie extension of $\mathbb{Q}$. Then we define the Selmer group $\mathrm{Sel}(\mathcal{T}/F_\infty)$ of $\mathcal{T}$ over F_∞ by taking the obvious generalization of Greenberg's definition discussed in §2, namely

$$\mathrm{Sel}(\mathcal{T}/F_\infty) = \mathrm{Ker}\left(H^1(F_\infty, \mathcal{A}) \to \prod_{w \nmid p} H^1(F_{\infty,w}, \mathcal{A}) \times \prod_{w \mid p} H^1(F_{\infty,w}, \mathcal{B}_p)\right). \tag{61}$$

This is a discrete p-primary R-module endowed with a natural left action of G. We will also consider its Pontryagin dual

$$X(\mathcal{T}/F_\infty) = \mathrm{Hom}\,(\mathrm{Sel}(\mathcal{T}/F_\infty), \mathbb{Q}_p/\mathbb{Z}_p), \tag{62}$$

which inherits natural left actions of R and G in the usual manner. These actions extend by continuity to a left action of the Iwasawa algebra

$$\Lambda_R(G) = \varprojlim_U R[G/U], \tag{63}$$

where U runs over all open normal subgroups of G. For each finite place v of $\mathbb{Q}$, the groups $J_v(\mathcal{T}/F_\infty)$ can be defined using the same definition as that given before Corollary 2.3, replacing T by $\mathcal{T}$. Assume from now on that Σ is any finite set of rational primes such that both the extension $F_\infty/\mathbb{Q}$ and the representation $\rho_\mathcal{T}$ are unramified outside Σ and the prime at ∞. Then the obvious analogue of Lemma 2.2 holds for the module $\mathcal{A}$, and we have the exact sequence

$$0 \to \mathrm{Sel}(\mathcal{T}/F_\infty) \to H^1(\mathcal{G}_\Sigma(F_\infty), \mathcal{A}) \to \bigoplus_{v \in \Sigma} J_v(\mathcal{T}/F_\infty). \tag{64}$$

For each integer $k \geq 2$, put

$$p_k = (1 + W) - (1 + p)^k, \tag{65}$$

so that $P_k = p_k R$. If M is any R-module, define

$$M[p_k] = \mathrm{Ker}\,(M \xrightarrow{p_k} M). \tag{66}$$

Recalling that T_k is given by (56), we also define

$$A_k = T_k \otimes_O \hat{O}, \qquad \text{where } \hat{O} = \text{Hom}\,(O, \mathbb{Q}_p/\mathbb{Z}_p). \tag{67}$$

Lemma 4.1 $A_k \simeq \mathcal{A}[p_k]$ *for all* $k \geq 2$.

Proof Taking Pontryagin duals, we see that

$$\widehat{\mathcal{A}[p_k]} = \text{Hom}_R(\mathcal{A}[p_k], \hat{R}) \simeq \text{Hom}_R(\mathcal{A}, \hat{R})/P_k \simeq \text{Hom}_R(\mathcal{T}, R)/P_k.$$

Since $\mathcal{T}$ is R-free, we have

$$\text{Hom}_R(\mathcal{T}, R)/P_k \simeq \text{Hom}_R(\mathcal{T}, R/P_k) \simeq \text{Hom}_O(T_k, O) \simeq \widehat{A_k}. \qquad \square$$

Now we clearly have two exact sequences, namely

$$0 \to \text{Sel}(\mathcal{T}/F_\infty)[p_k] \to H^1(\mathcal{G}_\Sigma(F_\infty), \mathcal{A})[p_k] \to \bigoplus_{v \in \Sigma} J_v(\mathcal{T}/F_\infty)[p_k], \tag{68}$$

and

$$0 \to \text{Sel}(T_k/F_\infty) \to H^1(\mathcal{G}_\Sigma(F_\infty), A_k) \to \bigoplus_{v \in \Sigma} J_v(T_k/F_\infty), \tag{69}$$

which are related as follows. As $\mathcal{A}$ is a divisible R-module, we have an exact sequence of Galois modules

$$0 \to A_k = \mathcal{A}[p_k] \to \mathcal{A} \xrightarrow{p_k} \mathcal{A} \to 0 \tag{70}$$

and similarly there is an exact sequence of $\text{Gal}(\bar{\mathbb{Q}}_p/\mathbb{Q}_p)$-modules

$$0 \to B_{k,p} = \mathcal{B}_p[p_k] \to \mathcal{B}_p \xrightarrow{p_k} \mathcal{B}_p \to 0, \tag{71}$$

where

$$B_{k,p} = A_k/F^+A_k. \tag{72}$$

Taking Galois cohomology of these two exact sequences, we obtain a commutative diagram with exact rows and where the middle map is obviously surjective

$$
\begin{array}{ccccccc}
0 & \longrightarrow & \text{Sel}(T_k/F_\infty) & \longrightarrow & H^1(\mathcal{G}_\Sigma(F_\infty), A_k) & \longrightarrow & \bigoplus_{v \in \Sigma} J_v(T_k/F_\infty) \\
 & & \downarrow{\scriptstyle a_{F_\infty,k}} & & \downarrow{\scriptstyle b_{F_\infty,k}} & & \downarrow{\scriptstyle d_{F_\infty,k} = \bigoplus_{v \in \Sigma} d_{F_\infty,k,v}} \\
0 & \longrightarrow & \text{Sel}(\mathcal{T}/F_\infty)[p_k] & \longrightarrow & H^1(\mathcal{G}_\Sigma(F_\infty), \mathcal{A})[p_k] & \longrightarrow & \bigoplus_{v \in \Sigma} J_v(\mathcal{T}/F_\infty)[p_k].
\end{array}
\tag{73}
$$

Theorem 4.2 *For all* $k \geq 2$, $\text{Ker}(a_{F_\infty,k})$ *is a cofinitely generated O-module, and* $\text{Coker}(a_{F_\infty,k})$ *a cofinitely generated $\Lambda_O(H)$-module, where* $H = \text{Gal}(F_\infty/F^{\text{cyc}})$.

Proof We first show that $\mathrm{Ker}\,(a_{F_\infty,k})$ is a cofinitely generated O-module. In view of (73), it suffices to show that $\mathrm{Ker}\,(b_{F_\infty,k})$ is a cofinitely generated O-module. However, taking Galois cohomology of (70), we see immediately that

$$\mathrm{Ker}\,(b_{F_\infty,k}) = \mathcal{A}(F_\infty)/p_k\,\mathcal{A}(F_\infty),$$

which is dual to $\hat{\mathcal{A}}(F_\infty)[p_k]$, where $\hat{\mathcal{A}}$ is the Pontryagin dual of $\mathcal{A}$, and $\hat{\mathcal{A}}(F_\infty) = H^0(\mathcal{G}_\Sigma(F_\infty), \hat{\mathcal{A}})$. But

$$\hat{\mathcal{A}} = \mathrm{Hom}_R(\mathcal{T}, R),$$

and thus is a finitely generated R-module. Since $R/p_kR = O$, it follows that $\hat{\mathcal{A}}(F_\infty)[p_k]$ is a finitely generated O-module, as required. To prove that $\mathrm{Coker}\,(a_{F_\infty,k})$ is cofinitely generated over $\Lambda_O(H)$, it suffices, thanks to the diagram (73), to show that, for each v in Σ, $\mathrm{Ker}\,(d_{F_\infty,k,v})$ is cofinitely generated over $\Lambda_O(H)$. Now H contains a pro-p open subgroup which we denote by H'. We will prove that $(\mathrm{Ker}\,d_{F_\infty,k,v})$ is a cofinitely generated $\Lambda_O(H')$-module. By Nakayama's lemma, this latter assertion will follow if we can show that the group $(\mathrm{Ker}\,d_{F_\infty,k,v})^{H'}$ is a cofinitely generated O-module. Let $\mathcal{K}$ be the fixed field of H'. Now, making use of the exact sequences (70) and (71), one sees easily that $(\mathrm{Ker}\,d_{F_\infty,k,v})^{H'}$ is equal to

$$\underset{w|v}{\oplus}\,\mathcal{A}(\mathcal{K}_w)/p_k\mathcal{A}(\mathcal{K}_w) \quad \text{or} \quad \underset{w|v}{\oplus}\,\mathcal{B}_p(\mathcal{K}_w)/p_k\mathcal{B}_p(\mathcal{K}_w), \qquad (74)$$

according as v does not or does divide p; here w runs over the fnite set of primes of $\mathcal{K}$ above v. But $\mathcal{A}$ and $\mathcal{B}_p$ are cofinitely generated R-modules, whence both modules appearing in (74) are plainly cofinitely generated over O. This completes the proof of Theorem 4.2. $\qquad\square$

5 Analogue of the $\mathfrak{M}_H(G)$ conjecture for Hida families

We use the same notation as in §4. In particular, $\mathcal{F}$ will continue to be an R-adic cusp form of level N and character ψ, and $\mathcal{T}$ will denote an R-lattice in $V_{\mathcal{F}}$. Moreover, F_∞ will again be an arbitrary admissible p-adic Lie extension of $\mathbb{Q}$. Our goal in this section is to propose analogues of Conjecture M and Conjecture $\mathfrak{M}_H(G)$ of §2 for the Selmer group of the Hida family.

If B is any R-module, we define $B(R)$ to be the R-torsion submodule of B. Our proposed analogue of Conjecture M is as follows.

Conjecture MR For each finite extension F of $\mathbb{Q}$, the module

$$X(\mathcal{T}/F^{\mathrm{cyc}})/X(\mathcal{T}/F^{\mathrm{cyc}})(R)$$

is finitely generated over R.

Similarly, we propose the following analogue of Conjecture $\mathfrak{M}_H(G)$.

Conjecture $\mathfrak{M}_H^R(G)$ For every admissible p-adic Lie extension F_∞ of F, the $\Lambda_R(G)$-module $X(\mathcal{T}/F_\infty)/X(\mathcal{T}/F_\infty)(R)$ is finitely generated over $\Lambda_R(H)$.

First, note the following parallel result to Proposition 2.5.

Proposition 5.1   *Assume that Conjecture $\mathfrak{M}_H^R(G)$ holds for an admissible p-adic Lie extension F_∞ of $\mathbb{Q}$. Then, for each finite extension L of $\mathbb{Q}$ contained in F_∞, the module $X(\mathcal{T}/L^{\mathrm{cyc}})/X(\mathcal{T}/L^{\mathrm{cyc}})(R)$ is a finitely generated R-module.*

We shall need the following analogue of Lemma 2.4 to prove this proposition. As always, let F_∞ be an admissible p-adic Lie extension of $\mathbb{Q}$. For each finite extension L of $\mathbb{Q}$ contained in F_∞, the restriction map

$$\beta_L^R \colon H^1(\mathcal{G}_\Sigma(L^{\mathrm{cyc}}), \mathcal{A}) \to H^1(\mathcal{G}_\Sigma(F_\infty), \mathcal{A})^{H_L}$$

induces a canonical R-homomorphism

$$r_L^R \colon X(\mathcal{T}/F_\infty)_{H_L} \to X(\mathcal{T}/F^{\mathrm{cyc}}).$$

Lemma 5.2 $\mathrm{Ker}\,(r_L^R)$ *and* $\mathrm{Coker}\,(r_L^R)$ *are finitely generated R-modules.*

Proof   This is entirely parallel to the proof of Lemma 2.4. We have a commutative diagram with exact rows

$$
\begin{array}{ccccccc}
0 & \longrightarrow & \mathrm{Sel}(\mathcal{T}/F_\infty)^{H_L} & \longrightarrow & H^1(\mathcal{G}_\Sigma(F_\infty), \mathcal{A})^{H_L} & \longrightarrow & \bigoplus_{v \in \Sigma} J_v(\mathcal{T}/F_\infty)^{H_L} \\
& & \big\uparrow{\alpha_L^R} & & \big\uparrow{\beta_L^R} & & \big\uparrow{\delta_L^R = \oplus_{v \in \Sigma} \delta_{L,v}^R} \\
0 & \longrightarrow & \mathrm{Sel}(\mathcal{T}/L^{\mathrm{cyc}}) & \longrightarrow & H^1(\mathcal{G}_\Sigma(L^{\mathrm{cyc}}), \mathcal{A}) & \longrightarrow & \bigoplus_{v \in \Sigma} J_v(\mathcal{T}/L^{\mathrm{cyc}}).
\end{array}
$$

Thus, as Σ is a finite set, it suffices via the snake lemma to show that both $\mathrm{Ker}\,\beta_L^R$ and $\mathrm{Ker}\,\delta_{L,v}^R$ are cofinitely generated R-modules. But

$$\mathrm{Ker}\,\beta_L^R = H^1(H_L, \mathcal{A}(F_\infty)), \qquad \text{where } \mathcal{A}(F_\infty) = \mathcal{A}^{\mathcal{G}_\Sigma(F_\infty)}.$$

Moreover, when v does not divide p, we have

$$\mathrm{Ker}\,\delta_L^R = \bigoplus_{w|v} H^1(\mathrm{Gal}(F_{\infty,w}/L_w^{\mathrm{cyc}}), \mathcal{A}(F_{\infty,w})),$$

where w runs over the finite set of places of L^{cyc} above v, and we have also written w for some fixed place of F_∞ above w; when v divides p, a similar assertion holds with $\mathcal{A}$ replaced by $\mathcal{A}/F_v^+\mathcal{A}$. Now $\mathcal{A}(F_\infty)$ and $\mathcal{A}(F_{\infty,w})$ are cofinitely generated R-modules, because $\mathcal{A}$ is a cofinitely generated R-module. The conclusion of the lemma follows on noting that, for any p-adic Lie group $\mathcal{H}$ and any R- cofinitely generated $\mathcal{H}$-module W, all of the cohomology groups

$H^i(\mathcal{H}, W)$ are cofinitely generated R-modules. This last assertion is immediate on observing that the Pontryagin dual of W has a projective resolution by finitely generated $\Lambda_R(\mathcal{H})$-modules. This completes the proof of the lemma. □

We now prove Proposition 5.1. For every algebraic extension $\mathcal{L}$ of F^{cyc}, define

$$Y(\mathcal{T}/\mathcal{L}) = X(\mathcal{T}/\mathcal{L})/X(\mathcal{T}/\mathcal{L})(R). \tag{75}$$

Let L be any finite extension of F contained in F_∞. We have the commutative diagram with exact rows

$$
\begin{array}{ccccc}
X(\mathcal{T}/F_\infty)_{H_L} & \longrightarrow & Y(\mathcal{T}/F_\infty)_{H_L} & \longrightarrow & 0 \\
\downarrow r_L^R & & \downarrow s_L^R & & \\
X(\mathcal{T}/L^{\text{cyc}}) & \longrightarrow & Y(\mathcal{T}/L^{\text{cyc}}) & \longrightarrow & 0,
\end{array} \tag{76}
$$

where s_L^R is induced by r_L^R. By Lemma 5.2, Coker r_L^R is a finitely generated R-module, and so (76) shows that Coker s_L^R is finitely generated over R. Our hypothesis that $Y(\mathcal{T}/F_\infty)$ is finitely generated over $\Lambda_R(H)$, and the fact that H_L is of finite index in H, imply that $Y(\mathcal{T}/F_\infty)_{H_L}$ is also finitely generated over R. Thus $Y(\mathcal{T}/L^{\text{cyc}})$ is finitely generated over R, as required. This completes the proof of the proposition. □

We also have the following result parallel to Theorem 2.1.

Theorem 5.3 *Let F_∞ be an admissible p-adic Lie extension of $\mathbb{Q}$. Assume that there exists a finite extension L of $\mathbb{Q}$ contained in F_∞ such that $X(\mathcal{T}/L^{\text{cyc}})$ is a finitely generated R-module, and $\mathrm{Gal}(F_\infty/L)$ is pro-p. Then $X(\mathcal{T}/F_\infty)$ is finitely generated over $\Lambda_R(H)$. In particular, Conjecture $\mathfrak{M}_H^R(G)$ is valid for $X(\mathcal{T}/F_\infty)$.*

Proof Suppose that there does exist an L as in the theorem, and put $H_L = \mathrm{Gal}(F_\infty/L^{\text{cyc}})$. We conclude from Lemma 5.2 that the coinvariants $X(\mathcal{T}/F_\infty)_{H_L}$ is a finitely generated R-module. As H_L is pro-p and R is a local ring, it follows that $\Lambda_R(H_L)$ is itself a local ring. Hence, by Nakayama's lemma, one concludes that $X(\mathcal{T}/F_\infty)$ is finitely generated over $\Lambda_R(H_L)$, as required. □

Proposition 5.4 *Let F_∞ be an admissible p-adic Lie extension of $\mathbb{Q}$. Assume that there exists an integer $k \geq 2$ such that $X(T_k/F_\infty)$ is a finitely generated $\Lambda_O(H)$-module. Then $X(\mathcal{T}/F_\infty)$ is a finitely generated $\Lambda_R(H)$-module.*

Proof Assume that $X(T_k/F_\infty)$ is a finitely generated $\Lambda_O(H)$-module. It follows from Theorem 4.2 that $X(\mathcal{T}/F_\infty)/p_k X(\mathcal{T}/F_\infty)$ is also a finitely generated $\Lambda_O(H)$-module, whence, recalling that $R/p_k R = O$, Nakayama's lemma shows that $X(\mathcal{T}/F_\infty)$ is a finitely generated $\Lambda_R(H)$-module. □

Example   Take $p = 11$ and let $\mathcal{F}$ be the well-known Hida family of tame level 1 such that in the notation of (54),

$$F_2 = g, \quad F_{11} = \Delta,$$

where

$$g = q \prod_{n=1}^{\infty} (1 - q^n)^2 \prod_{n=1}^{\infty} (1 - q^{11n})^2,$$
$$\Delta = q \prod_{n=1}^{\infty} (1 - q^n)^{24};$$

here g is the cusp form of weight 2 and level 11 corresponding to the elliptic curve

$$X_1(11) : y^2 + y = x^3 - x^2,$$

and Δ is the discriminant cusp form of weight 12 and level 1. Let $\mathcal{T}$ be the corresponding R-lattice, where $R = \mathbb{Z}_{11}[[W]]$. A classical computation shows that $X(T_2/\mathbb{Q}(\mu_{11^\infty})) = 0$. Let $K_\infty = \mathbb{Q}(\mu_{11^\infty})$. Applying Proposition 5.4 to the admissible 11-adic extension K_∞ of $\mathbb{Q}$, and noting that $\text{Gal}(K_\infty/\mathbb{Q}^{\text{cyc}})$ is finite, it follows that $X(\mathcal{T}/K_\infty)$ is a finitely generated R-module. Next, fix an integer $m > 1$, and define

$$F_\infty = \mathbb{Q}(\mu_{11^\infty}, m^{1/11^n} : n = 1, 2, \ldots).$$

Thus F_∞ is an admissible 11-adic Lie extension of $\mathbb{Q}$ whose Galois group G is the semi-direct product of $\mathbb{Z}_{11}$ and $\mathbb{Z}_{11}^\times$. Since $\text{Gal}(F_\infty/K_\infty) = \mathbb{Z}_{11}$ is pro-11, it follows from Theorem 5.3 that $X(\mathcal{T}/F_\infty)$ is a finitely generated $\Lambda_R(H)$-module. In particular, Conjecture $\mathfrak{M}_H^R(G)$ holds for $X(\mathcal{T}/F_\infty)$. Also, by Theorem 4.2, we conclude that, for all integers $k \geq 2$, $X(T_k/F_\infty)$ is a finitely generated $\mathbb{Z}_{11}[[G]]$-module. This latter result was proven earlier by Chandrakant in [1].

We end this section by remarking that one of the main interests of the $\mathfrak{M}_H^R(G)$-conjecture is that it enables us to define a characteristic element of $X(\mathcal{T}/F_\infty)$. We briefly indicate how the arguments of [5] can easily be generalised to achieve this. Let

$$S_R = \{f \in \Lambda_R(G) : \Lambda_R(G)/\Lambda_R(G)f \text{ is a finitely generated } \Lambda_R(H)\text{-module}\}.$$

Entirely parallel arguments to those given in §2 of [5] show that (i) S_R is a left and right Ore set consisting of non-zero divisors in $\Lambda_R(G)$ (ii) a finitely generated $\Lambda_R(G)$-module is annihilated by S_R if and only if it is finitely generated over $\Lambda_R(H)$. Define

$$S_R^* = \bigcup_q q S_R,$$

where q runs over all finite products of integral powers of a fixed set of generators of the height-one prime ideals of R. Since the elements of R lie in the centre of $\Lambda_R(G)$, it is clear that (i) S_R^* is also a left and right Ore set of non-zero divisors in $\Lambda_R(G)$, and (ii) a finitely generated $\Lambda_R(G)$-module W is annihilated by S_R^* if and only if $W/W(R)$ is finitely generated over $\Lambda_R(H)$, where $W(R)$, as above, denotes the R-torsion submodule of W.

Since S_R and S_R^* are Ore sets in $\Lambda_R(G)$, we can form the localizations $\Lambda(G)_{S_R}$ and $\Lambda_R(G)_{S_R^*}$, and then take the associated K_1-groups , $K_1(\Lambda_R(G)_{S_R})$ and $K_1(\Lambda_R(G)_{S_R^*})$. Let $C_{S_R^*}$ be the category of all finitely generated $\Lambda_R(G)$-modules W such that (i) $W/W(R)$ is finitely generated over $\Lambda_R(H)$, and (ii) W has a finite resolution by projective $\Lambda_R(G)$-modules. Note that condition (ii) is automatically true if G has no elements of order p. Then classical K-theory, combined with an additional algebraic argument (see [5, prop. 3.4]), shows that there is an exact sequence

$$K_1(\Lambda_R(G)) \to K_1(\Lambda_R(G)_{S_R^*}) \xrightarrow{\partial} K_0(C_{S_R^*}) \to 0. \qquad (77)$$

where $K_0(C_{S_R^*})$ denotes the Grothendieck group of the category $C_{S_R^*}$. If W is any module in $C_{S_R^*}$, its characteristic element is then defined to be any element ξ of $K_1(\Lambda_R(G)_{S_R^*})$ such that $\partial(\xi) = [W]$, where $[W]$ denotes the class of W in $K_0(C_{S_R^*})$. In particular, if Conjecture $\mathfrak{M}_H^R(G)$ is valid for $X(\mathcal{T}/F_\infty)$, and if we also assume that $X(\mathcal{T}/F_\infty)$ has a finite resolution by projective $\Lambda_R(G)$-modules (no condition at all if G has no element of order p), we can define in this way a characteristic element for $X(\mathcal{T}/F_\infty)$. There must be a "main conjecture" for the module $X(\mathcal{T}/F_\infty)$, in the spirit of [5], for every admissible p-adic Lie extension F_∞ of $\mathbb{Q}$, asserting that a characteristic element can be chosen which is a p-adic L-function interpolating certain critical values of the twists of the modular forms $F_k (k \geq 2)$ by Artin characters of G. We hope to formulate such a precise non-commutative "main conjecture" for $X(\mathcal{T}/F_\infty)$ in a subsequent paper (see [19] for the special case when $F_\infty = \mathbb{Q}^{cyc}$).

6 Vanishing of the R-torsion

Our goal in this section is to prove a result about the vanishing of the R-torsion submodule of $X(\mathcal{T}/F_\infty)$. Here again F_∞ will denote an arbitrary admissible p-adic Lie extension of $\mathbb{Q}$, and $\mathcal{F}$ an ordinary primitive R-adic Hida form of tame level N.

Fix a finite extension L of $\mathbb{Q}$, contained in F_∞, such that

$$G_L = \text{Gal}(F_\infty/L)$$

is pro-p, and has no element of order p. Let π denote a local parameter of O. If W is any finitely generated $\Lambda_O(G)$-module, which is annihilated by a power of π, the structure theory (see [21]) shows that there is an exact sequence of $\Lambda_O(G_L)$-modules

$$0 \to \bigoplus_{i=1}^{r} \Lambda_O(G_L)/\pi^{m_i}\Lambda_O(G_L) \to W \to D \to 0,$$

where the m_i are integers ≥ 1, and D is pseudo-null. As usual, we define $\mu_{G_L}(W) = \sum_{i=1}^{r} m_i$.

We are grateful to K. Ardakov for pointing out to us that the following algebraic lemma is a special case of a more general result of T. Levasseur [16, §4.5]. If M is a module over a ring A, we put

$$E_A^i(M) := \text{Ext}_A^i(M, A) \quad (i \geq 0).$$

Recall also that $p_k = (1 + W) - (1 + p)^k$, $k \geq 2$.

Lemma 6.1 *Let M be a finitely generated torsion $\Lambda_R(G_L)$-module and assume that there exists an integer $k \geq 2$, such that multiplication by p_k is injective on M and $M/p_k M$ is pseudo-null as a $\Lambda_O(G_L)$-module. Then M is pseudo-null as a $\Lambda_R(G_L)$-module.*

Proof A well-known spectral sequence argument (see [21, §3.4]) shows that, for all $i \geq 0$,

$$E_{\Lambda_O(G_L)}^i (M/p_k M) \simeq E_{\Lambda_R(G_L)}^{i+1} (M/p_k M).$$

As $M/p_k M$ is a pseudo-null $\Lambda_O(G_L)$-module, we have

$$E_{\Lambda_O(G_L)}^i (M/p_k M) = 0 \quad \text{for } i = 0, 1, \tag{78}$$

whence

$$E_{\Lambda_R(G_L)}^i (M/p_k M) = 0 \quad \text{for } i = 0, 1, 2. \tag{79}$$

Using the long exact Ext sequence arising from the short exact sequence

$$0 \to M \xrightarrow{p_k} M \to M/p_k M \to 0,$$

we conclude from (79) that multiplication by p_k is an isomorphism on $E_{\Lambda_R(G_L)}^1(M)$. In other words,

$$E_{\Lambda_R(G_L)}^1(M)/p_k E_{\Lambda_R(G_L)}^1(M) = 0.$$

Hence, as p_k is contained in the maximal ideal of $\Lambda_R(G_L)$, we conclude from Nakayama's lemma that $E_{\Lambda_R(G_L)}^1(M) = 0$. Thus M is pseudo-null as a $\Lambda_R(G_L)$-module. $\square$

Theorem 6.2 *Assume that there exists a integer $k \geq 2$, such that (i) $X(\mathcal{T}/F_\infty)$ has no p_k-torsion, and (ii) $\mu_{G_L}(X(T_k/F_\infty)) = 0$. Then $X(\mathcal{T}/F_\infty)(R)$ is pseudo-null.*

Proof Put

$$Z = X(\mathcal{T}/F_\infty)(R), \quad X = X(\mathcal{T}/F_\infty), \quad Y = Y(\mathcal{T}/F_\infty).$$

The hypothesis that Z has no p_k-torsion allows us to choose an annihilator $\xi \in R$ of Z with the property that $(\xi, p_k) = 1$. As the image of ξ in R/p_kR is a non-zero element of O, it follows that Z/p_kZ is annihilated by a power of π, where π is a uniformizer of O. We conclude that there is a commutative diagram with exact rows

$$
\begin{array}{ccccccccc}
0 & \longrightarrow & Z/p_kZ & \longrightarrow & X/p_kX & \longrightarrow & Y/p_kY & \longrightarrow & 0 \\
 & & {\scriptstyle \alpha'_{F_\infty,k}}\downarrow & & {\scriptstyle \alpha_{F_\infty,k}}\downarrow & & {\scriptstyle \alpha''_{F_\infty,k}}\downarrow & & \\
0 & \longrightarrow & X(T_k/F_\infty)(\pi) & \longrightarrow & X(T_k/F_\infty) & \longrightarrow & Y(T_k/F_\infty) & \longrightarrow & 0,
\end{array}
\tag{80}
$$

where $\alpha_{F_\infty,k}$ is the dual of the map $a_{F_\infty,k}$ appearing in (73). By Theorem 4.2, Ker $\alpha_{F_\infty,k}$ is a finitely generated $\Lambda_O(H)$-module. Thus Ker $\alpha'_{F_\infty,k}$ is pseudo-null as a $\Lambda_O(G)$-module, because it is finitely generated and torsion as a $\Lambda_O(H)$-module. Invoking our hypothesis that $\mu_{G_L}(X(T_k/F_\infty)(\pi)) = 0$, we conclude from (80) that

$$\mu_{G_L}(Z/p_kZ) = 0. \tag{81}$$

Recalling that Z/p_kZ is annihilated by some power of π, it follows from (81) that Z/p_kZ is pseudo-null as a $\Lambda_O(G_L)$-module. Applying Lemma 6.1, we conclude that Z is pseudo-null as a $\Lambda_R(G_L)$-module, thereby completing the proof of Theorem 6.2. $\qquad\square$

Corollary 6.3 *Assume that $X(\mathcal{T}/\mathcal{F}_\infty)(\mathcal{R})$ has no non-zero pseudo-null $\Lambda_R(G)$-submodule. Then, under the hypotheses of the theorem, we have $X(\mathcal{T}/\mathcal{F}_\infty)(\mathcal{R})$ is equal to zero.*

We first remark that in [20], it is shown that, for a certain class of admissible p-adic Lie extensions, the module $X(\mathcal{T}/F_\infty)(R)$ indeed has no non-zero pseudo-null submodules.

Secondly, we point out that if we assume both that (i) $X(\mathcal{T}/\mathcal{F}_\infty)(\mathcal{R})$ has no non-zero pseudo-null $\Lambda_R(G)$-submodule, and also (ii) $X(\mathcal{T}/F_\infty)/X(\mathcal{T}/F_\infty)(R)$ is finitely generated over $\Lambda_R(H_L)$, where $H_L = \mathrm{Gal}(F_\infty/F^{\mathrm{cyc}})$, then the vanishing of $\mu_{G_L}(X(T_k/F_\infty))$ for one integer $k \geq 2$ implies that $\mu_{G_L}(X(T_m/F_\infty)) = 0$ for all integers $m \geq 2$. This is immediate from the above corollary, and the fact that, for all integers $m \geq 2$, $\mathrm{Coker}(\alpha_{F_\infty,m})$ is a finitely generated O-module. When $F_\infty = \mathbb{Q}^{\mathrm{cyc}}$, results of this kind have been proven earlier in [8] and [19].

References

[1] A. C. SHARMA, *Iwasawa invariants for the False-Tate extension and congruences between modular forms*, Jour. Number Theory **129**, (2009), 1893–1911.

[2] N. BOURBAKI, *Elements of Mathematics, Commutative Algebra*, Chapters 1–7, Springer (1989).

[3] J.COATES *Fragments of the Iwasawa theory of elliptic curves without complex multiplication. Arithmetic theory of elliptic curves*, (Cetraro, 1997), Lecture Notes in Math., 1716, Springer, Berlin (1999), 1–50.

[4] J. COATES, R. GREENBERG, *Kummer theory for abelian varieties over local fields*, Invent. Math. **124** (1996), 129–174.

[5] J. COATES, T. FUKAYA, K. KATO, R. SUJATHA, O. VENJAKOB, *The* GL_2 *main conjecture for elliptic curves without complex multiplication*, Publ. Math. Inst. Hautes Études Sci. **101** (2005), 163–208.

[6] J. COATES, P. SCHNEIDER, R. SUJATHA, *Links between cyclotomic and* GL_2 *Iwasawa theory*, Kazuya Kato's fiftieth birthday. Doc. Math. 2003, Extra Vol., 187–215.

[7] A. CUOCO, P. MONSKY, *Class numbers in* $\mathbb{Z}_p^d$-*extensions*, Math. Ann. **255** (1981), 235–258.

[8] M. EMERTON, R. POLLACK, T. WESTON, *Variation of Iwasawa invariants in Hida families*, Invent. Math. **163** (2006), 523–580.

[9] R. GREENBERG, *Iwasawa theory for p-adic representations*, Algebraic number theory, 97–137, Adv. Stud. Pure Math., 17, Academic Press, Boston, MA, 1989.

[10] R. GREENBERG, *Iwasawa theory and p-adic deformations of motives*. Motives (Seattle, WA, 1991), 193–223, Proc. Sympos. Pure Math., 55, Part 2, Amer. Math. Soc., Providence, RI, (1994).

[11] R. GROSS, *On the integrality of some Galois representations*, Proc. Amer. Math. Soc. **123** (1995), 299–301.

[12] K. KATO, *p-adic Hodge theory and values of zeta functions of modular forms*, Cohomologies p-adiques et applications arithmétiques. III. Astérisque **295** (2004), 117–290.

[13] Y. HACHIMORI, O. VENJAKOB, *Completely faithful Selmer groups over Kummer extensions*, Kazuya Kato's fiftieth birthday. Doc. Math. 2003, Extra Vol., 443–478.

[14] H. HIDA, *Iwasawa modules attached to congruences of cusp forms*, Ann. Sci. École Norm. Sup. **19** (1986), 231–273.

[15] H. HIDA, *Galois representations into* $GL_2(Z_p[[X]])$ *attached to ordinary cusp forms*, Invent. Math. **85** (1986), 545–613.

[16] T. LEVASSEUR, *Some properties of non-commutative regular rings*, Glasgow Journal of Math. **34** (1992), 277–300.

[17] B. MAZUR, *Rational points of abelian varieties with values in towers of number fields*, Invent. Math. **18** (1972), 183–266.

[18] B. MAZUR, A. WILES, *On p-adic analytic families of Galois representations*, Compositio Math. **59** (1986), 231–264.

[19] T. OCHIAI, *On the two-variable Iwasawa main conjecture*, Compositio Math. **142** (2006), 1157–1200.

[20] S. SUDHANSHU, R. SUJATHA, *On the structure of Selmer groups of Λ-adic deformations over p-adic Lie extensions*, preprint.

[21] O. VENJAKOB, *On the structure theory of the Iwasawa algebra of a p-adic Lie group*, J. Eur. Math. Soc. **4** (2002), 271–311.

[22] A. WILES, *On ordinary λ-adic representations associated to modular forms*, Invent. Math. **94** (1988), 529–573.

Galois theory and Diophantine geometry

Minhyong Kim

University College London

1 The deficiency of abelian motives

1.1 The author must confess to having contemplated for some years a diagram of the following sort.

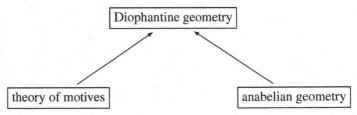

To a large extent, the investigations to be brought up today for discussion arise from a curious inadequacy having to do with the arrow on the left. On the one hand, it is widely acknowledged that the theory of motives finds a strong source of inspiration in Diophantine geometry, inasmuch as many of the structures, conjectures, and results therein have as model the conjecture of Birch and Swinnerton-Dyer, where the concern is with rational points on elliptic curves that can be as simple as

$$x^3 + y^3 = 1729.$$

Even in the general form discovered by Deligne, Beilinson, Bloch and Kato, (see, for example, [19]) it is clear that motivic L-functions are supposed, in an ideal world, to give access to invariants in arithmetic geometry of a *Diophantine nature*. The difficulty arises when we focus not on the highly sophisticated general conjectures, but rather on the very primitive concerns of Diophantine geometry, which might broadly be characterized as the study of maps between schemes of finite type over $\mathbb{Z}$ or $\mathbb{Q}$.

One might attempt, for example, to define the points of a motive M over $\mathbb{Q}$

Non-abelian Fundamental Groups and Iwasawa Theory, eds. John Coates, Minhyong Kim, Florian Pop, Mohamed Saïdi and Peter Schneider. Published by Cambridge University Press. ©Cambridge University Press 2012.

using a formula like

$$\mathrm{Ext}^1(\mathbb{Q}(0), M)$$

or even

$$\mathrm{RHom}(\mathbb{Q}(0), M),$$

hoping it eventually to be adequate in sufficiently many situations of interest. However, even in the best of all worlds, this formula will never provide direct access to the points of a scheme, except in very special cases like

$$M = H_1(A)$$

with A an abelian variety. This is a critical limitation of the abelian nature of motives, rendering it quite difficult to find direct applications to any mildly non-abelian Diophantine problem, say that posed by a curve of genus 2. It is worth remarking that this limitation is essentially by design, since the whole point of the motivic category is to *linearize* by increasing the number of morphisms.[1]

Even with such reservation in mind, we would still do well to acknowledge the role of technology that is more or less motivic in two of the most celebrated Diophantine results of our times, namely the theorems of Faltings and of Wiles [9, 45]. But there, the idea is to constrain points on a non-abelian variety by forcing them to *parametrize* motives of very special types. The method of achieving this is highly ingenious in each case and, hence, underscores our concern that it is rather unlikely to be part of a general system, and certainly not of the motivic philosophy as it stands.

1.2 Much has been written about the meaning of anabelian geometry, with a general tendency to retreat to the realm of curves as the only firm ground on which to venture real assertions or conjectures. We will also proceed to use X to denote a smooth projective curve of genus at least two over an algebraic number field K. The basic anabelian proposal then is to replace the Ext group that appeared above by the topological space

$$H^1(G_K, \pi_1^{et}(\bar{X}, b)),$$

the non-abelian continuous cohomology [37] of the absolute Galois group $G = \mathrm{Gal}(\bar{K}/K)$ of K with coefficients in the profinite étale fundamental group of X. The notation will suggest that a rational basepoint $b \in X(K)$ has been introduced. Many anabelian results do not require it [31], but the Diophantine issues discussed today will gain in clarity by having it at the outset, even if the

---

[1] Even then, we complain that there are not enough.

resulting restriction may appear as serious to many. An immediate relation to the full set of points is established by way of a non-abelian Albanese map

$$X(K) \xrightarrow{\kappa^{na}} H^1(G_K, \pi_1^{et}(\bar{X}, b));$$

$$x \mapsto [\pi_1^{et}(\bar{X}; b, x)].$$

We remind ourselves that the definition of fundamental groups in the style of Grothendieck [41] typically starts from a suitable category over X, in this case that of finite étale covers of

$$\bar{X} = X \times_{\text{Spec}(K)} \text{Spec}(\bar{K})$$

that we might denote by

$$\text{Cov}(\bar{X}).$$

The choice of any point $y \in \bar{X}(\bar{K})$ determines a fiber functor

$$F_y \colon \text{Cov}(\bar{X}) \longrightarrow \text{Finite Sets},$$

using which the fundamental group is defined to be

$$\pi_1^{\text{et}}(\bar{X}, y) := \text{Aut}(F_y),$$

in the sense of invertible natural transformations familiar from category theory.[2] Given two points y and z, there is also the set of étale paths

$$\pi_1^{\text{et}}(\bar{X}; y, z) := \text{Isom}(F_y, F_z)$$

from y to z that the bare definitions equip with a right action of $\pi_1^{\text{et}}(\bar{X}, y)$, turning it thereby into a torsor for the fundamental group. When y and z are rational points, the naturality of the constructions equips all objects with a compatible action of G_K, appearing in the non-abelian cohomology set and the definition of the map κ^{na}.

The context should make it clear that $H^1(G_K, \pi_1^{et}(\bar{X}, b))$ can be understood

---

[2] The reader unfamiliar with such notions would do well to think about the case of a functor

$$F \colon \mathbb{N}^{\text{op}} \to C$$

whose source is the category of natural numbers with a single morphism from n to m for each pair $m \leq n$. Of course this is just a sequence

$$\to F(3) \to F(2) \to F(1) \to F(0)$$

of objects in C, and an automorphism of F is a compatible sequence $(g_i)_{i \in \mathbb{N}}$ of automorphisms

$$g_i \colon F(i) \simeq F(i).$$

For a general $F \colon \mathcal{B} \to C$, it is sensible to think of $\mathcal{B}$ as a complicated indexing set for things in C.

as a non-abelian Jacobian in an étale profinite realization, where the analogy might be strengthened by the interpretation of the G-action as defining a sheaf on $\mathrm{Spec}(K)$ and $H^1(G_K, \pi_1^{et}(\bar{X}, b))$ as the moduli space of torsors for $\pi_1^{et}(\bar{X}, b)$ in the étale topos of $\mathrm{Spec}(K)$. It is instructive to compare this space with the moduli space $Bun_n(X)$ of rank n vectors bundles on X for $n \geq 2$. Their study was initiated in a famous paper of André Weil [44] whose title suggests the intention of the author to regard them also as non-abelian Jacobians. Perhaps less well known is the main motivation of the paper, which the introduction essentially states to be the study of rational points on curves of higher genus. Weil had at that point already expected non-abelian fundamental groups to intervene somehow in a proof of the Mordell conjecture, except that a reasonable arithmetic theory of π_1 was not available at the time. In order to make the connection to fields of definition, Weil proceeded to interpret the representations of the fundamental group in terms of algebraic vector bundles, whose moduli would then have the same field of definition as the curve. In this sense, the paper is very much a continuation of Weil's thesis [43], where an algebraic interpretation of the Jacobian is carried out with the same goal in mind, however with only the partial success noted by Hadamard. The spaces $Bun_n(X)$ of course fared no better, and it might be a good idea to ask why. One possibility was suggested by Serre [38] in his summary of Weil's mathematical contributions, where he calls attention to the lack of the geometric technology requisite to a full construction of Bun_n, which was subsequently developed only in the 1960s by Mumford, Narasimhan, Seshadri, and others [29, 32]. However, even with geometric invariant theory and its relation to π_1 completed in the remarkable work of Carlos Simpson [40], there has never been any direct applications of these moduli spaces (or their cotangent bundles) to Diophantine problems. It is for this reason that the author locates the difficulty in a far more elementary source, namely, *the lack of an Albanese map to go with Bun_n*. Unless $n = 1$, there is no obvious relation between Bun_n and the points on X. It is fortunate then that the étale topology manages to provide us with two valuable tools, namely, topological fundamental groups that come with fields of definition; and topological classifying spaces with extremely canonical Albanese maps. We owe this to a distinguished feature of Grothendieck's theory: the flexible use of basepoints, which are allowed to be any geometric point at all. The idea that Galois groups of a certain sort should be regarded as fundamental groups is likely to be very old, as Takagi [18] refers to Hilbert's preoccupation with Riemann surfaces as inspiration for class field theory. Indeed, it is true that that the fundamental group of a smooth variety V will be isomorphic to the Galois group $\mathrm{Gal}(k(V)^{nr}/k(V))$ of a maximal unramified extension $k(V)^{nr}$ of its function field $k(V)$. However, this isomorphism will be *canonical* only when the basepoint is taken to be a separable closure of $k(V)$ that contains $k(V)^{nr}$:

$$b : \mathrm{Spec}(k(V)^s) \to \mathrm{Spec}(k(V)^{nr}) \to \mathrm{Spec}(k(V)) \to V.$$

Within the Galois group approach, there is little room for small basepoints that come through rational points, or a study of variation. In fact, there seems to be no reasonable way to fit path spaces at all into the field picture. This could then be described as the precise ingredient missing in the arithmetic theory of fundamental groups at the time of Weil's paper. Even after the introduction of moving basepoints, appreciation of their genuine usefulness appears to have taken some time to develop. A rather common response is to pass quickly to invariants or situations where the basepoint can be safely ignored. This author, for example, came to appreciate the basepoint as a variable only after reading Deligne's paper written in the 1980s [8] as well as the papers of Hodge-theorists like Hain [17].

1.3 One convenient way to visualize path spaces is to consider a universal (pro-)cover

$$\tilde{\tilde{X}} \longrightarrow \tilde{X}.$$

The choice of a lifting $\tilde{b} \in \tilde{\tilde{X}}_b$ turns the pair into a *universal pointed covering space*. The uniqueness then allows us to descend to $\mathbb{Q}$, while the universal property determines canonical isomorphisms

$$\tilde{\tilde{X}}_x \simeq \pi_1^{et}(\tilde{X}; b, x),$$

so that the Galois action can be interpreted using the action on fibers.[3] This is one way to see that the map κ^{na} will never send $x \neq b$ to the trivial torsor, that is, a torsor with an element fixed by G_K, since, by the Mordell–Weil theorem, nothing but the basepoint will lift rationally even up to the maximal abelian quotient of $\tilde{\tilde{X}}$. A change of basepoint[4] then shows that the map must in fact be injective. That is, we have arrived at the striking fact that points can really be distinguished through the associated torsors.[5] In elementary topology, one

---

[3] The difficult problem of coming to actual grips with this is that of constructing a cofinal system making up $\tilde{\tilde{X}}$ in a manner that makes the action maximally visible. Consider $\mathbb{G}_m$ or an elliptic curve.

[4] One needs here the elementary fact that an isomorphism of torsors

$$\pi_1(\tilde{X}; b, x) \simeq \pi_1(\tilde{X}; b, y)$$

is necessarily induced by a path $F_x \simeq F_y$.

[5] It is, however, quite interesting to work out injectivity or its failure for quotients of fundamental groups corresponding to other natural systems, like modular towers. Alternatively, one could use the full fundamental group for a variety where the answer is much less obvious, like a moduli space of curves.

already encounters the warning that such path spaces are isomorphic, but not in a canonical fashion. The distinction may appear pedantic until one meets such enriched situations as to endow the torsors with the extra structure necessary to make them genuinely different.[6]

The remarkable *section conjecture* of Grothendieck [16] proposes that κ^{na} is even surjective:

$$X(K) \simeq H^1(G_K, \pi_1^{et}(\bar{X}, b)),$$

that is,

every torsor should be a path torsor.

The reader is urged to compare this conjecture with the assertion that the map

$$\widehat{E(K)} \simeq H_f^1(G_K, \pi_1^{et}(\bar{E}, e)),$$

from Kummer theory is supposed to be bijective for an elliptic curve (E, e). A small difference has to do with the local 'Selmer' conditions on cohomology indicated by the subscript 'f', which the complexity of the non-abelian fundamental group is supposed to render unnecessary. This is a subtle point on which the experts seem not to offer a consensus. Nevertheless, the comparison should make it clear to the newcomer that a resolution of the section conjecture is quite unlikely to be straightforward, being, as it is, a deep non-abelian incarnation of the principle that suitable conditions on a Galois-theoretic construction should force it to 'come from geometry'.[7] And then, the role of this bijection in the descent algorithm for elliptic curves might suggest a useful Diophantine context for the section conjecture [24]. Yet another reason for thinking the analogy through is a hope that the few decades' worth of effort that went into the study of Selmer groups of elliptic curves might illuminate certain aspects of the section conjecture as well, even at the level of concrete tools that might lead to its (partial) resolution.

2 Motivic fundamental groups and Selmer varieties

2.1 Our main concern today is with a version of these ideas where the parallel with elliptic curves is especially compelling, in that a good deal of unity

---

[6] For the author, this elementary fact has been a source of some satisfaction while working on this topic, since considerable good fortune seems to be necessary before such pedantry can lead to substantial results.

[7] This notion in abelian settings coincides roughly with 'motivic.'

between the abelian and non-abelian realms is substantially realized. This is when the profinite fundamental group is replaced by the motivic one [8]:

$$\pi_1^M(\bar{X}, b).$$

The motivic fundamental group lies between the profinite π_1 and homology in complexity:

$$\hat{\pi}_1(\bar{X}, b)$$
$$|$$
$$\pi_1^M(\bar{X}, b)$$
$$|$$
$$H_1(\bar{X})$$

although it should be acknowledged right away that it is much closer to the bottom of the hierarchy. The precise meaning of 'motivic' should not worry us here more than in other semi-formal expositions on the subject, since we will regress quickly to the rather precise use of realizations. But still, some inspiration may be gathered from a vague awareness of a classifying space

$$H_M^1(G_K, \pi_1^M(\bar{X}, b))$$

of motivic torsors as well that of a motivic Albanese map

$$\kappa^M : X(K) \longrightarrow H_M^1(G_K, \pi_1^M(\bar{X}, b))$$

that associates to points motivic torsors

$$\pi_1^M(\bar{X}; b, x)$$

of paths. The astute reader will object that we are again using the points of X to parametrize motives as in the subtle constructions of Parshin and Frey, to which we reply that the current family is entirely intrinsic to the curve X, and requires no particular ingenuity to consider.

When it comes to precise definitions [22, 23] that we must inflict upon the reader in a rapid succession of mildly technical paragraphs, the most important (Tannakian) category

$$\mathrm{Un}(\bar{X}, \mathbb{Q}_p)$$

consists of locally constant unipotent $\mathbb{Q}_p$-sheaves on $\bar{X}$, where a sheaf is unipotent if it can be constructed using successive extensions starting from the constant sheaf $[\mathbb{Q}_p]_{\bar{X}}$. As in the profinite theory, we have a fiber functor

$$F_b : \mathrm{Un}(\bar{X}, \mathbb{Q}_p) \to \mathrm{Vect}_{\mathbb{Q}_p}$$

that associates to a sheaf $\mathcal{V}$ its stalk $\mathcal{V}_b$, which has now acquired a linear nature. The $\mathbb{Q}_p$-pro-unipotent étale fundamental group is defined to be

$$U := \pi_1^{u,\mathbb{Q}_p}(\bar{X}, b) := \mathrm{Aut}^{\otimes}(F_b),$$

the tensor-compatible[8] automorphisms of the fiber functor, which the linearity equips with the added structure of a pro-algebraic pro-unipotent group over $\mathbb{Q}_p$. In fact, the descending central series filtration

$$U = U^1 \supset U^2 \supset U^3 \supset \cdots$$

yields the finite-dimensional algebraic quotients

$$U_n = U^{n+1}\backslash U,$$

at the very bottom of which is an identification

$$U_1 = H_1^{\mathrm{et}}(\bar{X}, \mathbb{Q}_p) = V_p J := T_p J \otimes \mathbb{Q}_p$$

with the $\mathbb{Q}_p$-Tate module of the (abelian) Jacobian J of X. The different levels are connected by exact sequences

$$0 \to U^{n+1}\backslash U^n \to U_n \to U_{n-1} \to 0$$

that add the extra term $U^{n+1}\backslash U^n$ at each stage, which, moreover, is a vector group that can be approached with techniques that are more or less conventional.

In fact, the G_K-action on U lifts the well-studied one on $V = V_p J$, and repeated commutators come together to a quotient map

$$V^{\otimes n} \longrightarrow U^{n+1}\backslash U^n,$$

placing the associated graded pieces into the category of motives generated by J. The inductive pattern of these exact sequences is instrumental in making the unipotent completions considerably more tractable than their profinite ancestors.

We will again denote by $H^1(G_K, U_n)$ continuous Galois cohomology with values in the points of U_n. For $n \geq 2$, this is still non-abelian cohomology, and hence, lacks the structure of a group. Nevertheless, the proximity to homology is evidenced in the presence of a remarkable subspace

$$H_f^1(G_K, U_n) \subset H^1(G_K, U_n)$$

---

[8] To see the significance of this notion, one should consider the group algebra $\mathbb{C}[G]$ of a finite group G. On the category $Rep_G(\mathbb{C})$ of G-representations on complex vector spaces, we have the fiber functor that forgets the G-action. Any unit in $\mathbb{C}[G]$ defines an automorphism of this functor, while the elements of G will then be picked out by the condition of being tensor-compatible.

defined by local 'Selmer' conditions. Starting at this point, one should choose p to be split in K with the property that X has good reduction at all $v|p$. The conditions in question require the classes to be

(a) unramified outside $T = S \cup \{v : v|p\}$, where S is the set of primes of bad reduction;
(b) and *crystalline* at the primes dividing p, a condition coming from p-adic Hodge theory.

The locality of the conditions refers to their focus on the pull-back of a torsor for U to the completed fields $\mathrm{Spec}(F_w)$. For $w \notin T$, (a) requires the torsor to trivialize over an unramified extension of F_w, while condition (b) requires it to trivialize over Fontaine's ring B_{cr} of crystalline periods [10]. One could equivalently describe the relevant torsors as having coordinate rings that are unramified or crystalline as representations of the local Galois groups.

Quite important to our purposes is the *algebraicity* of the system

$$\cdots \to H^1_f(G_K, U_{n+1}) \to H^1_f(G_K, U_n) \to H^1_f(G_K, U_{n-1}) \to \cdots .$$

This is the *Selmer variety* of X. That is, each $H^1_f(G_K, U_n)$ is an algebraic variety over $\mathbb{Q}_p$ and the transition maps are algebraic, so that

$$H^1_f(G_K, U) = \{H^1_f(G_K, U_n)\}$$

is now a moduli space very similar to the ones that come up in the study of Riemann surfaces [13], in that it parametrizes crystalline principal bundles for U in the étale topology of $\mathrm{Spec}(O_K[1/S])$. By comparison $H^1(G_K, \pi_1^{et}(\bar{X}, b))$ has no apparent structure but that of a pro-finite space: the motivic context has restored some geometry[9] to the moduli spaces of interest.

The algebraic structure is best understood in terms of $G_T = \mathrm{Gal}(K_T/K)$, where K_T is the maximal extension of K unramified outside T. Our moduli space $H^1_f(G_K, U_n)$ sits inside $H^1(G_T, U_n)$ as a subvariety defined by the additional crystalline condition. For the latter, there are sequences

$$0 \to H^1(G_T, U^{n+1} \backslash U^n) \to H^1(G_T, U_n) \to H^1(G_T, U_{n-1}) \overset{\delta_{n-1}}{\to} H^2(G_T, U^{n+1} \backslash U^n)$$

exact in a natural sense, and the algebraic structures are built up iteratively from the $\mathbb{Q}_p$-linear structure on the

$$H^i(G_T, U^{n+1} \backslash U^n)$$

using the fact that the boundary maps δ_{n-1} are algebraic.[10] That is, $H^1(G_T, U_n)$ is inductively realized as a torsor for the vector group $H^1(G_T, U^{n+1} \backslash U^n)$ lying over the kernel of δ_{n-1}.

---

[9] 'Coefficient geometry,' one might say, in contrast to Bun_n, which carries the algebraic geometry of the field of definition.

[10] The reader is warned that it is non-linear in general.

2.2 It should come as no surprise at this point that there is a map

$$\kappa^u = \{\kappa^u_n\}: X(K) \longrightarrow H^1_f(G_K, U)$$

associating to a point x the principal U-bundle

$$P(x) = \pi^{u,\mathbb{Q}_p}_1(\bar{X}; b, x) := \mathrm{Isom}^\otimes(F_b, F_x)$$

of tensor-compatible isomorphisms from F_b to F_x, that is, the $\mathbb{Q}_p$-pro-unipotent étale paths from b to x. This map is best viewed as a tower:

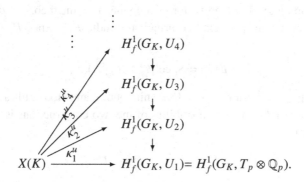

For $n = 1$,

$$\kappa^u_1 : X(K) \to H^1_f(G_K, U_1) = H^1_f(G_K, T_p J \otimes \mathbb{Q}_p)$$

reduces to the map from Kummer theory. But the maps κ^u_n for $n \geq 2$, much weaker as they are than the κ^{na} discussed in the profinite context, still do not extend to cycles in any natural way, and hence, retain the possibility of separating the structure[11] of $X(K)$ from that of $J_X(K)$.

Restricting U to the étale site of K_v for a fixed $v|p$, there are local analogues

$$\kappa^u_{v,n} : X(K_v) \to H^1_f(G_v, U_n)$$

(with $G_v = \mathrm{Gal}(\bar{K}_v/K_v)$) that can be described explicitly (and rather surprisingly) using non-abelian p-adic Hodge theory. More precisely, there is a compatible family of isomorphisms

$$D: H^1_f(G_v, U_n) \simeq U^{\mathrm{DR}}_n/F^0$$

to homogeneous spaces for the *De Rham fundamental group*

$$U^{\mathrm{DR}} = \pi^{\mathrm{DR}}_1(X \otimes K_v, b)$$

---

[11] It might be suggested, only half in jest, that the Jacobian, introduced by Weil to aid in the Diophantine study of a curve, has been getting in the way ever since.

of $X \otimes K_v$. Here, U^{DR} classifies unipotent vector bundles with flat connections on $X \otimes K_v$, while

$$U^{DR}/F^0$$

is a moduli space for U^{DR}-torsor that carry compatible Hodge filtrations and Frobenius actions, the latter being obtained from a comparison isomorphism[12] with the crystalline fundamental group and path torsors associated to a reduction modulo v. The advantage of the De Rham realization is its expression as a p-adic homogenous space whose form is far more transparent than that of Galois cohomology. The map D (for Dieudonné, as in the theory of p-divisible groups) associates to a crystalline principal bundle $P = \mathrm{Spec}(\mathcal{P})$ for U, the space

$$D(P) = \mathrm{Spec}([\mathcal{P} \otimes B_{cr}]^{G_p}).$$

This ends up as a U^{DR}-torsor with Frobenius action and Hodge filtration inherited from that of B_{cr}. The compatibility of the two constructions is expressed by a diagram

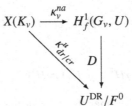

whose commutativity amounts to the non-abelian comparison isomorphism [33]

$$\pi_1^{DR}(X \otimes K_v; b, x) \otimes B_{cr} \simeq \pi_1^{u, \mathbb{Q}_p}(\bar{X}; b, x) \otimes B_{cr}.$$

The explicit nature of the map

$$\kappa^u_{dr/cr} : X(K_v) \to U^{DR}/F^0,$$

is a consequence of the p-adic iterated integrals[13] [12]

$$\int_b^z \alpha_1 \alpha_2 \cdots \alpha_n$$

---

[12] That is to say, if X denotes a smooth and proper O_{K_v}-model of $X \otimes K_v$, the category of unipotent vector bundles with flat connections on $X \otimes K_v$ is equivalent to the category of unipotent convergent isocrystals on $X \otimes O_{K_v}/m_v$. This comparison is the crucial ingredient in defining p-adic iterated integrals [12].

[13] Special values of such integrals have attracted attention because of the connection to values of L-functions. Here we are interested primarily in the integrals themselves as analytic functions, and in their zeros.

that appear in its coordinates. This expression endows the map with a highly transcendental nature: for any residue disk $]y[\subset X(K_v)$,

$$\kappa^{\mu}_{dr/cr,n}(]y[) \subset U_n^{\mathrm{DR}}/F^0$$

is Zariski dense for each n, and is made up of non-zero convergent power series that are obtained explicitly as repeated anti-derivatives starting from differential forms on X.

Finally, the local and global constructions fit into a family of commutative diagrams

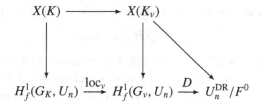

where the bottom horizontal maps are algebraic and the vertical maps transcendental. Thus, the difficult inclusion $X(K) \subset X(K_v)$ has been replaced by the map[14] $\log_v := D \circ \mathrm{loc}_v$, whose algebraicity gives a glimmer of hope that the arithmetic geometry of the global Diophantine set can be understood and controlled.

3 Diophantine finiteness

3.1 The following result is basic to the theory.

Theorem 1 *Suppose*

$$\log_p(H^1_f(G_K, U_n)) \subset U_n^{\mathrm{DR}}/F^0$$

is not Zariski dense for some n. Then $X(K)$ is finite.

---

[14] The strange notation reflects the view that D is itself a log map, according to Bloch–Kato [2].

Proof The proof of this assertion in its entirety is captured by the diagram

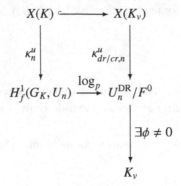

indicating the existence of a non-zero algebraic function ϕ vanishing on

$$\log_p(H^1_f(G_K, U_n)).$$

Hence, the function $\phi \circ \kappa^u_{dr/cr,n}$ on $X(K_v)$ vanishes on $X(K)$. But this function is a non-vanishing convergent power series on each residue disk, which therefore can have only finitely many zeros. (Recall that $K_v = \mathbb{Q}_p$ under our assumption on p.)   □

A slightly more geometric account of the proof might point to the fact that the image of $X(K_v)$ in U^{DR}_n/F^0 is a space-filling curve, with no portion contained in a proper subspace. Hence, its intersection with any proper subvariety must be discrete. Being compact as well, it must then be finite.[15] Serge Lang once proposed a strategy for proving the Mordell conjecture by deducing it from a purely geometric hope that the complex points on a curve of higher genus might intersect a finitely generated subgroup of the Jacobian in finitely many points. While that idea turned out to be very difficult to realize, here we have a non-Archimedean analog, with U^{DR}_n/F^0 playing the role of the complex Jacobian, and the Selmer variety that of the Mordell–Weil group.

When $K = \mathbb{Q}$, the hypothesis of the theorem on non-denseness of the global Selmer variety is expected always to hold for n large, in that we should have [23]

$$\dim H^1_f(G_K, U_n) << \dim U^{DR}_n/F^0.$$

(Recall that the map $\log_p$ is algebraic.) Such an inequality follows, for example, from the reasonable folklore conjecture that

$$H^1_f(G_K, M) = 0$$

---

[15] This proof, involving a straightforward interplay of denseness, non-denseness, and compactness, is a curious avatar of some ideas of Professor Deligne relating the section conjecture to Diophantine finiteness.

for a motivic Galois representation[16] M of weight > 0. This, in turn, might be deduced from the conjecture of Fontaine and Mazur on Galois representations of geometric origin [11], or from portions of the Bloch–Kato conjecture[17] [2]. The point is that if we recognized the elements of $H^1_f(G_K, M)$ themselves to be motivic, then the vanishing would follow from the existence of a weight filtration.

Through these considerations, we see that instead of the implication

Non-abelian 'finiteness of Sha' (= *section conjecture*) $\Rightarrow$ finiteness of $X(\mathbb{Q})$.

expected by Grothendieck, we have

'Higher abelian finiteness of Sha' (that $H^1_f(G, M)$ is generated by motives)

$$\Rightarrow \text{finiteness of } X(\mathbb{Q}).$$

This is not the only place that our discussion revolve around pale shadows of the section conjecture. One notes, for example, the critical use of the dense image of $\kappa^u_{dr/cr}$, which could itself be thought of as an 'approximate local section conjecture.'

3.2 In spite of all such lucubrations (that fascinate the author and quite likely no one else), we must now face the plain and painful fact that an unconditional proof of the hypothesis for large n (and hence, a new proof of finiteness) can be given only when $K = \mathbb{Q}$, and in situations where the image of G inside $\text{Aut}(H_1(\bar{X}, \mathbb{Z}_p))$ is *essentially abelian*. That is, when

- X is an affine hyperbolic of genus zero (say $\mathbf{P}^1 \setminus \{0, 1, \infty\}$) [22];
- $X = E \setminus \{e\}$ for an elliptic curve E with complex multiplication [25];
- (with John Coates) X is compact of genus ≥ 2 and the Jacobian J factors into abelian varieties with potential complex multiplication [4].

The first two cases require a rather obvious modification tailored to the study of integral points, while the two CM cases require p to be split inside the CM fields.

Given the half-completed state of the purported application, the reason for persevering in an abstruse investigation of known results might seem obscure indeed. We will return to this point towards the end of the lecture, side-stepping the issue for now in favor of a brief sketch of the methodology, confining our attention to the third class of curves.

---

[16] It suffices here to take M to be among the motives generated by $H^1(X)$.

[17] We thus have reason, in the manner of physicists, to regard Theorem 1 as good news for mixed motives, in that highly non-trivial real phenomena are among the corollaries of their theory.

3.3 For the remainder of this section, we assume that $K = \mathbb{Q}$ and write G for G_K. There is a pleasant quotient[18]

$$U \longrightarrow W := U/[[U, U], [U, U]]$$

of U that allows us to extend the key diagrams:

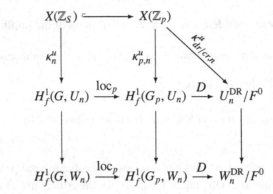

The structure of W turns out to be much simpler than that of U, and we obtain the following result.

Theorem 2 (with John Coates)   *Suppose J is isogenous to a product of abelian varieties having potential complex multiplication. Choose the prime p to split in all the CM fields that occur. Then*

$$dim\, H^1_f(G, W_n) < dim\, W_n^{DR}/F^0$$

for n sufficiently large.

The non-denseness of $\log_p(H^1_f(G, U))$ is an obvious corollary.

Proof   We give an outline of the proof assuming J is simple. Since

$$dim\, H^1_f(G, W_n) \leq dim\, H^1(G_T, W_n),$$

it suffices to estimate the dimension of cohomology with restricted ramification. Via the exact sequences

$$0 \to H^1(G_T, W^{n+1}\backslash W^n) \to H^1(G_T, W_n) \to H^1(G_T, W_{n-1})$$

the estimate can be reduced to a sum of abelian ones:

$$dim\, H^1(G_T, W_n) \leq \sum_{i=1}^{n} dim\, H^1(G_T, W^{i+1}\backslash W^i).$$

---

[18] For $\mathbf{P}^1 \setminus \{0, 1, \infty\}$, such quotients arise in the process of isolating (simple-)polylogarithms [1].

The linear representations $W^{i+1}\backslash W^i$ come with Euler characteristic formulas[19] [30]:

$$\dim H^0(G_T, W^{i+1}\backslash W^i) - \dim H^1(G_T, W^{i+1}\backslash W^i)$$
$$+ \dim H^2(G_T, W^{i+1}\backslash W^i) = -\dim[W^{i+1}\backslash W^i]^-$$

out of which the H^0 term always vanishes, leaving

$$\dim H^1(G_T, W^{i+1}\backslash W^i) = \dim[W^{i+1}\backslash W^i]^- + \dim H^2(G_T, W^{i+1}\backslash W^i).$$

The comparison with the topological fundamental group of $X(\mathbb{C})$ reveals U to be the unipotent completion of a free group on $2g$ generators modulo a single relation. This fact can applied to construct a Hall basis for the Lie algebra of W [35], from which we get an elementary estimate

$$\sum_{i=1}^n \dim[W^{i+1}\backslash W^i]^- \le [(2g-1)/2]\frac{n^{2g}}{(2g)!} + O(n^{2g-1}).$$

Similarly, on the De Rham side the dimension

$$\dim W_n^{DR}/F^0 = W_2/F^0 + \sum_{i=3}^n \dim[W^{DR,i+1}\backslash W^{DR,i}]$$

can easily be bounded below by

$$(2g-2)\frac{n^{2g}}{(2g)!} + O(n^{2g-1}).$$

Hence, since $g \ge 2$, we have

$$\sum_{i=1}^n \dim[W^{i+1}\backslash W^i]^- << \dim W_n^{DR}/F^0.$$

Therefore, it remains to show that

$$\sum_{i=1}^n \dim H^2(G_T, W^{i+1}\backslash W^i) = O(n^{2g-1}).$$

Standard arguments with Poitou–Tate duality[20] [30] eventually reduces the problem to the study of

$$\mathrm{Hom}_\Gamma\left[M(-1), \sum_{i=1}^n [W^{i+1}\backslash W^i]^*\right],$$

---

[19] The minus sign in the superscript refers to the negative eigenspace of complex conjugation. This has roughly half the dimension of the total space, and ends up unduly important to our estimates.

[20] This switches the focus from H^2 to H^1 at the cost of dealing with some insignificant local terms.

where

- F contains $\mathbb{Q}(J[p])$ and is a field of definition for all the complex multiplication;
- $\Gamma = \mathrm{Gal}(F_\infty/F)$ for the field $F_\infty = F(J[p^\infty])$ generated by the p-power torsion of J (we can assume $\Gamma \simeq \mathbb{Z}_p^r$); and
- $M = \mathrm{Gal}(H/F_\infty)$ is the Galois group of the p-Hilbert class field H of F_∞.

Choosing an annihilator[21]

$$\mathcal{L} \in \Lambda := \mathbb{Z}_p[[\Gamma]] \simeq \mathbb{Z}_p[[T_1, T_2, \ldots, T_r]]$$

for $M(-1)$ in the Iwasawa algebra, we need to count its zeros among the characters that appear in

$$\sum_{i=1}^{n} [W^{i+1} \backslash W^i]^*.$$

After a change of variables, a lemma of Greenberg [15] allows us to assume a form

$$\mathcal{L} = a_0(T_1, \ldots, T_{r-1}) + a_1(T_1, \ldots, T_{r-1})T_r + \cdots$$
$$+ a_{l-1}(T_1, \ldots, T_{r-1})T_r^{l-1} + T_r^l,$$

a polynomial in T_r. An elementary combinatorial argument then shows that the number of zeros is of the form $O(n^{2g-1})$. $\square$

We have now set up the first genuine occasion to motivate our constructions. The annihilator $\mathcal{L}$ is a version of an *algebraic p-adic L-function* controlling the situation. It is therefore of non-trivial interest that the sparseness of its zeros is responsible for the finiteness of points. The parallel with the case of elliptic curves [6, 20, 28, 36] might be seen clearly by comparing the implications

non-vanishing of $L \Rightarrow$ control of Selmer groups $\Rightarrow$ finiteness of points

familiar from the arithmetic of elliptic curves to the one given:

sparseness of L-zeros $\Rightarrow$ control of Selmer *varieties* $\Rightarrow$ finiteness of points.

As promised, the motivic fundamental group has provided a natural thread linking abelian and non-abelian Diophantine problems.

We remark that the non-CM case could proceed along the same lines, except that the group Γ and hence, the corresponding Iwasawa algebra is non-abelian. But the fact remains that the estimate

$$\dim \mathrm{Hom}_\Lambda(M, \oplus_{i=1}^{n} W^{i+1} \backslash W^i) = O(n^{2g-1})$$

---

[21] Provided by a theorem of Greenberg [14].

is sufficient for the analog of Theorem 2, and hence, for the finiteness of points. The representation $W^{i+1}\backslash W^i$ is a subquotient of the more familiar one

$$(\Lambda^2 V_p) \otimes (\text{Sym}^{i-2} V_p)$$

and the difference in dimensions is likely to count for very little in the coarse estimates. It might therefore be easier to work with

$$\text{Hom}_\Lambda[M, \oplus_{i=1}^{n-2}(\Lambda^2 V_p) \otimes (\text{Sym}^i V_p)].$$

Otmar Venjakob [42] has shown that M is locally torsion, so that a generating set $\{m_1, m_2, \ldots, m_d\}$ for M determines for each i a non-commutative power series $f_i \in \Lambda$ annihilating m_i. We must then count the *non-abelian zeros*[22] of f_i, that is, the representations containing vectors annihilated by f_i among the irreducible factors of $\oplus_{i=1}^{n-2}(\Lambda^2 V_p) \otimes (\text{Sym}^i V_p)$.

John Coates has stressed the role played by the ideal class group M in this picture, which is a priori smaller than the Iwasawa module relevant to elliptic curves. The reason that ramification at p can be ignored for now is that the local contribution at p is also of lower order as a function of n. For the Diophantine geometry of abelian fundamental groups, however, the option of passing to large n is absent. One is tempted to offer this as a kind of explanation for the infinitely many rational points that can live on an elliptic curve.

4 An explicit formula and speculations

4.1 Some preliminary evidence at present suggests another reason to pursue a π_1 approach to finiteness [26, 3]. This is the possibility that the function ϕ occurring in the proof of theorem 1 can be made explicit, leading to analytic defining equations for

$$X(\mathbb{Q}) \subset X(\mathbb{Q}_p).$$

For one thing, the map

$$\log_p : H_f^1(G, U_n) \to U_n^{\text{DR}}/F^0$$

occurs in the category of algebraic varieties over $\mathbb{Q}_p$, and is therefore amenable (in principle) to computation [7, 34]. Whenever the map itself can be presented, the computation of the image is then a matter of applying standard algorithms.

---

[22] As noted by Mahesh Kakde, it would be nice to know enough to formulate this in terms of a characteristic element $f \in K_1(\Lambda_{S^*})$ for M, whereby the count will be of irreducible representations $\rho : \Gamma \to N$ for which $f(\rho) = 0$ [5].

A genuinely *feasible* approach, however, should be effected by the *cohomological construction* of a function ψ as below that vanishes on global classes:

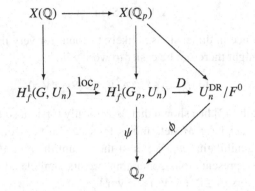

That is, once we have ψ, we can put

$$\phi = \psi \circ D^{-1},$$

a function whose precise computation might be regarded as a 'non-abelian explicit reciprocity law.' The vanishing itself should be explained by a local-to-global reciprocity, as in the work of Kolyvagin, Rubin, and Kato on the conjecture of Birch and Swinnerton-Dyer [28, 36, 20].

4.2 These speculations are best given substance with an example, albeit in an affine setting. Let $X = E \setminus \{e\}$, where E is an elliptic curve satisying the following hypotheses:

- rank $E(\mathbb{Q}) = 1$;
- Sha$(E)[p^\infty] < \infty$;
- the local Tamagawa numbers c_l of E are one for each prime l.

Among the consequences of the hypotheses is that the $\mathbb{Q}_p$-localization map is bijective on points,

$$\mathrm{loc}_p \colon E(\mathbb{Q}) \otimes \mathbb{Q}_p \simeq H^1_f(G_p, V_p(E)),$$

and the second cohomology with restricted ramification vanishes:

$$H^2(G_T, V_p(E)) = 0.$$

We will construct a diagram

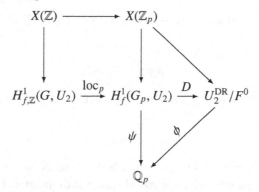

using just the first non-abelian level U_2 of the unipotent fundamental group. We have introduced here a refined Selmer variety $H^1_{f,\mathbb{Z}}(G, U_2)$ consisting of classes that are actually trivial at all places $l \neq p$. It is a relatively straightforward matter to show that the integral points land in this subspace.

The relevant structure now is a Heisenberg group

$$0 \to \mathbb{Q}_p(1) \to U_2 \to V_p \to 0,$$

that we will analyze in terms of the corresponding extension of Lie algebras

$$0 \to \mathbb{Q}_p(1) \to L_2 \to V_p \to 0.$$

Conveniently, at this level, the Galois action on L_2 splits:[23]

$$L_2 = V_p \oplus \mathbb{Q}_p(1),$$

provided we use a tangential base-point at the missing point e. With the identification[24] of U_2 and L_2, non-abelian cochains can be thought of as maps

$$\xi \colon G_p \longrightarrow L_2$$

and expressed in terms of components $\xi = (\xi_1, \xi_2)$ with respect to the decomposition. The cocycle condition in these coordinates reads[25]

$$d\xi_1 = 0, \quad d\xi_2 = (-1/2)[\xi_1, \xi_1].$$

---

[23] This uses the multiplication by $[-1]$, as in Mumford's theory of theta functions.

[24] For unipotent groups, the power series for the log map stops after finitely many terms, defining an algebraic isomorphism. The group then can be thought of as the Lie algebra itself with a twisted binary operation given by the Baker–Campbell–Hausdorff formula [39].

[25] In his book on gerbes, Breen emphasizes the importance of a familiarity with the 'calculus of cochains.' Indeed, the typical number-theorist will be quite anxious about non-closed cochains like ξ_2. Unfortunately, they are as unavoidable as the components of connection forms in non-abelian gauge theory, which obey complicated equations even when the connections themselves are closed in a suitable sense.

Define

$$\psi(\xi) := [\mathrm{loc}_p(x), \xi_1] - 2\log\chi_p \cup \xi_2 \in H^2(G_p, \mathbb{Q}_p(1)) \simeq \mathbb{Q}_p,$$

where

$$\log\chi_p \colon G_p \to \mathbb{Q}_p$$

is the logarithm of the $\mathbb{Q}_p$-cyclotomic character and x is a *global* solution, to the equation

$$dx = \log\chi_p \cup \xi_1.$$

The equation makes sense on G_T since both χ_p and ξ_1 have natural extensions to global classes, while the non-trivial existence of the global solution

$$x \colon G_T \to V_p$$

is guaranteed by the aforementioned vanishing of H^2. One checks readily that $\psi(\xi)$ is indeed a 2-cocycle whose class is independent of the choice of x.

Theorem 3 *ψ vanishes on the image of*

$$\mathrm{loc}_p \colon H^1_{f,\mathbb{Z}}(G, U_2) \to H^1_f(G_p, U_2).$$

The proof is a simple consequence of the standard reciprocity sequence

$$0 \to H^2(G_T, \mathbb{Q}_p(1)) \to \oplus_{v \in T} H^2(G_v, \mathbb{Q}_p(1)) \to \mathbb{Q}_p \to 0.$$

The point is that if ξ is global then so is $\psi(\xi)$. But this class has been constructed to vanish at all places $l \neq p$. Hence, it must also vanish at p.

An explicit formula on the De Rham side in this case is rather easily obtained. Choose a Weierstrass equation for E and let

$$\alpha = dx/y, \quad \beta = xdx/y.$$

Define

$$\log_\alpha(z) := \int_b^z \alpha, \quad \log_\beta(z) := \int_b^z \beta,$$

$$D_2(z) := \int_b^z \alpha\beta,$$

via (iterated) Coleman integration.

Corollary 4 *For any two points $y, z \in X(\mathbb{Z}) \subset X(\mathbb{Z}_p)$, we have*

$$\log_\alpha^2(y)D_2(z) = \log_\alpha^2(z)D_2(y).$$

The proof uses an action of the multiplicative monoid $\mathbb{Q}_p$ on $H^1_f(G, U_2)$ that covers the scalar multiplication on $E(\mathbb{Q}) \otimes \mathbb{Q}_p$. That is,

$$\lambda \cdot (\xi_1, \xi_2) = (\lambda \xi_1, \lambda^2 \xi_2).$$

Evaluating ψ on the class

$$\log_\alpha(x)\kappa_2^u(y) - \log_\alpha(y)\kappa_2^u(x) \in H^1_f(G_p, U^3 \backslash U^2)$$

leads directly to the formula displayed.

The harmonious form of the resulting constraint is perhaps an excuse for some general optimism. Of course, as it stands, the formula is useful only if there is a point y of infinite order already at hand. One can then look for the other integrals points in the zero set of the function

$$D_2(z) - \left(\frac{D_2(y)}{\log_\alpha^2(y)} \right) \log_\alpha^2(z)$$

in the coordinate z.

The *meaning* of the construction given is not yet clear to the author, even as some tentative avenues of interpretation are opening up quite recently. If the analogy with the abelian case is to be taken seriously, ψ should be a small fragment of *non-abelian duality* in Galois cohomology.[26] For the abelian quotient, one has the usual duality

$$H^1(G_p, V) \times H^1(G_p, V^*(1)) \longrightarrow H^2(G_p, \mathbb{Q}_p(1)) \simeq \mathbb{Q}_p$$

with respect to which $H^1_f(G_p, V)$ and $H^1_f(G_p, V^*(1))$ are mutual annihilators. We take the view that

$$H^1(G_p, V^*(1))/H^1_f(G_p, V^*(1))$$

is thereby a systematic source of functions on $H^1_f(G_p, V)$, which can then be used to annihilate global classes when the function itself comes from a suitable class[27] in $H^1(G, V^*(1))$. After a minimal amount of non-commutativity has been introduced, our ψ is exactly such a global function on the local cohomology $H^1_f(G_p, U_2)$ that ends up thereby *annihilating the Selmer variety*. The main difficulty is that we know not yet a suitable space in which ψ lives. Allowing ourselves a further flight of fancy, the elusive function in general might

---

[26] Kazuya Kato's immediate reaction to the idea of non-abelian duality was that it should have an 'automorphic' nature. Such a suggestion might be highly relevant if the *reductive completion* of fundamental groups could somehow be employed in an arithmetic setting. For the unipotent completions under discussion, the author's inclination is to look for duality that is a relatively straightforward lift of the abelian phenomenon.

[27] The author is not competent to review here the laborious procedure for producing such classes as was developed in the work of Kolyvagin and Kato. The guiding concept in the abelian case is that of a *zeta element*.

eventually be the subject of an Iwawasa theory rising out of a landscape radically more non-abelian and non-linear than we have dared to dream of thus far [21].

4.3 It has been remarked that the title of this lecture was chosen to be maximally ambiguous. Notice, however, that Galois theory in dimension zero, according to Galois, proposes groups as structures encoding the Diophantine geometry of equations in one variable. The proper subject of Galois theory in dimension one should then be a unified network of structures relevant to the Diophantine geometry of polynomials in two variables. Included therein one may find the arithmetic fundamental groups, motivic L-functions of weight one, and moduli spaces of torsors that have already proved their scattered usefulness to the trade.[28] The picture as a whole is blatantly far from clear, coherent, or complete at this stage. [29]

---

[28] The section conjecture says the set of points on a curve of higher genus *is* a moduli space of torsors. One might take this to be a categorical structure that generalizes the abelian groups that come up in elliptic curves.

[29] It has been an enduring source of amazement to the author that true number-theorists employ philosophies that never work in practice as planned at the outset. The numerous subtle twists and turns that one may find, for example, in the beautiful theorems of Richard Taylor, that adhere nevertheless to the overall form of a grand plan, are hallmarks of the kind of artistry that a mere generalist could never aspire to. It is essentially for this reason that the author has avoided thus far the question of applying the techniques of this paper to varieties of higher dimension, for example, those with a strong degree of hyperbolicity. A theory whose end product is a single function applies immediately only in dimension one. It is not inconceivable that an arsenal of clever tricks will strengthen the machinery shown here to make it more broadly serviceable. A robust strategy that makes minimal demands on the user's ingenuity, however, should expect the requisite structures to evolve as one climbs up the dimension ladder, perhaps in a manner reminiscent of Grothendieck's *poursuite*.

References

[1] Beilinson, A.; Deligne, P. Interpretation motivique de la conjecture de Zagier reliant polylogarithmes et régulateurs. Motives (Seattle, WA, 1991), 97–121, Proc. Sympos. Pure Math., 55, Part 2, Amer. Math. Soc., Providence, RI, 1994.

[2] Bloch, Spencer; Kato, Kazuya. *L*-functions and Tamagawa numbers of motives. The Grothendieck Festschrift, Vol. I, 333–400, Progr. Math., 86, Birkhauser Boston, Boston, MA, 1990.

[3] Balakrishnan, Jennifer S.; Kedlaya, Kiran S.; Kim, Minhyong Appendix and erratum: 'Massey products for elliptic curves of rank 1.' Jour. Amer. Math. Soc. (to be published).

[4] Coates, John; Kim, Minhyong. Selmer varieties for curves with CM Jacobians. To be published, Kyoto Mathematical Journal. Available at the mathematics archive, arXiv:0810.3354 .

[5] Coates, John; Fukaya, Takako; Kato, Kazuya; Sujatha, Ramdorai; Venjakob, Otmar. The GL_2 main conjecture for elliptic curves without complex multiplication. Publ. Math. Inst. Hautes Études Sci. No. 101 (2005), 163–208.

[6] Coates, J.; Wiles, A. On the conjecture of Birch and Swinnerton-Dyer. Invent. Math. 39 (1977), no. 3, 223–251.

[7] Coleman, Robert F. Effective Chabauty. Duke Math. J. 52 (1985), no. 3, 765–770.

[8] Deligne, Pierre. Le groupe fondamental de la droite projective moins trois points. Galois groups over $\mathbb{Q}$ (Berkeley, CA, 1987), 79–297, Math. Sci. Res. Inst. Publ., 16, Springer, New York, 1989.

[9] Faltings, G. Endlichkeitssätze für abelsche Varietäten über Zahlkörpern. Invent. Math. 73 (1983), no. 3, 349–366.

[10] Fontaine, Jean-Marc. Sur certains types de représentations p-adiques du groupe de Galois d'un corps local; construction d'un anneau de Barsotti-Tate. Ann. of Math. (2) 115 (1982), no. 3, 529–577.

[11] Fontaine, Jean-Marc; Mazur, Barry. Geometric Galois representations. Elliptic curves, modular forms, & Fermat's last theorem (Hong Kong, 1993), 41–78, Ser. Number Theory, I, Int. Press, Cambridge, MA, 1995.

[12] Furusho, Hidekazu. p-adic multiple zeta values. I. p-adic multiple polylogarithms and the p-adic KZ equation. Invent. Math. 155 (2004), no. 2, 253–286.

[13] Goldman, William M.; Millson, John J. The deformation theory of representations of fundamental groups of compact Kähler manifolds. Inst. Hautes Études Sci. Publ. Math. No. 67 (1988), 43–96.

[14] Greenberg, Ralph. The Iwasawa invariants of Γ-extensions of a fixed number field. Amer. J. Math. 95 (1973), 204–214.

[15] Greenberg, Ralph. On the structure of certain Galois groups. Invent. Math. 47 (1978), no. 1, 85–99.

[16] Grothendieck, Alexander. Brief an G. Faltings. London Math. Soc. Lecture Note Ser., 242, Geometric Galois actions, 1, 49–58, Cambridge University Press, Cambridge, 1997.

[17] Hain, Richard M. The de Rham homotopy theory of complex algebraic varieties. I. *K*-Theory 1 (1987), no. 3, 271–324.

[18] Iyanaga, Shokichi. Memories of Professor Teiji Takagi [1875–1960]. Class field theory—its centenary and prospect (Tokyo, 1998), 1–11, Adv. Stud. Pure Math., 30, Math. Soc. Japan, Tokyo, 2001.

[19] Motives. Proceedings of the AMS-IMS-SIAM Joint Summer Research Conference held at the University of Washington, Seattle, Washington, July 20–August 2, 1991. Edited by Uwe Jannsen, Steven Kleiman and Jean-Pierre Serre. Proceedings of Symposia in Pure Mathematics, 55, Part 1. American Mathematical Society, Providence, RI, 1994. xiv+747 pp. ISBN: 0-8218-1636-5

[20] Kato, Kazuya. p-adic Hodge theory and values of zeta functions of modular forms. Cohomologies p-adiques et applications arithmétiques. III. Astérisque No. 295 (2004), ix, 117–290.

[21] Kato, Kazuya. Lectures on the approach to Iwasawa theory for Hasse-Weil L-functions via B_{dR}. I. Arithmetic algebraic geometry (Trento, 1991), 50–163, Lecture Notes in Math., 1553, Springer, Berlin, 1993.

[22] Kim, Minhyong. The motivic fundamental group of $\mathbb{P}^1 \setminus \{0, 1, \infty\}$ and the theorem of Siegel. Invent. Math. 161 (2005), no. 3, 629–656.

[23] Kim, Minhyong. The unipotent Albanese map and Selmer varieties for curves. Publ. Res. Inst. Math. Sci. 45 (2009), no. 1, 89–133. (Proceedings of special semester on arithmetic geometry, Fall, 2006.)

[24] Kim, Minhyong. Remark on fundamental groups and effective Diophantine methods for hyperbolic curves. To be published in Serge Lang memorial volume. Available at mathematics archive, arXiv:0708.1115.

[25] Kim, Minhyong. p-adic L-functions and Selmer varieties associated to elliptic curves with complex multiplication. Annals of Mathematics, 172 (2010), no. 1, 751–759.

[26] Kim, Minhyong. Massey products for elliptic curves of rank 1. Jour. Amer. Math. Soc. 23 (2010), no. 3, 725–748.

[27] Kim, Minhyong, and Tamagawa, Akio. The l-component of the unipotent Albanese map. Math. Ann. 340 (2008), no. 1, 223–235.

[28] Kolyvagin, Victor A. On the Mordell–Weil group and the Shafarevich–Tate group of modular elliptic curves. Proceedings of the International Congress of Mathematicians, Vol. I, II (Kyoto, 1990), 429–436, Math. Soc. Japan, Tokyo, 1991.

[29] Mumford, David; Fogarty, John. Geometric invariant theory. Second edition. Ergebnisse der Mathematik und ihrer Grenzgebiete, 34. Springer-Verlag, Berlin, 1982. xii+220 pp.

[30] Milne, J. S. Arithmetic duality theorems. Perspectives in Mathematics, 1. Academic Press, Inc., Boston, MA, 1986.

[31] Nakamura, Hiroaki; Tamagawa, Akio; Mochizuki, Shinichi. The Grothendieck conjecture on the fundamental groups of algebraic curves [translation of Su-gaku 50 (1998), no. 2, 113–129; MR1648427 (2000e:14038)]. Sugaku Expositions. Sugaku Expositions 14 (2001), no. 1, 31–53.

[32] Narasimhan, M. S.; Seshadri, C. S. Stable and unitary vector bundles on a compact Riemann surface. Ann. of Math. (2) 82 1965 540–567.

[33] Olsson, Martin. The bar construction and affine stacks. Preprint. Available at http://math.berkeley.edu/ molsson/.

[34] Poonen, Bjorn. Computing rational points on curves. Number theory for the millennium, III (Urbana, IL, 2000), 149–172, A K Peters, Natick, MA, 2002.

[35] Reutenauer, Christophe. Free Lie algebras. London Mathematical Society Monographs. New Series, 7. Oxford Science Publications. The Clarendon Press, Oxford University Press, New York, 1993.

[36] Rubin, Karl. The "main conjectures" of Iwasawa theory for imaginary quadratic fields. Invent. Math. 103 (1991), no. 1, 25–68.

[37] Serre, Jean-Pierre. Galois cohomology. Translated from the French by Patrick Ion and revised by the author. Springer-Verlag, Berlin, 1997. x+210 pp.

[38] Serre, Jean-Pierre. Andr Weil 6 May 1906-6 August 1998 Biographical Memoirs of Fellows of the Royal Society, Vol. 45, (Nov., 1999), pp. 521–529

[39] Serre, Jean-Pierre. Lie algebras and Lie groups. 1964 lectures given at Harvard University. Second edition. Lecture Notes in Mathematics, 1500. Springer-Verlag, Berlin, 1992. viii+168 pp.

[40] Simpson, Carlos T. Higgs bundles and local systems. Inst. Hautes Études Sci. Publ. Math. No. 75 (1992), 5–95.

[41] Szamuely, Tamas. Galois Groups and Fundamental Groups. Cambridge Studies in Advanced Mathematics, vol. 117, Cambridge University Press, 2009.

[42] Venjakob, Otmar. On the Iwasawa theory of p-adic Lie extensions. Compositio Math. 138 (2003), no. 1, 1–54.

[43] Weil, André. L'arithmétique sur les courbes algébriques. Acta Math. 52 (1929), no. 1, 281–315.

[44] Weil, André. Généralisation des fonctions abéliennes. J. Math Pur. Appl. 17 (1938), no. 9, 47–87.

[45] Wiles, Andrew. Modular elliptic curves and Fermat's last theorem. Ann. of Math. (2) 141 (1995), no. 3, 443–551.

Potential modularity – a survey

Kevin Buzzard

Imperial College London

1 Introduction

Our main goal in this article is to talk about recent theorems of Taylor and his co-workers on modularity and potential modularity of Galois representations, particularly those attached to elliptic curves. However, so as to not bog down the exposition unnecessarily with technical definitions right from the off, we will build up to these results by starting our story with Wiles' breakthrough paper [Wil95], and working towards the more recent results. We will however assume some familiarity with the general area – for example we will assume the reader is familiar with the notion of an elliptic curve over a number field, and a Galois representation, and what it means for such things to be modular (when such a notion makes sense). Let us stress now that, because of this chronological approach, some theorems stated in this paper will be superseded by others (for example Theorem 1 gets superseded by Theorem 6 which gets superseded by Theorem 7), and similarly some conjectures (for example Serre's conjecture) will become theorems as the story progresses. The author hopes that this slightly non-standard style nevertheless gives the reader the feeling of seeing how the theory evolved.

We thank Toby Gee for reading through a preliminary draft of this article and making several helpful comments, and we also thank Matthew Emerton and Jan Nekovář for pointing out various other inaccuracies and ambiguities.

2 Semistable elliptic curves over Q are modular

The story, of course, starts with the following well-known result proved in [Wil95] and [TW95].

Non-abelian Fundamental Groups and Iwasawa Theory, eds. John Coates, Minhyong Kim, Florian Pop, Mohamed Saïdi and Peter Schneider. Published by Cambridge University Press. ©Cambridge University Press 2012.

Theorem 1 (Wiles, Taylor–Wiles) *Any semistable elliptic curve over the rationals is modular.*

This result, together with work of Ribet and others on Serre's conjecture, implies Fermat's Last Theorem. This meant that the work of Wiles and Taylor captured the imagination of the public. But this article is not about Fermat's Last Theorem, it is about how the *modularity theorem* above has been vastly generalised. Perhaps we should note here though that that there is still a long way to go! For example, at the time of writing, it is still an open problem as to whether an arbitrary elliptic curve over an arbitrary totally real field is modular, and over an general number field, where we cannot fall back on the theory of Hilbert modular forms, the situation is even worse (we still do not have a satisfactory theorem attaching elliptic curves to modular forms in this generality, let alone a result in the other direction).

Before we go on to explain the generalisations that this article is mainly concerned with, we take some time to remind the reader of some of the details of the strategy of the Wiles/Taylor–Wiles proof. The ingredients are as follows. For the first, main, ingredient, we need to make some definitions. Let p be a prime number, let O denote the integers in a finite extension of $\mathbf{Q}_p$, let $\rho \colon \mathrm{Gal}(\overline{\mathbf{Q}}/\mathbf{Q}) \to \mathrm{GL}_2(O)$ denote a continuous odd (by which we mean $\det(\rho(c)) = -1$, for c a complex conjugation) irreducible representation, unramified outside a finite set of primes, and let $\overline{\rho}$ denote its reduction modulo the maximal ideal of O. Recall that there is a general theorem due to Deligne and others, which attaches p-adic Galois representations to modular eigenforms; we say a p-adic Galois representation is *modular* if it arises in this way, and that a mod p Galois representation is *modular* if it arises as the semisimplification of the mod p reduction of a modular p-adic Galois representation. Note that we will allow myself the standard abuse of notation here, and talk about "mod p reduction" when we really mean "reduction modulo the maximal ideal of O".

Note that if a p-adic Galois representation ρ is modular, then its reduction $\overline{\rho}$ (semisimplified if necessary) is trivially modular. Wiles' insight is that one could sometimes go the other way.

Theorem 2 (Modularity lifting theorem) *If $p > 2$, if $\overline{\rho}$ is irreducible and modular, and if furthermore ρ is semistable and has cyclotomic determinant, then ρ is modular.*

This is Corollary 3.46 of [DDT97] (the aforementioned paper is an overview of the Wiles/Taylor–Wiles work; the theorem is essentially due to Wiles and Taylor–Wiles). The proof is some hard work, but is now regarded as "standard"

– many mathematicians have read and verified the proof. We shall say a few words about the proof later on. Semistability is a slightly technical condition (see op. cit. for more details) but we shall be removing it soon so we do not go into details. Rest assured that if $E/\mathbf{Q}$ is a semistable elliptic curve then its Tate module is semistable.

The next ingredient is a very special case of the following conjecture of Serre.

Conjecture 3 (Serre, 1987)   If k is a finite field and $\overline{\rho} : \mathrm{Gal}(\overline{\mathbf{Q}}/\mathbf{Q}) \to \mathrm{GL}_2(k)$ is a continuous odd absolutely irreducible representation, then $\overline{\rho}$ is modular.

We will say more about this conjecture and its generalisations later. Note that this conjecture is now a theorem of Khare and Wintenberger, but we are taking a chronological approach so will leave it as a conjecture for now. In the early 1990s this conjecture was wide open, but one special case had been proved in 1981 (although perhaps it was not stated in this form in 1981; see for example prop. 11 of [Ser87] for the statement we need).

Theorem 4 (Langlands, Tunnell)   *If $\overline{\rho} : \mathrm{Gal}(\overline{\mathbf{Q}}/\mathbf{Q}) \to \mathrm{GL}_2(\mathbf{F}_3)$ is continuous, odd, and irreducible, then $\overline{\rho}$ is modular.*

The original proof of Theorem 4 is a huge amount of delicate analysis: let it not be underestimated! One needs (amongst other things) the full force of the trace formula in a non-compact case to prove this result, and hence a lot of delicate analysis. We freely confess to not having checked the details of this proof ourselves. Note also that the point is *not* just that the image of $\overline{\rho}$ is solvable, it is that the image is very small (just small enough to be manageable, in fact). Note that because we are in odd characteristic, the notions of irreducibility and absolute irreducibility coincide for an odd representation (complex conjugation has two distinct eigenvalues, both defined over the ground field). Langlands' book [Lan80] proves much of what is needed; the proof was finished by Tunnell in [Tun81] using the non-solvable cubic base change results of [JPSS81]. This result is of course also now regarded as standard – many mathematicians have read and verified this proof too. The author remarks however that due to the rather different techniques involved in the proofs of the two results, he has the impression that the number of mathematicians who have read and verified all the details of the proofs of *both* the preceding theorems is rather smaller!

Let us see how much of Theorem 1 we can prove so far, given Theorems 2 and 4. Let E be a semistable elliptic curve, set $p = 3$ and let $\rho : \mathrm{Gal}(\overline{\mathbf{Q}}/\mathbf{Q}) \to \mathrm{GL}_2(\mathbf{Z}_3)$ be the 3-adic Tate module of E. Then ρ is continuous, odd, unramified outside a finite set of primes, and semistable, with cyclotomic determinant. Furthermore, *if $\overline{\rho}$, the Galois representation on $E[3]$, is irreducible, then $\overline{\rho}$ is* modular by the Langlands–Tunnell theorem 4 and so ρ, and hence E, is modular, by the modularity lifting theorem. Of course the problem is that $E[3]$ may be reducible – for example if E has a $\mathbf{Q}$-point of order 3 (or more generally a subgroup of order 3 defined over $\mathbf{Q}$). To deal with this situation, Wiles developed a technique known as the "3–5 trick".

Lemma 5 (The 3–5 trick) *If $E/\mathbf{Q}$ is a semistable elliptic curve with $E[3]$ reducible, then $E[5]$ is irreducible, and there is another semistable elliptic curve $A/\mathbf{Q}$ with $E[5] \cong A[5]$ and $A[3]$ irreducible.*

The 3–5 trick is all we need to finish the proof of Theorem 1. For if $E/\mathbf{Q}$ is semistable but $E[3]$ is reducible, choose A as in the lemma, and note that $A[3]$ is irreducible, hence $A[3]$ is modular (by the Langlands–Tunnell Theorem 4), hence A is modular (by the Modularity Lifting Theorem 2), so $A[5]$ is modular, so $E[5]$ is modular, and irreducible, so E is modular by the Modularity Lifting Theorem 2 applied to the 5-adic Tate module of E.

Let us say a few words about the proof of the last lemma. The reason that reducibility of $E[3]$ implies irreducibility of $E[5]$ is that reducibility of both would imply that E had a rational subgroup of order 15, and would hence give rise to a point on the modular curve $Y_0(15)$, whose compactification $X_0(15)$ is a curve of genus 1 with finitely many rational points, and it turns out that the points on this curve are known, and one can check that none of them can come from semistable elliptic curves (and all of them are modular anyway). So what is left is that given E as in the lemma, we need to produce A. Again we use a moduli space trick. We consider the moduli space over $\mathbf{Q}$ parametrising elliptic curves B equipped with an isomorphism $B[5] \cong E[5]$ that preserves the Weil pairing. This moduli space has a natural smooth compactification X over $\mathbf{Q}$, obtained by adding cusps. Over the complexes the resulting compactified curve is isomorphic to the modular curve $X(5)$, which has genus zero. Hence X is a genus zero curve over $\mathbf{Q}$. Moreover, X has a rational point (coming from E) and hence X is itself isomorphic to the projective line over $\mathbf{Q}$ (rather than a twist of the projective line). In particular, X has infinitely many rational points. Now using Hilbert's Irreducibility Theorem, which in this setting can be viewed as some sort of refinement of the Chinese Remainder Theorem, it is possible to find a point on X which is 5-adically very close to E, and 3-adically very far away from E (far enough so that the Galois representation on the 3-torsion of the corresponding elliptic curve is irreducible: this is the crux of the Hilbert Irreducibility Theorem, and this is where we are using more than the naive Chinese Remainder Theorem). Such a point corresponds to the elliptic curve A we seek, and the lemma, and hence Theorem 1, is proved.

The reason we have broken up the proof of Theorem 1 into these pieces is that we would like to discuss generalisations of Theorem 1, and this will entail discussing generalisations of the pieces that we have broken it into.

3 Why the semistability assumption?

All semistable elliptic curves were known to be modular by 1995, but of course one very natural question was whether the results could be extended to all elliptic curves. Let us try and highlight the issues involved with trying to extend the proof; we will do this by briefly reminding readers of the strategy of the *proof* of a modularity lifting theorem such as Theorem 2. The strategy is that given a modular $\bar{\rho}$, one considers two kinds of lifting to characteristic zero. The first is a "universal deformation" $\rho^{\text{univ}} \colon \text{Gal}(\overline{\mathbf{Q}}/\mathbf{Q}) \to \text{GL}_2(R^{\text{univ}})$, where one considers *all* deformations satisfying certain properties fixed beforehand (in Wiles' case these properties were typically "unramified outside S and semistable at all the primes in S" for some fixed finite set of primes S), and uses the result of Mazur in [Maz89] that says that there is a *universal* such deformation, taking values in a ring R^{univ}. The second is a "universal modular deformation" $\rho_{\mathbf{T}} \colon \text{Gal}(\overline{\mathbf{Q}}/\mathbf{Q}) \to \text{GL}_2(\mathbf{T})$ comprising of a lift of $\bar{\rho}$ to a representation taking values in a Hecke algebra over $\mathbf{Z}_p$ built from modular forms of a certain level, weight and character (or perhaps satisfying some more refined local properties). Theorems about modular forms (typically local-global theorems) tell us that the deformation to $\text{GL}_2(\mathbf{T})$ has the properties used in the definition of R^{univ}, and there is hence a map

$$R^{\text{univ}} \to \mathbf{T}.$$

The game is to prove that this map is an isomorphism; then all deformations will be modular, and in particular ρ, the representation we started with, will be modular. The insight that the map may be an isomorphism seems to be due to Mazur: see conjecture (*) of [MT] and the comments preceding it. One underlying miracle is that this procedure can only work if R^{univ} has no p-torsion, something which is not at all evident, but which came out of the Wiles/Taylor–Wiles proof as a consequence.

The original proof that the map $R^{\text{univ}} \to \mathbf{T}$ is an isomorphism breaks up into two steps: the first one, referred to as the minimal case, deals with situations where ρ is "no more ramified than $\bar{\rho}$", and is proved by a patching argument via the construction of what is now known as a Taylor–Wiles system: one checks that certain projective limits of Rs and Ts (using weaker and weaker deformation conditions, and more and more modular forms) are power series rings, and that the natural map between them is an isomorphism for commutative algebra reasons (for dimension reasons, really), and then one descends back to the case of interest. The second is how to deduce the general case from the minimal case – this is an inductive procedure (which relies on a result on Jacobians of modular curves known as Ihara's Lemma, the analogue of which still appears to be open for GL_n, $n > 2$; this provided a serious stumbling-block in generalising the theory to higher dimensions for many years). Details of both of these arguments can be found in [DDT97], especially §5 (as well, of course, as in the original sources).

To see why the case of semistable elliptic curves was treated first historically, we need to look more closely at the nitty-gritty of the details behind a deformation problem.

Wiles had a semistable irreducible representation $\bar{\rho}\colon \mathrm{Gal}(\overline{\mathbf{Q}}/\mathbf{Q}) \to GL_2(k)$ with k a finite field of characteristic p. Say $\bar{\rho}$ is unramified outside some finite set $S \ni p$ of primes, and has cyclotomic determinant. Crucially, Wiles knew what it meant (at least when $p > 2$) for a deformation $\rho\colon \mathrm{Gal}(\overline{\mathbf{Q}}/\mathbf{Q}) \to GL_2(A)$ to be semistable and unramified outside S, where A is now a general Artin local ring with residue field k (or even a projective limit of such rings). For a prime $q \notin S$ it of course means ρ is unramified at q. For $q \neq p$, $q \in S$, it means that the image of an inertia group at q under ρ can be conjugated into the upper triangular unipotent matrices. For $q = p$ one needs more theory. The observation is that an elliptic curve with semistable reduction either has good reduction, or multiplicative reduction. The crucial point is that for a general Artin local A with finite residue field one can make sense of the notion that $\rho\colon \mathrm{Gal}(\overline{\mathbf{Q}}/\mathbf{Q}) \to GL_2(A)$ has "good reduction" – one demands that it is the Galois action on the generic fibre of a finite flat group scheme with good reduction, and work of Fontaine [Fon77] and Fontaine–Laffaille [FL82] shows that one can translate this notion into "linear algebra" which is much easier to work with (this is where the assumption $p > 2$ is needed). Also crucial are the results of Raynaud [Ray74], which show that the category of Galois representations with these properties is very well behaved. Similarly one can make sense of the notion that ρ has "multiplicative reduction": one can demand that ρ on a decomposition group at p is upper triangular.

We stress again that the crucial point is that the notions of "good reduction" and "multiplicative reduction" above make sense for an *arbitrary* Artin local A, and patch together well to give well-behaved local deformation conditions which are locally representable (by which we mean the deformations of the Galois representation $\bar{\rho}|_{G_{\mathbf{Q}_p}}$ are represented by some universal ring). So we get a nicely behaved local deformation ring – in particular we get a ring for which we can compute the tangent space $\mathfrak{m}/\mathfrak{m}^2$ of its mod p reduction. If this tangent space has dimension at most 1 then the dimension calculations work out in the patching argument and the modularity lifting theorem follows.

If one is prepared to take these observations on board, then it becomes manifestly clear what the one of the main problems will be in proving that an arbitrary elliptic curve over the rationals is modular: we will have to come up with deformation conditions that are small enough to make the dimension calculations work, but big enough to encompass Galois representations that are not semistable. At primes $q \neq p$ this turned out to be an accessible problem; careful calculations by Fred Diamond in [Dia96] basically resolved these issues completely. Diamond's main theorem had as a consequence the result that if $E/\mathbf{Q}$ had semistable reduction at both 3 and 5 then E was modular. The reason that both 3 and 5 occur is of course because he has to use the 3–5 trick if $E[3]$ is reducible. A year or so later, Diamond, and independently Fujiwara, had another insight: instead of taking limits of Hecke algebras to prove that a deformation ring equalled a Hecke ring, one can instead take limits of modules that these algebras act on naturally. The resulting commutative algebra is more delicate, and one does not get modularity of any more elliptic curves in this way, but the result is of importance because it enables one to apply the machinery in situations where certain "mod p multiplicity one" hypotheses are not known. These multiplicity one hypotheses were known in the situations that Wiles initially dealt with but were not known in certain more general situations; the consequence was that Wiles' method could now be applied more generally. Diamond's paper [Dia97] illustrated the point by showing that the methods could now be applied in the case of Shimura curves over $\mathbf{Q}$ (where new multiplicity one results could be deduced as a byproduct), and Fujiwara (in [Fuj], an article which remains unpublished, for reasons unknown to this author) illustrated that the method enabled one to generalise Wiles' methods to the Hilbert modular case, on which more later.

Getting back to elliptic curves over the rationals, the situation in the late 1990s, as we just indicated, was that any elliptic curve with semistable reduction at 3 and 5 was now proven to be modular. To get further, new ideas were needed, because in the 1990s the only source of modular mod p Galois representations were those induced from a character, and those coming from the Langlands–Tunnell theorem. Hence in the 1990s one was forced to ultimately work with the prime $p = 3$ (the prime $p = 2$ was another possibility; see for example [Dic], but here other technical issues arise). Hence, even with the 3–5 trick, it was clear that if one wanted to prove that all elliptic curves over $\mathbf{Q}$ were modular using these methods then one was going to have to deal with elliptic curves that have rather nasty non-semistable reduction at 3 (one cannot use the 3–5 trick to get around this because if E has very bad reduction at 3 (e.g.

if its conductor is divisible by a large power of 3) then this will be reflected in the 5-torsion, which will also have a large power of 3 in its conductor, so any curve A with $A[5] \cong E[5]$ will also be badly behaved at 3; one can make certain simplifications this way but one cannot remove the problem entirely). The main problem is then deformation-theoretic: given some elliptic curve E which is highly ramified at some odd prime p, how does one write down a reasonable deformation problem for $E[p]$ at p, which is big enough to see the Tate module of the curve, but is still sufficiently small for the Taylor–Wiles method to work? By this we mean that the tangent space of the mod p local deformation problem at p has to have dimension at most 1. This thorny issue explains the five year gap between the proof of the modularity of all semistable elliptic curves, and the proof for all elliptic curves.

4 All elliptic curves over Q are modular

As explained in the previous section, one of the main obstacles in proving that all elliptic curves over the rationals are modular is that we are forced, by Langlands–Tunnell, to work with $p = 3$, so elliptic curves with conductor a multiple of a high power of 3 are going to be difficult to deal with. Let us review the situation at hand. Let $\bar{\rho}$: $\mathrm{Gal}(\overline{\mathbf{Q}}_p/\mathbf{Q}_p) \to \mathrm{GL}_2(k)$ be an irreducible Galois representation, where here k is a finite field of characteristic p. Such a $\bar{\rho}$ has a universal deformation to ρ^{univ}: $\mathrm{Gal}(\overline{\mathbf{Q}}_p/\mathbf{Q}_p) \to \mathrm{GL}_2(R^{\mathrm{univ}})$. This ring R^{univ} is a quotient of a power series ring $W(k)[[x_1, x_2, \ldots, x_n]]$ in finitely many variables, where here $W(k)$ denotes the Witt vectors of the field k. But this universal deformation ring is too big for our purposes – a general lifting of $\bar{\rho}$ to the integers O of a finite extension of $\mathrm{Frac}(W(k))$ will not look anything like the Tate module of an elliptic curve (it will probably not even be Hodge–Tate, for example). The trick that Wiles used was to not look at such a big ring as R^{univ}, but to look at more stringent deformation problems, such as deforming $\bar{\rho}$ to representations which came from finite flat group schemes over $\mathbf{Z}_p$. This more restricted space of deformations is represented by a smaller deformation ring $R^\flat$, a quotient of R^{univ}, and it is rings such as $R^\flat$ that Wiles could work with (the relevant computations in this case were done in Ravi Ramakrishna's thesis [Ram93]).

In trying to generalise this idea we run into a fundamental problem. The kind of deformation problems that one might want to look at are problems of the form "ρ that become finite and flat when restricted to $\mathrm{Gal}(\overline{\mathbf{Q}}_p/K)$ for K this fixed finite extension of $\mathbf{Q}_p$". However, for K a wildly ramified extension of $\mathbf{Q}_p$ the linear algebra methods alluded to earlier on become much more complex,

and indeed at this point historically there was no theorem classifying finite
flat group schemes over the integers of such p-adic fields which was concrete
enough to enable people to check that the resulting deformation problems were
representable, and represented by rings whose tangent spaces were sufficiently
small enough to enable the methods to work.

A great new idea, however, was introduced in the paper [CDT99]. Instead
of trying to write down a complicated deformation problem that made sense
for all Artin local rings and then to analyse the resulting representing ring,
Conrad, Diamond and Taylor construct "deformation rings" in the following
manner. First, they consider deformations $\rho\colon \mathrm{Gal}(\overline{\mathbf{Q}}_p/\mathbf{Q}_p) \to \mathrm{GL}_2(O)$ of $\overline{\rho}$ in
the case that O is the integers of a finite extension of $\mathbf{Q}_p$. In this special set-
ting there is a lot of extra theory available: one can ask if ρ is Hodge–Tate,
de Rham, potentially semistable, crystalline and so on (these words *do not
make sense* when applied to a general deformation $\rho\colon \mathrm{Gal}(\overline{\mathbf{Q}}_p/\mathbf{Q}_p) \to \mathrm{GL}_2(A)$
of $\overline{\rho}$, they only make sense when applied to a deformation to $\mathrm{GL}_2(O)$), and
furthermore if ρ is potentially semistable then the associated Fontaine module $D_{\mathrm{pst}}(\rho)$ is a 2-dimensional vector space with an action of the inertia subgroup of $\mathrm{Gal}(\overline{\mathbf{Q}}_p/\mathbf{Q}_p)$ which factors through a finite quotient. This finite image 2-dimensional representation of inertia is called the *type* of the potentially
semistable representation ρ, and so we can fix a type τ and then ask that a
deformation $\rho\colon \mathrm{Gal}(\overline{\mathbf{Q}}_p/\mathbf{Q}_p) \to \mathrm{GL}_2(O)$ be potentially semistable of a given
type.

Again we stress that this notion of being potentially semistable of a given
type certainly does *not* make sense for a deformation of $\overline{\rho}$ to an arbitrary Artin
local $W(k)$-algebra, so in particular this notion is *not* a deformation problem
and we cannot speak of its representability. One of the insights of [CDT99]
however, is that we can construct a "universal ring" for this problem anyway! Here is the trick, which is really rather simple. We have $\overline{\rho}$ and its universal formal deformation ρ^{univ} to $GL_2(R^{\mathrm{univ}})$. Now let us consider all maps
$s\colon R^{\mathrm{univ}} \to O$, where O is as above. Given such a map s, we can compose ρ^{univ}
with s to get a map $\rho_s\colon \mathrm{Gal}(\overline{\mathbf{Q}}_p/\mathbf{Q}_p) \to \mathrm{GL}_2(O)$. Let us say that the kernel of s
is *of type τ* if ρ_s is of type τ, and if furthermore ρ_s is potentially Barsotti–Tate
(that is, comes from a p-divisible group over the integers of a finite extension
of $\mathbf{Q}_p$) and has determinant equal to the cyclotomic character. A good example
of a potentially Barsotti–Tate representation is the representation coming from
the Tate module of an elliptic curve with potentially good reduction at p, and
such things will give rise to points of type τ for an appropriate choice of τ.

Let R_τ denote the quotient of R^{univ} by the intersection of all the prime ideals
of R^{univ} which are of type τ (with the convention that $R_\tau = 0$ if there are no
such prime ideals). Geometrically, what is happening is that the kernel of s is a

prime ideal and hence a point in $\mathrm{Spec}(R^{\mathrm{univ}})$, and we are considering the closed subscheme of $\mathrm{Spec}(R^{\mathrm{univ}})$ obtained as the Zariski-closure of all the prime ideals of type τ. So, whilst R_τ does not represent the moduli problem of being "of type τ" (because this is not even a moduli problem, as mentioned above), it is a very natural candidate for a ring to look at if one wants to consider deformations of type τ. It also raises the question as to whether the set of points which are of type τ actually form a closed set in, say, the rigid space generic fibre of R^{univ}. If they were to not form a closed set then R_τ would have quotients corresponding to points which were not of type τ, but which were "close" to being of type τ (more precisely, whose reductions modulo p^n were also reductions of type τ representations). The paper [CDT99] calls the points in the closure "weakly of type τ" and conjectures that being weakly of type τ is equivalent to being of type τ. This conjecture was proved not long afterwards for tame types by David Savitt in [Sav05].

Now of course, one hopes that for certain types τ, the corresponding rings R_τ are small enough for the Taylor–Wiles method to work (subject to the restriction that $p > 2$ and that $\overline{\rho}$ is absolutely irreducible even when restricted to the absolute Galois group of $\mathbf{Q}(\sqrt{(-1)^{(p-1)/2}p})$, an assumption needed to make the Taylor–Wiles machine work, and big enough to capture some new elliptic curves). Even though the definition of R_τ is in some sense a little convoluted, one can still hope to write down a surjection $W(k)[[t]] \to R_\tau$ in some cases (and thus control the tangent space of R_τ), for example by writing down a deformation problem which is known to be representable by a ring isomorphic to $W(k)[[t]]$, and showing that it contains all the points of type τ (geometrically, we are writing down a closed subset of $\mathrm{Spec}(R^{\mathrm{univ}})$ with sufficiently small tangent space, checking it contains all the points of type τ and concluding that it contains all of $\mathrm{Spec}(R_\tau)$). The problem with such a strategy is that it requires a good understanding of finite flat group schemes over the integers of the p-adic field K corresponding to the kernel of τ. In 1998 the only fields for which enough was known were those extensions K of $\mathbf{Q}_p$ which were tamely ramified. For such extensions, some explicit calculations were done in [CDT99] at the primes 3 and 5, where certain explicit R_τ were checked to have small enough tangent space. There is a general modularity lifting theorem announced in [CDT99] but it includes, in the non-ordinary case, an assumption the statement that R_τ is small enough for the method to work, and this is difficult to check in practice, so the result has limited applicability. However the authors did manage to check this assumption in several explicit cases when $p \in \{3, 5\}$, and deduced the following.

Theorem 6 *If $E/\mathbf{Q}$ is an elliptic curve which becomes semistable at 3 over a tamely ramified extension of $\mathbf{Q}_3$, then E is modular.*

This is the main theorem of [CDT99] (see the second page of loc. cit.). The proof is as follows: if $E[3]$ is irreducible when restricted to the absolute Galois group of $\mathbf{Q}(\sqrt{-3})$ then they verify by an explicit calculation that either E is semistable at 3, or some appropriate R_τ is small enough, and in either case this is enough. If $E[5]$ is irreducible when restricted to the absolute Galois group of $\mathbf{Q}(\sqrt{-5})$ then $E[5]$ can be checked to be modular via the 3–5 trick, and E can be proven modular as a consequence, although again the argument relies on computing enough about an explicit R_τ to check that it is small enough. Finally Noam Elkies checked for the authors that the number of j-invariants of elliptic curves over $\mathbf{Q}$ for which neither assertion holds is finite and worked them out explicitly; each j-invariant was individually checked to be modular.

At around the same time, Christophe Breuil had proven the breakthrough theorem [Bre00], giving a "linear algebra" description of the category of finite flat group schemes over the integers of an *arbitrary* p-adic field. Armed with this, Conrad, Diamond and Taylor knew that there was a chance that further calculations of the sort done in [CDT99] had a chance of proving the full Taniyama-Shimura conjecture. The main problem was that the rings R_τ were expected to be small enough for quite a large class of tame types τ, but were rarely expected to be small enough if τ was wild. After much study, Breuil, Conrad, Diamond and Taylor found a list of triples $(p, \overline{\rho}, \tau)$ for which R_τ could be proved to be small enough (p was always 3 in this list, and in one extreme case they had to use a mild generalisation of a type called an "extended type" in a case where R_τ was just too big; the extended type cut it down enough), and this list and the 3–5 trick was enough to prove the following.

Theorem 7 (Breuil, Conrad, Diamond, Taylor (2001)) *Any elliptic curve $E/\mathbf{Q}$ is modular.*

5 Kisin's modularity lifting theorems

In this section we briefly mention some important work of Kisin that takes the ideas above much further.

As we have just explained, the Breuil–Conrad–Diamond–Taylor strategy for proving a modularity lifting theorem was to write down subtle local conditions at p which were representable by a ring which was "not too big" (that is, its tangent space is at most 1-dimensional). The main problem with this approach was that the rings that this method needs to use in cases where the representation is coming from a curve of large conductor at p are (a) difficult to control, and (b) very rarely small enough in practice. The authors of [BCDT01] only just got away with proving modularity of all elliptic curves because of some coincidences specific to the prime 3, where the rings turned out to be computable using Breuil's ideas, and just manageable enough for the method to work. These calculations inspired conjectures of Breuil and Mézard ([BM02]) relating an invariant of R_τ (the Hilbert–Samuel multiplicity of the mod p reduction of this ring) to a representation-theoretic invariant (which is much easier to compute).

Kisin in the breakthrough paper [Kis09] (note that this paper was published in 2009 but the preprint had been available since 2004) gave a revolutionary new way to approach the problem of proving modularity lifting theorems. Kisin realised that rather than doing the commutative algebra in the world of $\mathbf{Z}_p$-algebras, one could instead just carry around the awkward rings R_τ introduced in [CDT99], and instead do all the dimension-counting in the world of R_τ-algebras (that is, count relative dimensions instead). This insight turns out to seriously reduce the amount of information one needs about R_τ; rather than it having to have a 1-dimensional tangent space, it now basically only needs to be an integral domain of Krull dimension 2. In fact one can get away with even less (which is good because R_τ is not always an integral domain); one can even argue using only an irreducible component of $\mathrm{Spec}(R_\tau[1/p])$, as long as one can check that the deformations one is interested in live on this component.

There is one problem inherent in this method, as it stands: the resulting modularity lifting theorems have a form containing a condition which might be tough to verify in practice. For example, they might say something like this: "Say $\overline{\rho}$ is modular, coming from a modular form f. Say ρ lifts $\overline{\rho}$. Assume furthermore that ρ_f and ρ both correspond to points on the same component of some $\mathrm{Spec}(R_\tau[1/p])$. Then ρ is modular." The problem here is that one now needs either to be able to check which component various deformations of $\overline{\rho}$ are on, or to be able somehow to jump between components (more precisely, one needs to prove theorems of the form "if $\overline{\rho}$ is modular coming from some modular form, then it is modular coming from some modular form whose associated local Galois representation lies on a given component of $\mathrm{Spec}(R_\tau[1/p])$". Kisin managed to prove that certain R_τ only had one component, and others had two components but that sometimes one could move from one to the other, and as a result of these "component-hopping" tricks ended up proving the following much cleaner theorem ([Kis09]).

Theorem 8 (Kisin) *Let $p > 2$ be a prime, let ρ be a 2-dimensional p-adic* *representation of* $\mathrm{Gal}(\overline{\mathbf{Q}}/\mathbf{Q})$ *unramified outside a finite set of primes, with reduction* $\overline{\rho}$, *and assume that* $\overline{\rho}$ *is modular and* $\overline{\rho}|\,\mathrm{Gal}(\overline{\mathbf{Q}}/K)$ *is absolutely irreducible, where* $K = \mathbf{Q}(\sqrt{(-1)^{(p-1)/2}p})$. *Assume furthermore that* ρ *is potentially Barsotti–Tate and has determinant equal to a finite order character times the cyclotomic character. Then* ρ *is modular.*

Note that we do not make any assumption on the type of ρ; this is why the theorem is so strong. This result gives another proof of the modularity of all elliptic curves, because one can argue at 3 and 5 as in [CDT99] and [BCDT01] but is spared the hard computations of R_τ in [BCDT01]: the point is that Kisin's machine can often deal with them even if their tangent space has dimension greater than one by doing the commutative algebra in this different and more powerful way. These arguments ultimately led to a proof of the Breuil–Mézard conjectures: see for example Kisin's recent ICM talk.

The next part of the story in the case of 2-dimensional representations of $\mathrm{Gal}(\overline{\mathbf{Q}}/\mathbf{Q})$ would be the amazing work of Khare and Wintenberger ([KW09a], [KW09b]), proving the following result.

Theorem 9 (Khare–Wintenberger) *Serre's conjecture (Conjecture 3) is true.*

We have to stop somewhere however, so simply refer the interested reader to the very readable papers [KW09a] and [Kha07] for an overview of the proof of this breakthrough result.

The work of Khare and Wintenberger means that nowadays we do not have to rely on the Langlands–Tunnell theorem to "get us going", and indeed we now get two more proofs of Fermat's Last theorem and of the modularity of all elliptic curves: firstly, given an elliptic curve E, we can just choose a random large prime, apply Khare–Wintenberger to $E[p]$ and then apply the theorem of Kisin above. Secondly, given an elliptic curve, we can apply the Khare–Wintenberger theorem to $E[p]$ for all p at once, and then use known results about level optimisation in Serre's conjecture to conclude again that E is modular. In particular we get a proof of FLT that avoids non-Galois cubic base change. However it seems to the author that things like cyclic base change and the Jacquet–Langlands theorem will still be essentially used in this proof, and hence even now it seems that to understand a full proof of FLT one still needs to understand both a huge amount of algebraic geometry and algebra, and also a lot of hard analysis.

6 Generalisations to totally real fields

So far we have restricted our discussion to modularity lifting theorems that applied to representations of the absolute Galois group of $\mathbf{Q}$. It has long been realised that even if one is mainly interested in these sorts of questions over $\mathbf{Q}$, it is definitely worthwhile to prove as much as one can for a general totally real field, because then one can use base change tricks (the proofs of which are in [Lan80] and use a lot of hard analysis) to get more information about the situation over $\mathbf{Q}$. One of the first examples of this phenomenon, historically, was the result in §0.8 of [Car83], where Carayol proves that the conductor of a modular elliptic curve over $\mathbf{Q}$ was equal to the level of the newform giving rise to the curve – even though this is a statement about forms over $\mathbf{Q}$, the proof uses Hilbert modular forms over totally real fields.

Some of what we have said above goes through to the totally real setting. Let us summarise the current state of play. We fix a totally real number field F. The role of modular forms in the previous sections is now played by Hilbert modular forms; to a Hilbert modular eigenform there is an associated 2-dimensional p-adic representation of the absolute Galois group of F, and this representation is totally odd, in the sense that the determinant of $\rho(c)$ is -1 for all complex conjugations c (there is more than one conjugacy class of such things if $F \neq \mathbf{Q}$, corresponding to the embeddings $F \rightarrow \mathbf{R}$). So formally the situation is quite similar to the case of $F = \mathbf{Q}$. "Under the hood" there are some subtle differences, because there is more than one analogue of the theory of modular curves in this setting (Hilbert modular varieties and Shimura curves, both of which play a role, as do certain 0-dimensional "Shimura varieties"), but we will not go any further into these issues; the point is that one can *formulate* the notion of modularity, and hence of modularity lifting theorems in this setting. But can one prove anything? One thing we certainly cannot prove, at the time of writing, is the following.

Conjecture 10 ("Serre")   Any continuous totally odd irreducible representation $\overline{\rho}\colon \operatorname{Gal}(\overline{F}/F) \rightarrow \operatorname{GL}_2(\overline{\mathbf{F}}_p)$ is modular.

Serre did not (as far as we know) formulate his conjecture in this generality, but it has become part of the folklore and his name seems now to be attached to it. We mention this conjecture because of its importance in the theory. If one could prove this sort of conjecture then modularity of all elliptic curves over all totally real fields would follow.

Just as in the case of $F = \mathbf{Q}$, the conjecture can be refined – for example one can predict the weight and level of a modular form that should give rise to $\overline{\rho}$, as Serre did for $F = \mathbf{Q}$ in [Ser87]. These refinements, in the case of the weight, can be rather subtle: we refer the reader to [BDJ10] when F is unramified at p, and to work of Michael Schein (for example [Sch08]) and Gee (conjecture 4.2.1 of [Gee]).

Once these refinements are made, one can ask two types of questions. The first is of the form "given $\bar{\rho}$, is it modular?". These sorts of questions seem to be wide open for a general totally real field. The second type is of the form "given $\bar{\rho}$ which is assumed modular, is it modular of the conjectured weight and level?". This is the sort of question which was answered by Ribet (the level: see [Rib90]) and Edixhoven (the weight: see [Edi92]), following work of many many others, for $F = \mathbf{Q}$. Much progress has also been made in the totally real case. For example see work of Jarvis [Jar99] and Rajaei [Raj01] on the level, and Gee [Gee07] on the weight, so the situation for Serre's conjecture on Hilbert modular forms now is becoming basically the same as it was in the classical case before Khare–Wintenberger: various forms of the conjecture are known to be equivalent, but all are open.

Given that there is a notion of modularity, one can formulate modularity lifting conjectures in this setting. But what can one prove? The first serious results in this setting were produced by Skinner and Wiles in [SW99] and [SW01], which applied in the setting of *ordinary* representations (that is, basically, to representations ρ which were upper triangular when restricted to a decomposition group for each prime above p). Here is an example of a modularity lifting theorem that they prove ([SW01] and correction in [Ski]).

Theorem 11 (Skinner–Wiles, 2001) *Suppose p is an odd prime and F is a totally real field. Suppose $\rho \colon \operatorname{Gal}(\overline{F}/F) \to \operatorname{GL}_2(\overline{\mathbf{Q}}_p)$ is continuous, irreducible, and unramified outside a finite set of places. Suppose that $\det(\rho)$ is a finite order character times some positive integer power of the cyclotomic character, that $\rho|D_{\mathfrak{p}}$ is upper triangular with an unramified quotient, for all $\mathfrak{p}|p$, and that the two characters on the diagonal are distinct modulo p. Finally, suppose that $\bar{\rho}|G_{F(\zeta_p)}$ is absolutely irreducible, and that $\bar{\rho}$ is modular, coming from an ordinary Hilbert modular form f of parallel weight such that, for all $\mathfrak{p}|p$, the unramified quotients of $\rho_f|D_{\mathfrak{p}}$ and $\rho|D_{\mathfrak{p}}$ are congruent mod p. Then ρ is modular.*

Note that Skinner and Wiles show that if $\bar{\rho}$ has an ordinary modular lift, then many of its ordinary lifts are modular. Their technique is rather more involved than the usual numerical criterion argument – they make crucial use of deformations to characteristic p rings, and in fact do not show that the natural map $R \to T$ is an isomorphism, using base change techniques to reduce to a case where they can prove that it is a surjection with nilpotent kernel.

Kisin's work on the rings R_τ of the previous section all generalised to the totally real setting, enabling Kisin to prove some stronger modularity lifting theorems which were not confined to the ordinary case. Kisin's original work required p to be totally split in F, but Gee proved something in the general case. We state Gee's theorem below.

Theorem 12 (Gee [Gee06], [Gee09]) *Suppose $p > 2$, F is totally real, and ρ is a continuous potentially Barsotti–Tate 2-dimensional p-adic Galois representation of the absolute Galois group of F, unramified outside a finite set of primes, and with determinant equal to a finite order character times the cyclotomic character. Suppose that its reduction $\overline{\rho}$ is modular, coming from a Hilbert modular form f, and suppose that for all $v|p$, if ρ is potentially ordinary at v then so is ρ_f. Finally suppose $\overline{\rho}$ is irreducible when restricted to the absolute Galois group of $F(\zeta_p)$, and if $p = 5$ and the projective image of $\overline{\rho}$ is isomorphic to $\mathrm{PGL}_2(\mathbf{F}_5)$ then assume furthermore than $[F(\zeta_5) : F] = 4$.*
   Then ρ is modular.

Note that, in contrast to Kisin's result Theorem 8, in this generality "component hopping" is not as easy, and the assumption in this theorem that if ρ is potentially ordinary then ρ_f is too, are precisely assumptions ensuring that ρ and ρ_f are giving points on the same components of the relevant spaces $\mathrm{Spec}(R_\tau[1/p])$.

It is also worth remarking here that the Langlands–Tunnell theorem, Theorem 4, is true for totally odd irreducible representations of any totally real field to $\mathrm{GL}_2(\mathbf{F}_3)$, so we can start to put together what we have to prove some modularity theorems for elliptic curves. Note that Gee's result above has the delicate assumption that not only is $\overline{\rho}$ modular, but it is modular coming from a Hilbert modular form whose behaviour at primes dividing p is similar to that of ρ. However Kisin's "component hopping" can be done if p is totally split in F, and Kisin can, using basically the same methods, generalise his Theorem 8 to the totally real case if p is totally split, giving the following powerful modularity result:

Theorem 13 (Kisin [Kis07]) *Let F be a totally real field in which a prime $p > 2$ is totally split, let ρ be a continuous irreducible 2-dimensional representation of $\mathrm{Gal}(\overline{F}/F)$, unramified outside a finite set of primes, and potentially Barsotti–Tate at the primes above p. Suppose that $\det(\rho)$ is a finite order character times the cyclotomic character, that $\overline{\rho}$ is modular coming from a Hilbert modular form of parallel weight 2, and that $\overline{\rho}|G_{F(\zeta_p)}$ is absolutely irreducible. Then ρ is modular.*

As a consequence, if F is a totally real field in which $p = 3$ is totally split, and if E/F is an elliptic curve with $E[3]|F_{F(\zeta_3)}$ absolutely irreducible, then E is modular.

Of course *all* elliptic curves over F are conjectured to be modular, but this conjecture still remains inaccessible. If one were to attempt to mimic the strategy of proof in the case $F = \mathbf{Q}$ then one problem would be that for a general totally real field, there may be infinitely many elliptic curves with subgroups of order 15 defined over F, and how can one deal with such curves? There are infinitely many, so one cannot knock them off one by one as Elkies did. Their mod 3 and mod 5 Galois representations are globally reducible, and the best modularity lifting theorems we have in this situation are in [SW99], where various hypotheses on F are needed (for example $F/\mathbf{Q}$ has to be abelian in Theorem A of [SW99]). On the other hand, because Serre's conjecture is still open for totally real fields one cannot use the p-torsion for any prime $p \geq 7$ either, in general. It is not clear how to proceed in this situation!

7 Potential modularity pre-Kisin and the $\mathfrak{p}$–λ trick

We have been daydreaming in the previous section about the possibilities of proving that a general elliptic curve E over a general totally real field F is modular, and observing that we are not ready to prove this result yet. Modularity is a wonderful thing to know for an elliptic curve; for example, the Birch–Swinnerton-Dyer conjecture is a statement about the behaviour of the L-function of an elliptic curve at the point $s = 1$, but the L-function of an elliptic curve is defined by an infinite sum which converges for $\text{Re}(s) > 3/2$, and it is only a conjecture that this L-function has an analytic continuation to the entire complex plane. One very natural way of analytically continuing the L-function is to prove that the curve is modular, because modular forms have nice analytic properties and the analytic continuation of their L-functions is well-known.

For an elliptic curve over a general number field though, the L-function is currently not known to have an analytic continuation, or even a meromorphic continuation! However, perhaps surprisingly, it turns out that the results above, plus one more good new idea due to Taylor, enabled him to prove *meromorphic* continuation for a huge class of elliptic curves over totally real fields, and the ideas have now been pushed sufficiently far to show that the L-function of every elliptic curve over every totally real field can be meromorphically continued. We want to say something about how this all happened.

The starting point was Taylor's paper [Tay02]. The basic idea behind this breakthrough paper is surprisingly easy to explain! Recall first Lemma 5, the

3–5 trick. We have an elliptic curve $E/\mathbf{Q}$ with $E[5]$ irreducible, and we want to prove that $E[5]$ is modular. We do this by writing down a second elliptic curve $A/\mathbf{Q}$ with $A[3]$ irreducible and $A[5] \cong E[5]$. Then the trick, broadly, was that $A[3]$ is modular by the Langlands–Tunnell theorem, so A is modular by a modularity lifting theorem, so $A[5]$ is modular, so $E[5]$ is modular. The proof crucially uses the fact that the genus of the modular curve $X(5)$ is zero so clearly does not generalise to much higher numbers.

However, if we are allowed to be more flexible with our base field, then this trick generalises very naturally and easily. Let us say that we have an elliptic curve E over a totally real field F, and we want to prove E is potentially modular (that is, E becomes modular over a finite extension field F' of F, also assumed totally real). Here is a strategy. Say p is a large prime such that $E[p]$ is irreducible. Let us write down a random odd 2-dimensional mod ℓ Galois representation $\rho_\ell : \mathrm{Gal}(\overline{F}/F) \to \mathrm{GL}_2(\overline{\mathbf{F}}_\ell)$ which is induced from a character; because this representation is induced it is known to be modular. Now let us consider the moduli space parametrising elliptic curves A equipped with

(1) an isomorphism $A[p] \cong E[p]$,
(2) an isomorphism $A[\ell] \cong \rho_\ell$.

This moduli problem will be represented by some modular curve, whose connected components will be twists of $X(p\ell)$ and hence, if p and ℓ are large, will typically have large genus. However, such a curve may well still have lots of rational points, as long as I am allowed to look for such things over an arbitrary finite extension F' of F! So here is the plan: first, consider this moduli problem. Second, find a point on this moduli space defined over F', for F' some finite extension of F. Next, ensure that F' is totally real, and that our modularity lifting theorems are robust enough to apply in the two situations we'll need them. More precisely, we need one modularity lifting theorem of the form "ρ_ℓ is modular over F' and hence A/F' is modular", so $A[p] = E[p]$ is modular, and then another one which says "$E[p]$ is modular over F', and hence E is modular".

In 2001, our knowledge of modularity lifting theorems was poorer than it is now (because, for example, this was the era of [BCDT01] and before Kisin's work on local deformation problems), so this idea would not run as far as it naturally wanted to. Let us sketch some of the issues that arise here. Firstly, if our moduli space has no real points at all for some embedding $F \to \mathbf{R}$ then we cannot find a point over *any* totally real extension. So we need to check our moduli problem has real points. Secondly, the modularity lifting theorems available to Taylor in the totally real case were those of Skinner and Wiles so only applied in the ordinary setting (which is not much of a restriction because the curve will be ordinary at infinitely many places) but furthermore only applied in the "distinguished" setting (that is, the characters on the diagonal of the mod p local representation have to be distinct), so the completions of F' at the primes above p had better not be too big, and similarly the completions of F' at the primes above ℓ must not be too big either. This results in more local conditions on F', and so we need to ensure two things: firstly that our moduli problems have points defined over reasonably small extensions of $F_\mathfrak{p}$ and F_λ for $\mathfrak{p} \mid p$ and $\lambda \mid \ell$, and secondly that there is no local-global obstruction to the existence of a well-behaved F'-point (that is: given that our moduli problem has points over certain "small" local fields, we need to ensure it has a point over a totally real number field whose completions at the primes above p and ℓ are equally "small"). Fortunately, such a local-global theorem was already a result of Moret-Bailly (see [MB89]).

Theorem 14 *Let K be a number field and let S be a finite set of places of K. Let X be a geometrically irreducible smooth quasi-projective variety over K. Let L run through the finite field extensions of K in which all the primes of S split completely. Then the union of $X(L)$, as L runs through these extensions, is Zariski-dense in X.*

Let us see how our plan looks so far. The idea now is that given an elliptic curve E/F, we find a prime p such that the Skinner–Wiles theorem applies to $E[p]$ (so p is an ordinary prime for which $E[p]$ locally at p has two distinct characters on the diagonal) and then write down a random odd prime ℓ and an induced representation ρ_ℓ of $\mathrm{Gal}(\overline{F}/F)$. Because Moret-Bailly's theorem needs X geometrically irreducible, let us ensure that ρ_ℓ has cyclotomic determinant, and fix an alternating pairing on the underlying vector space (to be thought of as a Weil pairing). Consider now the moduli space of elliptic curves A equipped with isomorphisms $A[p] \cong E[p]$ and $A[l] \cong \rho_\ell$ both of which preserve the Weil pairing. If we can find points on this curve defined over the completions of F at all primes above p, ℓ and ∞, then we might hope to conclude. But there are obstacles to finding such points. For example, if $\lambda \mid \ell$ is a prime of F and $A[p]$ is unramified at λ, and b_λ is the number of points on A mod λ, then one can read off b_λ mod p from $A[p]$, and one also has the Weil bounds on the integer b_λ, and these two constraints on b_λ might not be simultaneously satisfiable. Because of obstructions of this form, Taylor finds it easier to work not moduli spaces of elliptic curves over F, but with moduli spaces of so-called Hilbert–Blumenthal abelian varieties over F, that is, of higher-dimensional abelian varieties (say g-dimensional) equipped with a certain kind of polarization and endomorphisms by the integers in a second totally real field M of degree g over $\mathbf{Q}$. As is becoming clear, this generalisation of the 3–5 trick is one of these ideas where the great inspiration now has to be offset by the large amount of perspiration that has to get the idea to work. On the other hand, the idea is certainly not restricted to elliptic curves E/F, and applies to a large class of ordinary Galois representations. Taylor managed to put everything together, even in the "pre-Kisin" modularity world, and managed in [Tay02] to prove that a wide class of ordinary 2-dimensional Galois representations of $\mathrm{Gal}(\overline{\mathbf{Q}}/\mathbf{Q})$ were potentially modular. Let us state his result here.

Theorem 15 (Taylor)   *Let p be an odd prime, let $\rho\colon \mathrm{Gal}(\overline{\mathbf{Q}}/\mathbf{Q}) \to \mathrm{GL}_2(\overline{\mathbf{Q}}_p)$ be a continuous odd irreducible representation unramified outside a finite set of primes, and assume*

$$\rho|D_p = \begin{pmatrix} \chi^n\psi_1 & * \\ 0 & \psi_2 \end{pmatrix}$$

with χ the cyclotomic character, $n \geq 1$, and ψ_1 and ψ_2 two finitely ramified characters, such that the mod p reductions of $\chi^n\psi_1$ and ψ_2 are not equal on the inertia subgroup of D_p. Then ρ becomes modular over some totally real number field.

As we said, the reason for the ordinarity assumption is that the result was proved in 2000 before the more recent breakthroughs in modularity lifting theorems. Taylor went on in 2001 in [Tay06] to prove an analogous theorem in the low weight crystalline case.

8 Potential modularity after Kisin

In this last section we put together Kisin's modularity lifting theorem methods with Taylor's potential modularity ideas. Together, the methods can be used to prove much stronger results such as the following.

Theorem 16   *Let E/F be an elliptic curve over a totally real field. Then there is some totally real extension F'/F such that E/F' is modular.*

In particular, the L-function of E has meromorphic continuation to the whole complex plane. It is difficult to give a precise attribution to this theorem – the history is a little complicated. Perhaps soon after Taylor saw Kisin's work on local deformation rings he realised that this theorem was accessible, but he could also see that perhaps the Sato–Tate conjecture was accessible too, and turned his attention to this problem instead. Whatever the history, it seems that it was clear to the experts around 2006 and possibly even earlier that the theorem above was accessible. The first published proof that we are aware of is in the appendix by Wintenberger to [Nek], published in 2009. Much has happened recently in higher-dimensional generalisations of modularity lifting theorems – so-called automorphy lifting theorems, proving that various n-dimensional Galois representations are automorphic or potentially automorphic, and as a result one could also point to, for example, Theorem 8.7 of [BLGHT], where a far stronger (n-dimensional) result is proven from which the theorem follows. Another place to read about the details of the proof of this potential modularity theorem would be the survey article of Snowden [Sno], who sticks to the 2-dimensional situation and does a very good job of explaining what is needed. Snowden fills in various gaps in the literature in order to make his paper relatively self-contained modulo some key ideas of Kisin; if the reader looks at Snowden's paper then they will see that the crucial ideas are basically due to Taylor and Kisin. One of the problems that needs to be solved is how to do the "component-hopping": a general modularity lifting theorem might look like "if ρ_1 is modular and ρ_2 is congruent to ρ_1, and ρ_1, ρ_2 give points on the same components of certain local deformation spaces, then ρ_2 is modular". As a result one needs very fine control on the abelian variety that is employed to do the $\mathfrak{p}{-}\lambda$ trick; however, this fine control can be obtained by Moret-Bailly's theorem. The interested reader should read these references, each of which gives the details of the argument.

9 Some final remarks

This survey has to stop somewhere so we just thought we would mention a few things that we have not touched on. In the 2-dimensional case there is of course the work of Khare and Wintenberger, which we have mentioned several times but not really touched on more seriously. As mentioned already, Khare and Wintenberger have done a good job of summarising their strategy in their papers, and the interested reader should start there. Another major area which we have left completely untouched is the higher-dimensional $R = T$ theorems in the literature, concerned with modularity of n-dimensional Galois representations. Here one uses automorphic forms on certain unitary groups to construct Galois representations, and new techniques are needed. The first big result is [CHT08], proving an $R = T$ result at minimal level. After Kisin's rings are factored into the equation and issues with components are resolved one can prove much stronger theorems; the state of the art at the time of writing seems to be the preprint [BLGGT], which is again very clearly written and might serve as a good introduction to the area. We apologise for not saying more about these recent fabulous works.

References

[BCDT01] Christophe Breuil, Brian Conrad, Fred Diamond, and Richard Taylor, *On the modularity of elliptic curves over* **Q**: *wild 3-adic exercises*, J. Amer. Math. Soc. **14** (2001), no. 4, 843–939 (electronic). MR1839918(2002d: 11058)

[BDJ10] Kevin Buzzard, Fred Diamond, and Frazer Jarvis, *On Serre's conjecture for mod ℓ galois representations over totally real fields*, Duke Math. J. **155** (2010), no. 1, 105–161.

[BLGGT] Tom Barnet-Lamb, Toby Gee, David Geraghty, and Richard Taylor, *Potential automorphy and change of weight*, Preprint.

[BLGHT] Tom Barnet-Lamb, David Geraghty, Michael Harris, and Richard Taylor, *A family of Calabi-Yau varieties and potential automorphy ii*, to appear in P.R.I.M.S.

[BM02] Christophe Breuil and Ariane Mézard, *Multiplicités modulaires et représentations de* $GL_2(\mathbf{Z}_p)$ *et de* $Gal(\overline{\mathbf{Q}}_p/\mathbf{Q}_p)$ *en* $l = p$, Duke Math. J. **115** (2002), no. 2, 205–310, With an appendix by Guy Henniart. MR1944572(2004i:11052)

[Bre00] Christophe Breuil, *Groupes p-divisibles, groupes finis et modules filtrés*, Ann. of Math. (2) **152** (2000), no. 2, 489–549. MR1804530(2001k:14087)

[Car83] Henri Carayol, *Sur les représentations l-adiques attachées aux formes modulaires de Hilbert*, C. R. Acad. Sci. Paris Sér. I Math. **296** (1983), no. 15, 629–632. MR705677(85e:11039)

[CDT99] Brian Conrad, Fred Diamond, and Richard Taylor, *Modularity of certain potentially Barsotti-Tate Galois representations*, J. Amer. Math. Soc. **12** (1999), no. 2, 521–567. MR1639612(99i:11037)

[CHT08] Laurent Clozel, Michael Harris, and Richard Taylor, *Automorphy for some l-adic lifts of automorphic mod l Galois representations*, Publ. Math. Inst. Hautes Études Sci. (2008), no. 108, 1–181, With Appendix A, summarizing unpublished work of Russ Mann, and Appendix B by Marie-France Vignéras. MR2470687(2010j:11082)

[DDT97] Henri Darmon, Fred Diamond, and Richard Taylor, *Fermat's last theorem*, Elliptic curves, modular forms & Fermat's last theorem (Hong Kong, 1993), Int. Press, Cambridge, MA, 1997, pp. 2–140. MR1605752(99d: 11067b)

[Dia96] Fred Diamond, *On deformation rings and Hecke rings*, Ann. of Math. (2) **144** (1996), no. 1, 137–166. MR1405946(97d:11172)

[Dia97] Fred Diamond, *The Taylor-Wiles construction and multiplicity one*, Invent. Math. **128** (1997), no. 2, 379–391. MR1440309(98c:11047)

[Dic] Mark Dickinson, *On the modularity of certain 2-adic Galois representations*, Duke Math. J. **109** (2001), no. 2, 319–382. MR1845182(2002k: 11079)

[Edi92] Bas Edixhoven, *The weight in Serre's conjectures on modular forms*, Invent. Math. **109** (1992), no. 3, 563–594. MR1176206(93h:11124)

[FL82] Jean-Marc Fontaine and Guy Laffaille, *Construction de représentations p-adiques*, Ann. Sci. École Norm. Sup. (4) **15** (1982), no. 4, 547–608 (1983). MR707328(85c:14028)

[Fon77] Jean-Marc Fontaine, *Groupes p-divisibles sur les corps locaux*, Société Mathématique de France, Paris, 1977, Astérisque, No. 47-48. MR0498610(58\#16699)

[Fuj] K. Fujiwara, *Deformation rings and Hecke algebras in the totally real case*, preprint available at http://arxiv.org/abs/math/0602606.

[Gee] Toby Gee, *Automorphic lifts of prescribed types*, to appear in Math Annalen.

[Gee06] ——, *A modularity lifting theorem for weight two Hilbert modular forms*, Math. Res. Lett. **13** (2006), no. 5-6, 805–811. MR2280776(2007m:11065)

[Gee07] ——, *Companion forms over totally real fields. II*, Duke Math. J. **136** (2007), no. 2, 275–284. MR2286631(2008e:11053)

[Gee09] ——, *Erratum—a modularity lifting theorem for weight two Hilbert modular forms [mr2280776]*, Math. Res. Lett. **16** (2009), no. 1, 57–58. MR2480560(2010c:11057)

[Jar99] Frazer Jarvis, *Mazur's principle for totally real fields of odd degree*, Compositio Math. **116** (1999), no. 1, 39–79. MR1669444(2001a:11081)

[JPSS81] Hervé Jacquet, Ilja I. Piatetski-Shapiro, and Joseph Shalika, *Relèvement cubique non normal*, C. R. Acad. Sci. Paris Sér. I Math. **292** (1981), no. 12, 567–571. MR615450(82i:10035)

[Kha07] Chandrashekhar Khare, *Serre's modularity conjecture: a survey of the level one case*, L-functions and Galois representations, London Math. Soc. Lecture Note Ser., vol. 320, Cambridge Univ. Press, Cambridge, 2007, pp. 270–299. MR2392357(2009g:11066)

[Kis07] Mark Kisin, *Modularity for some geometric Galois representations*, L-functions and Galois representations, London Math. Soc. Lecture Note Ser., vol. 320, Cambridge Univ. Press, Cambridge, 2007, With an appendix by Ofer Gabber, pp. 438–470. MR2392362(2009j:11086)

[Kis09] ——, *Moduli of finite flat group schemes, and modularity*, Ann. of Math. (2) **170** (2009), no. 3, 1085–1180. MR2600871

[KW09a] Chandrashekhar Khare and Jean-Pierre Wintenberger, *Serre's modularity conjecture. I*, Invent. Math. **178** (2009), no. 3, 485–504. MR2551763(2010k:11087)

[KW09b] ——, *Serre's modularity conjecture. II*, Invent. Math. **178** (2009), no. 3, 505–586. MR2551764(2010k:11088)

[Lan80] Robert P. Langlands, *Base change for* GL(2), Annals of Mathematics Studies, vol. 96, Princeton University Press, Princeton, N.J., 1980. MR574808(82a:10032)

[Maz89] B. Mazur, *Deforming Galois representations*, Galois groups over **Q** (Berkeley, CA, 1987), Math. Sci. Res. Inst. Publ., vol. 16, Springer, New York, 1989, pp. 385–437. MR1012172(90k:11057)

[MT] B. Mazur and J. Tilouine, *Reprsentations galoisiennes, diffrentielles de Khler et "conjectures principales"*, Inst. Hautes tudes Sci. Publ. Math. **71** (1990), 65–103. MR1079644(92e:11060)

[MB89] Laurent Moret-Bailly, *Groupes de Picard et problèmes de Skolem. I, II*, Ann. Sci. École Norm. Sup. (4) **22** (1989), no. 2, 161–179, 181–194. MR1005158(90i:11065)

[Nek] Jan Nekovář, *On the parity of ranks of Selmer groups. IV* (with an appendix by Jean-Pierre Wintenberger), Compos. Math. **145**, no. 6, 1351–1359. MR2575086(2010j:11106)

[Raj01] Ali Rajaei, *On the levels of mod l Hilbert modular forms*, J. Reine Angew. Math. **537** (2001), 33–65. MR1856257(2002i:11041)

[Ram93] Ravi Ramakrishna, *On a variation of Mazur's deformation functor*, Compositio Math. **87** (1993), no. 3, 269–286. MR1227448(94h:11054)

[Ray74] Michel Raynaud, *Schémas en groupes de type $(p, \ldots, p)$*, Bull. Soc. Math. France **102** (1974), 241–280. MR0419467(54\#7488)

[Rib90] K. A. Ribet, *On modular representations of $\mathrm{Gal}(\overline{\mathbf{Q}}/\mathbf{Q})$ arising from modular forms*, Invent. Math. **100** (1990), no. 2, 431–476. MR1047143(91g:11066)

[Sav05] David Savitt, *On a conjecture of Conrad, Diamond, and Taylor*, Duke Math. J. **128** (2005), no. 1, 141–197. MR2137952(2006c:11060)

[Sch08] Michael M. Schein, *Weights in Serre's conjecture for Hilbert modular forms: the ramified case*, Israel J. Math. **166** (2008), 369–391. MR2430440(2009e:11090)

[Ser87] Jean-Pierre Serre, *Sur les représentations modulaires de degré 2 de $\mathrm{Gal}(\overline{\mathbf{Q}}/\mathbf{Q})$*, Duke Math. J. **54** (1987), no. 1, 179–230. MR885783(88g:11022)

[Ski] Christopher Skinner, *Nearly ordinary deformations of residually dihedral representations (draft)*, Preprint.

[Sno] Andrew Snowden, *On two dimensional weight two odd representations of totally real fields*, preprint available at http://arxiv.org/abs/0905.4266.

[SW99] C. M. Skinner and A. J. Wiles, *Residually reducible representations and modular forms*, Inst. Hautes Études Sci. Publ. Math. (1999), no. 89, 5–126 (2000). MR1793414(2002b:11072)

[SW01] C. M. Skinner and Andrew J. Wiles, *Nearly ordinary deformations of irreducible residual representations*, Ann. Fac. Sci. Toulouse Math. (6) **10** (2001), no. 1, 185–215. MR1928993(2004b:11073)

[Tay02] Richard Taylor, *Remarks on a conjecture of Fontaine and Mazur*, J. Inst. Math. Jussieu **1** (2002), no. 1, 125–143. MR1954941(2004c:11082)

[Tay06] ——, *On the meromorphic continuation of degree two L-functions*, Doc. Math. (2006), no. Extra Vol., 729–779 (electronic). MR2290604(2008c:11154)

[Tun81] Jerrold Tunnell, *Artin's conjecture for representations of octahedral type*, Bull. Amer. Math. Soc. (N.S.) **5** (1981), no. 2, 173–175. MR621884(82j:12015)

[TW95] Richard Taylor and Andrew Wiles, *Ring-theoretic properties of certain Hecke algebras*, Ann. of Math. (2) **141** (1995), no. 3, 553–572. MR1333036(96d:11072)

[Wil95] Andrew Wiles, *Modular elliptic curves and Fermat's last theorem*, Ann. of Math. (2) **141** (1995), no. 3, 443–551. MR1333035(96d:11071)

Remarks on some locally $\mathbb{Q}_p$-analytic representations of $GL_2(F)$ in the crystalline case

Christophe Breuil

C.N.R.S. et Université Paris-Sud

1 Introduction and notations

Let p be a prime number, let F be a finite extension of $\mathbb{Q}_p$ and pick a finite extension E of $\mathbb{Q}_p$ such that the set of field embeddings of F into E has cardinality $[F : \mathbb{Q}_p]$. This note fits into the local p-adic Langlands programme, whose aim is to attach locally $\mathbb{Q}_p$-analytic or continuous p-adic representations of $GL_n(F)$ over E to n-dimensional p-adic representations of $\mathrm{Gal}(\overline{\mathbb{Q}_p}/F)$ over E where $\mathrm{Gal}(\overline{\mathbb{Q}_p}/F)$ is the absolute Galois group of F. One of the most important cases is when the Galois representation is crystalline with distinct Hodge–Tate weights. When $n = 2$ and $F = \mathbb{Q}_p$, we completely understand the $GL_2(\mathbb{Q}_p)$-representations both from a p-adic and from a locally analytic point of view ([1], [7], [6], [15]). When $n = 2$ but $F \neq \mathbb{Q}_p$, several complications occur, all more or less related to the fact that one has to deal with the "mixture" of several embeddings of the base field F into the coefficient field E. This note focusses on the locally analytic point of view when $n = 2$. To most of the 2-dimensional crystalline representations V of $\mathrm{Gal}(\overline{\mathbb{Q}_p}/F)$ over E with distinct Hodge–Tate weights for each embedding $F \hookrightarrow E$, we attach and study a locally $\mathbb{Q}_p$-analytic representation $\Pi(V)$ of $GL_2(F)$. Before we sum up the results of this note, let us mention right away that we *don't expect* $\Pi(V)$ to be the "complete" locally $\mathbb{Q}_p$-analytic representation associated to V when $F \neq \mathbb{Q}_p$, but only a subrepresentation of it. For instance, one can't recover V from $\Pi(V)$ in general when $F \neq \mathbb{Q}_p$. However, the study of $\Pi(V)$ is not hard and already reveals interesting features. Moreover it already seems a non-trivial task to prove that $\Pi(V) \otimes_E V$ occurs for instance inside the completed H^1 of Hilbert Shimura curves.

We now explain the main results of this note. Let f be the residual index of F. To any rank 2 filtered φ-module D (not necessarily weakly admissible) with distinct Hodge–Tate weights and such that φ^f has two distinct eigenval-

Non-abelian Fundamental Groups and Iwasawa Theory, eds. John Coates, Minhyong Kim, Florian Pop, Mohamed Saïdi and Peter Schneider. Published by Cambridge University Press. ©Cambridge University Press 2012.

ues, we first associate a locally $\mathbb{Q}_p$-analytic representation $\Pi(D)$ of $\mathrm{GL}_2(F)$. The underlying idea for the definition of $\Pi(D)$ is the following. Since we are in dimension 2 with distinct Hodge–Tate weights, the Hodge filtration on the rank 2 module D_F is just the datum of a rank 1 submodule. If this submodule is completely generic, then $\Pi(D)$ is just the amalgamated sum of two natural locally $\mathbb{Q}_p$-analytic parabolic inductions associated to D relative to their common locally algebraic vectors. But it can happen that the Hodge filtration is in a special position (with respect to the eigenvectors of φ^f). In that case, one replaces each locally $\mathbb{Q}_p$-analytic parabolic induction in the previous amalgamated sum by a certain direct sum of some of its subquotients (depending on the position of the Hodge filtration) so that the final representation has the same Jordan–Hölder constituents. One uses here results of Frommer and Schraen (and others) on the Jordan–Hölder filtration of locally $\mathbb{Q}_p$-analytic parabolic inductions of characters (which are themselves based on foundational results of Schneider, Teitelbaum and Morita).

Let $\mathrm{soc}_{\mathrm{GL}_2(F)}\,\Pi(D)$ be the direct sum of the topologically irreducible subrepresentations of $\Pi(D)$. We then prove the following statements, all giving evidence that $\Pi(D)$ is (a piece of) the right representation to consider.

(i) If $\mathrm{soc}_{\mathrm{GL}_2(F)}\,\Pi(D)$ has a p-adic norm which is invariant under the action of $\mathrm{GL}_2(F)$ (for instance if $\Pi(D)$ itself has such an invariant norm), then D is weakly admissible (§5).

(ii) If $\mathrm{soc}_{\mathrm{GL}_2(F)}\,\Pi(D)$ has a p-adic invariant norm, then the completion of $\mathrm{soc}_{\mathrm{GL}_2(F)}\,\Pi(D)$ with respect to this norm automatically contains a locally $\mathbb{Q}_p$-analytic representation $\Pi(D)^{\mathrm{Amice}}$ of $\mathrm{GL}_2(F)$ which is larger than $\mathrm{soc}_{\mathrm{GL}_2(F)}\,\Pi(D)$ and such that $\Pi(D)^{\mathrm{Amice}} \subsetneq \Pi(D)$ (§7).

(iii) If D is weakly admissible and corresponds to a reducible crystalline Galois representation $V = \begin{pmatrix} \chi_2\varepsilon & * \\ 0 & \chi_1 \end{pmatrix}$ (where ε is the p-adic cyclotomic character and χ_1, χ_2 are crystalline characters), then $\Pi(D)$ has a natural increasing filtration by $\mathrm{GL}_2(F)$-subrepresentations:

$$0 = \mathrm{Fil}^0\Pi(D) \subsetneq \mathrm{Fil}^1\Pi(D) \subsetneq \cdots \subsetneq \mathrm{Fil}^{[F:\mathbb{Q}_p]}\Pi(D) \subsetneq \mathrm{Fil}^{[F:\mathbb{Q}_p]+1}\Pi(D) = \Pi(D)$$

such that, when $\chi_1\chi_2^{-1} \notin \{1, \varepsilon^2\}$, the graded pieces

$$\Pi(D)_j := \mathrm{Fil}^{j+1}\Pi(D)/\mathrm{Fil}^j\Pi(D)$$

satisfy

$$\Pi(D)_0 = \left(\mathrm{Ind}_{B(F)}^{\mathrm{GL}_2(F)}\chi_1 \otimes \chi_2\right)^{\mathbb{Q}_p-\mathrm{an}},$$

$$\Pi(D)_{[F:\mathbb{Q}_p]} = \left(\mathrm{Ind}_{B(F)}^{\mathrm{GL}_2(F)}\chi_2\varepsilon \otimes \chi_1\varepsilon^{-1}\right)^{\mathbb{Q}_p-\mathrm{an}}$$

(where $B(F)$ is the upper parabolic) and such that $\Pi(D) \simeq \oplus_{j=0}^{[F:\mathbb{Q}_p]} \Pi(D)_j$ if and only if V is split (§9).

Statement (i) follows from an easy necessary condition for a parabolic induction which is locally analytic in some "directions" and locally algebraic in the others to admit a p-adic invariant norm. Statement (ii) is based on well-known techniques of Amice-Vélu and Vishik which give that, if a unitary Banach space representation of $GL_2(F)$ contains $soc_{GL_2(F)} \Pi(D)$, then p-adic analysis forces it to contain a larger representation $\Pi(D)^{\text{Amice}}$, which turns out to be a subrepresentation of $\Pi(D)$. Note that one knows examples of unitary Banach spaces representations of $GL_2(F)$ containing $soc_{GL_2(F)} \Pi(D)$ (e.g. completed cohomology groups when $soc_{GL_2(F)} \Pi(D)$ is the locally algebraic vectors of $\Pi(D)$). The results of (i) and (ii) might be (very) special cases of some of the results of [10], [11]. Finally, (iii) is consistent with results in characteristic p when F is unramified ([3]) giving evidence that the smooth representation(s) of $GL_2(F)$ in characteristic p corresponding to a reducible non-split (resp. split) 2-dimensional representation of $\text{Gal}(\overline{\mathbb{Q}_p}/F)$ should generically be a successive extension (resp. a direct sum) of $[F : \mathbb{Q}_p]$ irreducible representations, the first and the last being two principal series analogous to the two parabolic inductions in (iii) above. We refer to the body of the text for more detailed and more precise statements.

For any finite extension L of $\mathbb{Q}_p$, we denote by O_L its ring of integers, ϖ_L a uniformizer and $k_L := O_L/(\varpi_L)$ its residue field.

Throughout the text, F and E are two fixed finite extensions of $\mathbb{Q}_p$ such that the set S of field embeddings of F into E has cardinality $[F : \mathbb{Q}_p]$ (F is the base field, E the coefficient field). We denote by F_0 the maximal unramified subfield in F, $f := [F_0 : \mathbb{Q}_p]$, $e := [F : F_0]$, $q := p^f$ and S_0 the set of embeddings of F_0 into E. We let φ be the arithmetic Frobenius on F_0 inducing $x \mapsto x^p$ on k_F.

The p-adic valuation val_F on F or on E is normalized by $\text{val}_F(p) := [F : \mathbb{Q}_p]$ and we set $|x|_F := p^{-\text{val}_F(x)}$ if $x \in F$ or $x \in E$. If $\lambda \in E^\times$, we denote by:

$$\text{unr}_F(\lambda) : F^\times \to E^\times \tag{1}$$

the unramified character sending $x \in F^\times$ to $\lambda^{\text{val}_F(x)}$.

We normalize local class field theory such that uniformizers are sent to geometric Frobeniuses. We view without comment a character of $\text{Gal}(\overline{\mathbb{Q}_p}/F)$ as a character of $F^\times$. We denote by ε the p-adic cyclotomic character. It corresponds to the character of $F^\times$ given by $x \mapsto |x|_F \prod_{\sigma \in S} \sigma(x)$.

A p-adic norm on an E-vector space V is a function $\| \cdot \| : V \to \mathbb{R}_{\geq 0}$ such that $\|v\| = 0$ if and only if $v = 0$, $\|\lambda v\| = |\lambda|_F \|v\|$ ($\lambda \in E$, $v \in V$) and $\|v + w\| \leq \sup(\|v\|, \|w\|)$ ($v, w \in V$). A p-adic Banach space over E is an E-vector space endowed with a topology coming from a p-adic norm and such that the underlying metric space is complete. An invariant norm on an E-vector space V endowed with an E-linear action of a group G is a p-adic norm $\|\cdot\|$ such that $\|gv\| = \|v\|$ for all $v \in V$ and $g \in G$. A unitary Banach space representation of a topological group G over E is a p-adic Banach space B over E endowed with an E-linear action of G such that the map $G \times B \to B$ is continuous and such that the topology on B can be defined by an invariant norm.

By (topologically) irreducible for a (continuous) representation of a (topological) group on an E-vector space, we always mean (topologically) absolutely irreducible.

If R_0 and R_1 are objects in an abelian category, we denote by $R_0 \relbar\joinrel\relbar R_1$ an arbitrary non-split extension of R_1 by R_0. If R and $(R_j)_j$ are objects of this category, $R \simeq R_0 \relbar\joinrel\relbar R_1 \relbar\joinrel\relbar R_2 \relbar\joinrel\relbar \cdots$ means that R contains a non-split extension of R_1 by R_0 such that the quotient R/R_0 contains a non-split extension of R_2 by R_1, etc.

I thank K. Buzzard, F. Diamond, M. Dimitrov, F. Herzig, M. Kisin, R. Liu, J. Newton, T. Schmidt and B. Schraen for conversations or remarks related to this note, and the organizers of the programme "Non-abelian Fundamental Groups in Arithmetic Geometry", especially J. Coates, M. Kim and P. Schneider, for giving me the opportunity to give talks containing the results of this note at the Isaac Newton Institute in September 2009.

2 Quick review of the $\mathrm{GL}_2(\mathbb{Q}_p)$-case

We review the locally analytic representations of $\mathrm{GL}_2(\mathbb{Q}_p)$ associated to 2-dimensional crystalline representations of $\mathrm{Gal}(\overline{\mathbb{Q}}_p/\mathbb{Q}_p)$ over E with distinct Hodge–Tate weights.

Recall that, when $F = \mathbb{Q}_p$, a filtered φ-module $(D, \varphi, \mathrm{Fil}^{\cdot} D)$ is a finite dimensional E-vector space equipped with a bijective automorphism φ and with a decreasing filtration by subvector spaces $\mathrm{Fil}^i D$, $i \in \mathbb{Z}$ which is exhaustive ($\mathrm{Fil}^i D = D$ for $i \ll 0$) and separated ($\mathrm{Fil}^i D = 0$ for $i \gg 0$).

Let V be a 2-dimensional crystalline representation of $\mathrm{Gal}(\overline{\mathbb{Q}}_p/\mathbb{Q}_p)$ over E with distinct Hodge–Tate weights. Twisting V if necessary, we can assume that its Hodge–Tate weights are $(0, k - 1)$ where $k \in \mathbb{Z}_{\geq 2}$. Then by [13] we have $V = V_{\mathrm{cris}}(D) := \mathrm{Fil}^0(\mathrm{B}_{\mathrm{cris}} \otimes_{\mathbb{Q}_p} D)^{\varphi=1}$ for a filtered φ-module which can be written as follows.

(i) If φ has distinct eigenvalues, then $D = Ee \oplus E\widetilde{e}$, $\varphi(e) = \alpha^{-1}e$, $\varphi(\widetilde{e}) = \widetilde{\alpha}^{-1}\widetilde{e}$

(with $\alpha, \widetilde{\alpha} \in E^\times$, $\alpha \neq \widetilde{\alpha}$), $\mathrm{Fil}^i D = D$ if $i \leq -(k-1)$, $\mathrm{Fil}^i D = E(ae + \widetilde{a}\widetilde{e})$ if $-(k-2) \leq i \leq 0$ (with $(a, \widetilde{a}) \in E \times E \backslash \{(0,0)\}$) and $\mathrm{Fil}^i D = 0$ if $1 \leq i$.

(ii) If the eigenvalues of φ are the same, then $D = Ee \oplus E\widetilde{e}$, $\varphi(e) = \alpha^{-1} e$, $\varphi(\widetilde{e}) = \alpha^{-1}(e + \widetilde{e})$ (with $\alpha \in E^\times$), $\mathrm{Fil}^i D = D$ if $i \leq -(k-1)$, $\mathrm{Fil}^i D = E(e + \widetilde{e})$ if $-(k-2) \leq i \leq 0$ and $\mathrm{Fil}^i D = 0$ if $1 \leq i$.

Moreover, the weak admissibility conditions of [13] imply in (i): $\mathrm{val}_{\mathbb{Q}_p}(\alpha) + \mathrm{val}_{\mathbb{Q}_p}(\widetilde{\alpha}) = k - 1$ with $0 \leq \mathrm{val}_{\mathbb{Q}_p}(\alpha) \leq k - 1$ and $\mathrm{val}_{\mathbb{Q}_p}(\alpha) = k - 1$ (resp. $\mathrm{val}_{\mathbb{Q}_p}(\widetilde{\alpha}) = k - 1$) if $a = 0$ (resp. $\widetilde{a} = 0$), and in (ii): $\mathrm{val}_{\mathbb{Q}_p}(\alpha) = \frac{k-1}{2}$. Note that φ can never be scalar when $F = \mathbb{Q}_p$.

Let $B(\mathbb{Q}_p) \subset \mathrm{GL}_2(\mathbb{Q}_p)$ be the subgroup of upper triangular matrices and $\chi_1, \chi_2 : \mathbb{Q}_p^\times \to E^\times$ two locally analytic characters. We define the locally analytic parabolic induction:

$$\left(\mathrm{Ind}_{B(\mathbb{Q}_p)}^{\mathrm{GL}_2(\mathbb{Q}_p)} \chi_1 \otimes \chi_2 \right)^{\mathrm{an}} := \{ f \colon \mathrm{GL}_2(\mathbb{Q}_p) \longrightarrow E, \ f \text{ is locally analytic and}$$

$$f(bg) = (\chi_1 \otimes \chi_2)(b) f(g), \ b \in B(\mathbb{Q}_p), \ g \in \mathrm{GL}_2(\mathbb{Q}_p) \},$$

where $\chi_1 \otimes \chi_2$ maps $\left(\begin{smallmatrix} a & * \\ 0 & d \end{smallmatrix} \right) \in B(\mathbb{Q}_p)$ to $\chi_1(a) \chi_2(d) \in E^\times$. We endow this parabolic induction with a left E-linear action of $\mathrm{GL}_2(\mathbb{Q}_p)$ via $(g \cdot f)(g') := f(g'g)$. This makes $\left(\mathrm{Ind}_{B(\mathbb{Q}_p)}^{\mathrm{GL}_2(\mathbb{Q}_p)} \chi_1 \otimes \chi_2 \right)^{\mathrm{an}}$ into a locally analytic admissible representation of $\mathrm{GL}_2(\mathbb{Q}_p)$ in the sense of [18], [19].

Let D be a filtered module as in (i) above such that φ has distinct eigenvalues. We define (see (1) for $\mathrm{unr}_{\mathbb{Q}_p}$):

$$\pi_D := \left(\mathrm{Ind}_{B(\mathbb{Q}_p)}^{\mathrm{GL}_2(\mathbb{Q}_p)} \mathrm{unr}_{\mathbb{Q}_p}(\alpha^{-1}) \otimes \mathrm{unr}_{\mathbb{Q}_p}(p\widetilde{\alpha}^{-1}) \sigma^{k-2} \right)^{\mathrm{an}}, \tag{2}$$

$$\widetilde{\pi}_D := \left(\mathrm{Ind}_{B(\mathbb{Q}_p)}^{\mathrm{GL}_2(\mathbb{Q}_p)} \mathrm{unr}_{\mathbb{Q}_p}(\widetilde{\alpha}^{-1}) \otimes \mathrm{unr}_{\mathbb{Q}_p}(p\alpha^{-1}) \sigma^{k-2} \right)^{\mathrm{an}},$$

$$\pi_D^\infty := \mathrm{Sym}^{k-2} E^2 \otimes_E \left(\mathrm{Ind}_{B(\mathbb{Q}_p)}^{\mathrm{GL}_2(\mathbb{Q}_p)} \mathrm{unr}_{\mathbb{Q}_p}(\alpha^{-1}) \otimes \mathrm{unr}_{\mathbb{Q}_p}(p\widetilde{\alpha}^{-1}) \right)^\infty,$$

$$\widetilde{\pi}_D^\infty := \mathrm{Sym}^{k-2} E^2 \otimes_E \left(\mathrm{Ind}_{B(\mathbb{Q}_p)}^{\mathrm{GL}_2(\mathbb{Q}_p)} \mathrm{unr}_{\mathbb{Q}_p}(\widetilde{\alpha}^{-1}) \otimes \mathrm{unr}_{\mathbb{Q}_p}(p\alpha^{-1}) \right)^\infty,$$

where the parabolic inductions in the last two tensor products are the classical smooth parabolic inductions and where σ denotes the unique embedding $\mathbb{Q}_p \hookrightarrow E$. Note that we have inclusions $\pi_D^\infty \subset \pi_D$ and $\widetilde{\pi}_D^\infty \subset \widetilde{\pi}_D$ (write $\mathrm{Sym}^{k-2} E^2$ as an algebraic parabolic induction and take the product of functions). If $\alpha\widetilde{\alpha}^{-1} \neq p^{\pm 1}$ we have the classical intertwining $\pi_D^\infty \simeq \widetilde{\pi}_D^\infty$. If $\alpha\widetilde{\alpha}^{-1} = p^{-1}$ (resp. $\alpha\widetilde{\alpha}^{-1} = p$), we let $F_D \subset \pi_D^\infty$ (resp. $\widetilde{F}_D \subset \widetilde{\pi}_D^\infty$) be the unique non-zero finite dimensional subrepresentation. Otherwise, we let $F_D := 0$ (resp. $\widetilde{F}_D := 0$). In all cases, we denote by $\pi(D)$ the unique non-zero irreducible subrepresentation of both π_D^∞ / F_D and $\widetilde{\pi}_D^\infty / \widetilde{F}_D$. Note that $\pi(D) = \pi_D^\infty / F_D$ (resp. $\pi(D) = \widetilde{\pi}_D^\infty / \widetilde{F}_D$) if $\alpha\widetilde{\alpha}^{-1} \neq p$ (resp. $\alpha\widetilde{\alpha}^{-1} \neq p^{-1}$).

If D is a filtered module as in (ii) above such that φ has twice the same eigenvalue, we just define π_D as in (2) with $\widetilde{\alpha} = \alpha$.

To any 2-dimensional continuous representation of $\mathrm{Gal}(\overline{\mathbb{Q}}_p/\mathbb{Q}_p)$ over E, the local p-adic Langlands correspondence as in [7] associates a unitary Banach space representation $B(V)$ of $\mathrm{GL}_2(\mathbb{Q}_p)$ over E. The following theorem describes the locally analytic vectors $B(V)^{\mathrm{an}}$ inside $B(V)$. It was conjectured (in the case $\alpha \neq \widetilde{\alpha}$) in [1] and proved independently by Colmez [6] and Liu [15].

Theorem 2.1 *We keep all of the above notations.*

(i) Assume φ *has distinct eigenvalues, we have:*

$$B(V)^{\mathrm{an}} = (\pi_D/F_D) \oplus_{\pi(D)} (\widetilde{\pi}_D/\widetilde{F}_D) \text{ if } a\widetilde{a} \neq 0,$$
$$B(V)^{\mathrm{an}} = (\pi_D/\pi_D^\infty) \oplus ((\pi_D^\infty/F_D) \oplus_{\pi(D)} (\widetilde{\pi}_D/\widetilde{F}_D)) \text{ if } a = 0, \widetilde{a} \neq 0,$$
$$B(V)^{\mathrm{an}} = ((\pi_D/F_D) \oplus_{\pi(D)} (\widetilde{\pi}_D^\infty/\widetilde{F}_D)) \oplus (\widetilde{\pi}_D/\widetilde{\pi}_D^\infty) \text{ if } a \neq 0, \widetilde{a} = 0.$$

(ii) Assume the eigenvalues of φ *are the same, we have* $B(V)^{\mathrm{an}} = \pi_D$.

Let us rename $B(V)^{\mathrm{an}}$ as $\Pi(D)$, and note that, by the same formulas as those in Theorem 2.1, one can define $\Pi(D)$ for any filtered φ-module as in (i) or (ii) before which is not necessarily weakly admissible. When $F \neq \mathbb{Q}_p$, it is not known at present how to define a reasonable $B(V)$, but one can easily extend and study the definition of $\Pi(D)$, as we will see.

Remark 2.2 When $\alpha\widetilde{\alpha}^{-1} \neq p^{\pm 1}$, one can rewrite $\Pi(D)$ in Theorem 2.1(i) in a simpler way as $\pi_D \oplus_{\pi(D)} \widetilde{\pi}_D$ if $a\widetilde{a} \neq 0$, $(\pi_D/\pi(D)) \oplus \widetilde{\pi}_D$ if $a = 0$, $\pi_D \oplus (\widetilde{\pi}_D/\pi(D))$ if $\widetilde{a} = 0$.

3 Quick review of weakly admissible filtered φ-modules

We list weakly admissible filtered φ-modules of rank 2 with distinct Hodge–Tate weights and such that φ^f has distinct eigenvalues.

When F is not necessarily $\mathbb{Q}_p$, a filtered φ-module $(D, \varphi, \mathrm{Fil}^{\cdot} D_F)$ is a free $F_0 \otimes_{\mathbb{Q}_p} E$-module D of finite rank equipped with a bijective F_0-semi-linear and E-linear endomorphism φ such that $D_F := F \otimes_{F_0} D$ is equipped with a decreasing exhaustive separated filtration by $F \otimes_{\mathbb{Q}_p} E$-submodules $\mathrm{Fil}^i D_F$, $i \in \mathbb{Z}$ (not necessarily free over $F \otimes_{\mathbb{Q}_p} E$). Using the isomorphism:

$$F_0 \otimes_{\mathbb{Q}_p} E \xrightarrow{\;\sim\;} \prod_{\sigma \in S_0} E$$
$$x \otimes y \longmapsto (\sigma(x)y)_{\sigma \in S_0}$$

one can write D as $\prod_{\sigma_0 \in S_0} D_{\sigma_0}$ where $D_{\sigma_0} := (0, \cdots, 0, 1, 0, \cdots, 0)D$ (1 being "at σ_0"). Likewise, one has $D_F = \prod_{\sigma \in S} D_\sigma$ and

$$F \otimes_{F_0, \sigma_0} D_{\sigma_0} = \prod_{\substack{\sigma \in S \\ \sigma|_{F_0} = \sigma_0}} D_\sigma$$

(viewing D_{σ_0} as an F_0-vector space via $\sigma_0 : F_0 \hookrightarrow E$).

In the rest of the text, we consider rank 2 filtered φ-modules

$$D = D(\alpha, \widetilde{\alpha}, (k_\sigma, a_\sigma, \widetilde{a}_\sigma)_{\sigma \in S})$$

with $\alpha, \widetilde{\alpha} \in E^\times$, $\alpha^f \neq \widetilde{\alpha}^f$, $k_\sigma \in \mathbb{Z}_{>1}$, $(a_\sigma, \widetilde{a}_\sigma) \in E \times E \backslash \{(0,0)\}$ ($\forall \sigma \in S$) and with

$$\begin{cases} D_{\sigma_0} &= E e_{\sigma_0} \oplus E \widetilde{e}_{\sigma_0} & (\sigma_0 \in S_0) \\ \varphi(e_{\sigma_0}) &= \alpha^{-1} e_{\sigma_0 \circ \varphi^{-1}} \\ \varphi(\widetilde{e}_{\sigma_0}) &= \widetilde{\alpha}^{-1} \widetilde{e}_{\sigma_0 \circ \varphi^{-1}} \end{cases}$$

$$\begin{cases} D_\sigma &= E e_\sigma \oplus E \widetilde{e}_\sigma & (\sigma \in S) \\ \mathrm{Fil}^i D_\sigma &= D_\sigma & i \leq -(k_\sigma - 1) \\ \mathrm{Fil}^i D_\sigma &= E(a_\sigma e_\sigma + \widetilde{a}_\sigma \widetilde{e}_\sigma) & -(k_\sigma - 2) \leq i \leq 0 \\ \mathrm{Fil}^i D_\sigma &= 0 & 1 \leq i \end{cases}$$

where $1 \otimes e_{\sigma_0} = (e_\sigma)$ and $1 \otimes \widetilde{e}_{\sigma_0} = (\widetilde{e}_\sigma)$ in $F \otimes_{F_0, \sigma_0} D_{\sigma_0} = \prod_{\sigma|_{F_0} = \sigma_0} D_\sigma$.

Remark 3.1 The natural filtered φ-modules coming from Hilbert eigenforms of weight $(k_\sigma)_{\sigma \in S}$ have Hodge–Tate weights $(0, k_\sigma - 1)_{\sigma \in S}$ after maybe twisting by a crystalline character (see §8).

The following lemma is straightforward and left to the reader.

Lemma 3.2 *One has $D(\alpha, \widetilde{\alpha}, (k_\sigma, a_\sigma, \widetilde{a}_\sigma)_{\sigma \in S}) \simeq D(\alpha', \widetilde{\alpha}', (k'_\sigma, a'_\sigma, \widetilde{a}'_\sigma)_{\sigma \in S})$ if and only if $k_\sigma = k'_\sigma$ for all $\sigma \in S$ and there exists $(\lambda_{\sigma_0}, \widetilde{\lambda}_{\sigma_0})_{\sigma_0 \in S_0} \in (E^\times \times E^\times)^{|S_0|}$ such that either*

$$\begin{cases} \alpha' &= \alpha \dfrac{\lambda_{\sigma|_{F_0}}}{\lambda_{\sigma|_{F_0} \circ \varphi}} \\ \widetilde{\alpha}' &= \widetilde{\alpha} \dfrac{\widetilde{\lambda}_{\sigma|_{F_0}}}{\widetilde{\lambda}_{\sigma|_{F_0} \circ \varphi}} & \text{for all } \sigma \in S \\ (a'_\sigma, \widetilde{a}'_\sigma) &= (\lambda_{\sigma|_{F_0}} a_\sigma, \widetilde{\lambda}_{\sigma|_{F_0}} \widetilde{a}_\sigma) \text{ in } \mathbb{P}^1(E) \end{cases}$$

or

$$\begin{cases} \widetilde{\alpha}' &= \alpha \dfrac{\lambda_{\sigma|_{F_0}}}{\lambda_{\sigma|_{F_0} \circ \varphi}} \\ \alpha' &= \widetilde{\alpha} \dfrac{\widetilde{\lambda}_{\sigma|_{F_0}}}{\widetilde{\lambda}_{\sigma|_{F_0} \circ \varphi}} & \text{for all } \sigma \in S. \\ (\widetilde{a}'_\sigma, a'_\sigma) &= (\lambda_{\sigma|_{F_0}} a_\sigma, \widetilde{\lambda}_{\sigma|_{F_0}} \widetilde{a}_\sigma) \text{ in } \mathbb{P}^1(E) \end{cases}$$

Remark 3.3   Note that if $D(\alpha, \widetilde{\alpha}, (k_\sigma, a_\sigma, \widetilde{a}_\sigma)_{\sigma \in S}) \simeq D(\alpha', \widetilde{\alpha}', (k'_\sigma, a'_\sigma, \widetilde{a}'_\sigma)_{\sigma \in S})$ then one has $\{\alpha^f, \widetilde{\alpha}^f\} = \{\alpha'^f, \widetilde{\alpha}'^f\}$. This also imposes some restriction on the possible $(\lambda_{\sigma_0}, \widetilde{\lambda}_{\sigma_0})_{\sigma_0 \in S_0}$ that can occur.

For $D = D(\alpha, \widetilde{\alpha}, (k_\sigma, a_\sigma, \widetilde{a}_\sigma)_{\sigma \in S})$ we let

$$Z_D := \{\sigma \in S, a_\sigma = 0\}, \quad \widetilde{Z}_D := \{\sigma \in S, \widetilde{a}_\sigma = 0\}. \tag{3}$$

One obviously always has $Z_D \cap \widetilde{Z}_D = \emptyset$.

Lemma 3.4   *The filtered φ-module D is weakly admissible (in the sense of [13]) if and only if the following hold:*

$$\mathrm{val}_F(\alpha) + \mathrm{val}_F(\widetilde{\alpha}) = \sum_{\sigma \in S}(k_\sigma - 1), \tag{4}$$

$$\sum_{\sigma \in Z_D}(k_\sigma - 1) \leq \mathrm{val}_F(\alpha) \leq \sum_{\sigma \notin \widetilde{Z}_D}(k_\sigma - 1). \tag{5}$$

The proof is straightforward and omitted.

Remark 3.5   (i) In the presence of (4), (5) is equivalent to

$$\sum_{\sigma \in \widetilde{Z}_D}(k_\sigma - 1) \leq \mathrm{val}_F(\widetilde{\alpha}) \leq \sum_{\sigma \notin Z_D}(k_\sigma - 1). \tag{6}$$

(ii) One could also consider the case $\alpha^f = \widetilde{\alpha}^f$. When $F = \mathbb{Q}_p$, the weak admissibility condition forces φ to be non-semi-simple (see §2). But this breaks down when $F \neq \mathbb{Q}_p$; that is, there exist plenty of weakly admissible filtered φ-modules D with Hodge–Tate weights $(0, k_\sigma - 1)_{\sigma \in S}$ as above such that φ^f is scalar (i.e. is the multiplication by $\alpha^{-f} = \widetilde{\alpha}^{-f}$). As I am not sure how to define a reasonable $\Pi(D)$ if φ^f is scalar (see §4 below for $\Pi(D)$ when $\alpha^f \neq \widetilde{\alpha}^f$), I prefer to ignore this case here. Note that it is easy to build such weakly admissible filtered φ-modules with φ^f scalar from global origin: consider e.g. an elliptic curve over $\mathbb{Q}$ with φ given by $\begin{pmatrix} 0 & -1 \\ p^{-1} & 0 \end{pmatrix}$ and take its base change over a real quadratic extension of $\mathbb{Q}$ where p is inert (I thank M. Kisin and the referee for independently providing me with this example).

By the main result of [8], when D runs along the weakly admissible modules $D(\alpha, \widetilde{\alpha}, (k_\sigma, a_\sigma, \widetilde{a}_\sigma)_{\sigma \in S})$ of Lemma 3.4, the E-vector spaces

$$V_{\mathrm{cris}}(D) := (\mathrm{B}_{\mathrm{cris}} \otimes_{F_0} D)^{\varphi=1} \bigcap \mathrm{Fil}^0(\mathrm{B}_{\mathrm{dR}} \otimes_F D_F)$$

endowed with the continuous E-linear action of $\mathrm{Gal}(\overline{\mathbb{Q}_p}/F)$ induced by that on $\mathrm{B}_{\mathrm{cris}}$ and B_{dR} exhaust the 2-dimensional crystalline representations of

$\text{Gal}(\overline{\mathbb{Q}}_p/F)$ over E with Hodge–Tate weights $(0, k_\sigma - 1)_{\sigma \in S}$ such that the crystalline Frobenius has distinct eigenvalues.

One easily checks that $V_{\text{cris}}(D)$ is reducible if and only if either $\text{val}_F(\alpha) = \sum_{\sigma \in Z_D}(k_\sigma - 1)$ or $\text{val}_F(\widetilde{\alpha}) = \sum_{\sigma \in \widetilde{Z}_D}(k_\sigma - 1)$, and that $V_{\text{cris}}(D)$ is reducible split if and only if both equalities hold, which, granting (4), (5) and (6), is equivalent to just $Z_D \amalg \widetilde{Z}_D = S$.

For more general explicit rank 2 filtered φ-modules, the reader can see [9].

4 Some locally $\mathbb{Q}_p$-analytic representations of $\text{GL}_2(F)$

To a filtered φ-module D as in §3 (not necessarily weakly admissible), we associate a locally $\mathbb{Q}_p$-analytic representation $\Pi(D)$ of $\text{GL}_2(F)$ over E.

For every p-adic analytic group G, we have the E-vector space $C^{\mathbb{Q}_p-\text{an}}(G, E)$ of locally $\mathbb{Q}_p$-analytic functions $f : G \to E$. Let $\mathfrak{g}$ be the Lie algebra of G and for $\mathfrak{x} \in \mathfrak{g}$ and $f \in C^{\mathbb{Q}_p-\text{an}}(G, E)$, define as usual $\mathfrak{x} \cdot f : G \to E$ by

$$(\mathfrak{x} \cdot f)(g) := \frac{d}{dt} f(g \exp(t\mathfrak{x}))|_{t=0}. \tag{7}$$

This endows $C^{\mathbb{Q}_p-\text{an}}(G, E)$ with a $\mathbb{Q}_p$-linear action of $\mathfrak{g}$ which extends linearly to an E-linear action of $\mathfrak{g} \otimes_{\mathbb{Q}_p} E$. If G is F-analytic, then $\mathfrak{g}$ is an F-vector space and we have the usual decomposition induced by $F \otimes_{\mathbb{Q}_p} E \simeq \prod_{\sigma \in S} E$:

$$\mathfrak{g} \otimes_{\mathbb{Q}_p} E \simeq \prod_{\sigma \in S} \mathfrak{g} \otimes_{F,\sigma} E.$$

Let J be any subset of S. Following [21, §1.3.1] we say that $f \in C^{\mathbb{Q}_p-\text{an}}(G, E)$ is locally J-analytic if the action of $\mathfrak{g} \otimes_{\mathbb{Q}_p} E$ on f in (7) factors through $\prod_{\sigma \in J} \mathfrak{g} \otimes_{F,\sigma} E$. Note the two extreme cases: when $J = S$, we rather say that f is locally $\mathbb{Q}_p$-analytic (instead of S-analytic) and write "$\mathbb{Q}_p$-an" (instead of "S-an") and when $J = \emptyset$, we rather say that f is locally constant or smooth (instead of $\emptyset$-analytic).

Let $J \subseteq S$, $\chi_1, \chi_2 : F^\times \to E^\times$ be two locally J-analytic multiplicative characters and $B(F) \subset \text{GL}_2(F)$ the Borel subgroup of upper triangular matrices. We set

$$\chi_1 \otimes \chi_2 : B(F) \longrightarrow E^\times$$

$$\begin{pmatrix} a & * \\ 0 & d \end{pmatrix} \longmapsto \chi_1(a)\chi_2(d)$$

and define as in §2 the locally J-analytic parabolic induction

$$\left(\mathrm{Ind}_{B(F)}^{\mathrm{GL}_2(F)}\chi_1\otimes\chi_2\right)^{J\text{-an}}:=\{f:\ \mathrm{GL}_2(F)\longrightarrow E,\ f\text{ is locally }J\text{-analytic and}$$

$$f(bg)=(\chi_1\otimes\chi_2)(b)f(g),\ b\in B(F),\ g\in\mathrm{GL}_2(F)\}.$$

As in §2 we endow this parabolic induction with a left E-linear action of GL$_2(F)$ by $(g\cdot f)(g'):=f(g'g)$. This makes $\left(\mathrm{Ind}_{B(F)}^{\mathrm{GL}_2(F)}\chi_1\otimes\chi_2\right)^{J\text{-an}}$ into a locally $\mathbb{Q}_p$-analytic admissible representation of GL$_2(F)$ in the sense of [18], [19].

For the rest of this section, we fix $D=D(\alpha,\widetilde{\alpha},(k_\sigma,a_\sigma,\widetilde{a}_\sigma)_{\sigma\in S})$ a rank 2 filtered φ-module as in §3 (not necessarily weakly admissible). For $r_\sigma\in\mathbb{Z}_{\geq0}$ ($\sigma\in S$), denote by $(\mathrm{Sym}^{r_\sigma}E^2)^\sigma$ the r_σ-symmetric product of the representation E^2 on which GL$_2(F)$ acts via the embedding σ. For $J\subseteq S$ and $r_\sigma\in\mathbb{Z}$ ($\sigma\in J$), denote by $\prod_{\sigma\in J}\sigma^{r_\sigma}:F^\times\to E^\times$ the locally J-analytic character sending x to $\prod_{\sigma\in J}\sigma(x)^{r_\sigma}$ (it is in fact "J-algebraic").

For $J_1\subseteq J_2\subseteq S$, we first define the following locally J_2-analytic parabolic inductions (see (1) for unr$_F$):

$$I_D(J_1,J_2)$$

$$:=\left(\mathrm{Ind}_{B(F)}^{\mathrm{GL}_2(F)}\mathrm{unr}_F(\alpha^{-1})\textstyle\prod_{\sigma\in J}\sigma^{k_\sigma-1}\otimes\mathrm{unr}_F(p\widetilde{\alpha}^{-1})\prod_{\sigma\in J_1}\sigma^{-1}\prod_{\sigma\in J_2\backslash J_1}\sigma^{k_\sigma-2}\right)^{J\text{-an}}.$$

Note that by definition the unramified characters unr$_F(\alpha^{-1})$ and unr$_F(p\widetilde{\alpha}^{-1})$ only depend on α^f and $\widetilde{\alpha}^f$. Note also that $I_D(\emptyset,\emptyset)$ is a smooth unramified parabolic induction which is irreducible unless $(\alpha\widetilde{\alpha}^{-1})^f=q$ (resp. $(\alpha\widetilde{\alpha}^{-1})^f=q^{-1}$) in which case it is the twist by unr$_F(\widetilde{\alpha}^{-1})\circ\det$ (resp. by unr$_F(\alpha^{-1})\circ\det$) of the unique non-split extension of the trivial representation by the Steinberg representation (resp. of the Steinberg representation by the trivial representation).

For $J_1\subseteq J_2\subseteq S$, we then define the following locally $\mathbb{Q}_p$-analytic representations of GL$_2(F)$:

$$\pi_D(J_1,J_2):=(\otimes_{\sigma\notin J_2}(\mathrm{Sym}^{k_\sigma-2}E^2)^\sigma)\otimes_E I_D(J_1,J_2). \tag{8}$$

Theorem 4.1 ([21]) *(i) The $\pi_D(J_1',J_2')$ for $J_1\subseteq J_1'\subseteq J_2'\subseteq J_2\subseteq S$ are all distinct and are subquotients of $\pi_D(J_1,J_2)$. Moreover, if $J_1'=J_1$ (resp. $J_2'=J_2$), then $\pi_D(J_1',J_2')$ is a subrepresentation (resp. a quotient) of $\pi_D(J_1,J_2)$.*

(ii) If $(\alpha\widetilde{\alpha}^{-1})^f\neq q^{\pm1}$ or $J_1\neq\emptyset$, the representations $\pi_D(J,J)$ for $J_1\subseteq J\subseteq J_2$ are all topologically irreducible and exhaust the irreducible constituents of $\pi_D(J_1,J_2)$.

*(iii) If $|J_2 \backslash J_1| = 1$, $\pi_D(J_1, J_2)$ is the unique non-split extension of $\pi_D(J_2, J_2)$
by $\pi_D(J_1, J_1)$ (in the abelian category of admissible locally $\mathbb{Q}_p$-analytic
representations of $\mathrm{GL}_2(F)$ over E [19]).*

The embedding $\pi_D(J_1, J_2') \hookrightarrow \pi_D(J_1, J_2)$ in (i) of Theorem 4.1 easily fol-
lows from writing $\otimes_{\sigma \in J_2 \backslash J_2'}(\mathrm{Sym}^{k_\sigma - 2} E^2)^\sigma$ as a parabolic induction. The surjection $\pi_D(J_1', J_2) \twoheadrightarrow \pi_D(J_1, J_2)$ is the derivation map sending a pair of functions
(f_1, f_2) as in §7 (see the proof of Theorem 7.1) to the pair $(\partial_{J_1 \backslash J_1'}(f_1), \partial_{J_1 \backslash J_1'}(f_2))$
where $\partial_{J_1 \backslash J_1'} := \prod_{\sigma \in J_1 \backslash J_1'} \frac{\partial^{k_\sigma - 1}}{\partial \sigma(z)^{k_\sigma - 1}}$.

Theorem 4.1 is proved in detail by Schraen in [21, §2.3] (see in particular Cor. 2.10, Prop. 2.11 and Th. 2.12 of *loc. cit.*; the proof relies on work
of Schneider–Teitelbaum [18], Frommer [14] and Orlik–Strauch [16]). It tells
us that the position of the constituents $\pi_D(J, J)$ inside $\pi_D(J_1, J_2)$ form a "hypercube" with $\pi_D(J_1, J_1)$ as "first vertex" and $\pi_D(J_2, J_2)$ as "last vertex". For
instance, when $F = \mathbb{Q}_p$, $\pi_D(\emptyset, S)$ is the representation denoted by π_D in §2 and
it strictly contains the representation $\pi_D(\emptyset, \emptyset) = \pi_D^\infty$ with cokernel $\pi_D(S, S) =
\pi_D / \pi_D^\infty$. Note that $\pi_D(\emptyset, \emptyset)$ is a locally algebraic representation of $\mathrm{GL}_2(F)$.

We define $\widetilde{I}_D(J_1, J_2)$ and $\widetilde{\pi}_D(J_1, J_2)$ exactly as $I_D(J_1, J_2)$ and $\pi_D(J_1, J_2)$ by
exchanging α and $\widetilde{\alpha}$.

As in §2, if $(\alpha\widetilde{\alpha}^{-1})^f \neq q^{\pm 1}$, there is a $\mathrm{GL}_2(F)$-equivariant isomorphism
$I_D(\emptyset, \emptyset) \simeq \widetilde{I}_D(\emptyset, \emptyset)$ which induces a $\mathrm{GL}_2(F)$-equivariant isomorphism $\pi_D(\emptyset, \emptyset) \simeq
\widetilde{\pi}_D(\emptyset, \emptyset)$. When $(\alpha\widetilde{\alpha}^{-1})^f = q^{-1}$ (resp. $(\alpha\widetilde{\alpha}^{-1})^f = q$), we let $F_D \subset \pi_D(\emptyset, \emptyset)$ (resp.
$\widetilde{F}_D \subset \widetilde{\pi}_D(\emptyset, \emptyset)$) be the unique non-zero finite dimensional subrepresentation.
Otherwise, we let $F_D := 0$ (resp. $\widetilde{F}_D := 0$). We denote by $\pi(D)$ the unique non-
zero irreducible subrepresentation of both $\pi_D(\emptyset, \emptyset)/F_D$ and $\widetilde{\pi}_D(\emptyset, \emptyset)/\widetilde{F}_D$ (note
that $\pi(D)$ is $\pi_D(\emptyset, \emptyset)/F_D$ or $\widetilde{\pi}_D(\emptyset, \emptyset)/\widetilde{F}_D$ or both).

We define

$$\Pi(D) := \left(\pi_D(\emptyset, S \backslash Z_D)/F_D \oplus_{\pi(D)} \widetilde{\pi}_D(\emptyset, S \backslash \widetilde{Z}_D)/\widetilde{F}_D \right) \tag{9}$$
$$\bigoplus \left(\oplus_{\emptyset \subsetneq J \subseteq Z_D} \pi_D(J, J \amalg (S \backslash Z_D)) \right) \bigoplus \left(\oplus_{\emptyset \subsetneq J \subseteq \widetilde{Z}_D} \widetilde{\pi}_D(J, J \amalg (S \backslash \widetilde{Z}_D)) \right).$$

When $(\alpha\widetilde{\alpha}^{-1})^f \neq q^{\pm 1}$, we can rewrite it more simply as

$$\Pi(D) = \left(\oplus_{\emptyset \subseteq J \subseteq Z_D} \pi_D(J, J \amalg (S \backslash Z_D)) \right) \bigoplus \left(\oplus_{\emptyset \subsetneq J \subseteq \widetilde{Z}_D} \widetilde{\pi}_D(J, J \amalg (S \backslash \widetilde{Z}_D)) \right).$$
$$\underset{\pi(D)}{}$$

The representation $\Pi(D)$ is locally $\mathbb{Q}_p$-analytic and admissible. If $(\alpha\widetilde{\alpha}^{-1})^f \neq
q^{\pm 1}$, Theorem 4.1(ii) implies that it has $2^{[F:\mathbb{Q}_p]+1} - 1$ topologically irreducible
constituents and that its socle $\mathrm{soc}_{\mathrm{GL}_2(F)} \Pi(D)$ is exactly

$$\pi(D) \oplus \bigoplus \left(\oplus_{\emptyset \subsetneq J \subseteq Z_D} \pi_D(J, J) \right) \bigoplus \left(\oplus_{\emptyset \subsetneq J \subseteq \widetilde{Z}_D} \widetilde{\pi}_D(J, J) \right).$$

For later use, we also define

$$\mathrm{soc}'_{\mathrm{GL}_2(F)}\,\Pi(D) := \left(\pi_D(\emptyset,\emptyset)/F_D \oplus_{\pi(D)} \widetilde{\pi}_D(\emptyset,\emptyset)/\widetilde{F}_D\right) \tag{10}$$

$$\bigoplus\left(\oplus_{\emptyset \subsetneq J \subseteq Z_D}\,\pi_D(J,J)\right)\bigoplus\left(\oplus_{\emptyset \subsetneq J \subseteq \widetilde{Z}_D}\,\widetilde{\pi}_D(J,J)\right),$$

which coincides with the above socle when $(\alpha\widetilde{\alpha}^{-1})^f \neq q^{\pm 1}$ ($\pi_D(\emptyset,\emptyset)/F_D \oplus_{\pi(D)}$ $\widetilde{\pi}_D(\emptyset,\emptyset)/\widetilde{F}_D$ is the locally agebraic vectors of $\Pi(D)$ and is a reducible representation when $(\alpha\widetilde{\alpha}^{-1})^f = q^{\pm 1}$).

Basically, in (9) we decompose each "hypercube" $\pi_D(\emptyset,S)$ and $\widetilde{\pi}_D(\emptyset,S)$ into a direct sum of smaller "hypercubes" of the same size according to where the parameters a_σ and $\widetilde{a}_\sigma$ (defining the Hodge filtration) vanish. Note also that if $F = \mathbb{Q}_p$ and D is weakly admissible, we exactly recover the locally analytic representation in Theorem 2.1(i) (we leave this as an exercise). One big difference if $F \neq \mathbb{Q}_p$ is that one obviously can't recover D from $\Pi(D)$ as we miss the exact values of the a_σ.

Remark 4.2 By twisting by a suitable crystalline character, one can extend in an obvious way the definition of $\Pi(D)$ to any filtered φ-module with distinct Hodge–Tate weights for each $\sigma \in S$ and such that φ^f has distinct eigenvalues. This can be useful in view of Remark 3.1.

5 Weak admissibility and $\mathrm{GL}_2(F)$-unitarity I

The most interesting locally $\mathbb{Q}_p$-analytic representations of a p-adic analytic group are those which occur inside continuous unitary Banach spaces representations of this group. Assuming that $\Pi(D)$ occurs inside such a unitary representation of $\mathrm{GL}_2(F)$, we show that D is weakly admissible.

Recall that an invariant lattice on a locally $\mathbb{Q}_p$-analytic representation over E of a p-adic analytic group G is a closed O_E-submodule that generates the underlying E-vector space of the representation, that doesn't contain non-zero E-lines and that is preserved by G. A locally $\mathbb{Q}_p$-analytic representation of G contains an invariant lattice if and only if it is continuously contained in a unitary Banach space representation of G over E (the p-adic completion of a lattice is a unit ball).

Proposition 5.1 *Let* $J \subseteq S$, $r_\sigma \in \mathbb{Z}_{\geq 0}$ ($\sigma \in S \setminus J$) *and* $\chi_1,\chi_2 \colon F^\times \to E^\times$ *be two locally* J-*analytic multiplicative characters. If*

$$\left(\otimes_{\sigma \notin J}(\mathrm{Sym}^{r_\sigma}E^2)^\sigma\right)\otimes_E\left(\mathrm{Ind}_{B(F)}^{\mathrm{GL}_2(F)}\chi_1\otimes\chi_2\right)^{J\text{-}an} \tag{11}$$

is contained in a unitary Banach space representation of $\text{GL}_2(F)$, *then one has*

$$\text{val}_{\mathbb{Q}_p}(\chi_1(p)) + \text{val}_{\mathbb{Q}_p}(\chi_2(p)) + \sum_{\sigma \notin J} r_\sigma = 0, \tag{12}$$

$$\text{val}_{\mathbb{Q}_p}(\chi_2(p)) + \sum_{\sigma \notin J} r_\sigma \geq 0. \tag{13}$$

Proof The equality (12) is just the integrality of the central character, so we are left to prove (13).

Viewing the representation (11) inside $\left(\text{Ind}_{B(F)}^{\text{GL}_2(F)} \chi_1 \otimes \chi_2 \prod_{\sigma \notin J} \sigma(z)^{r_\sigma} \right)^{\mathbb{Q}_p - \text{an}}$, we see it contains the functions $\mathbf{1}_{O_F} \colon \text{GL}_2(F) \to E$ (resp. $\mathbf{1}_{x + pO_F} \colon \text{GL}_2(F) \to E$ for $x \in O_F$) defined by $\mathbf{1}_{O_F}\left(\left(\begin{smallmatrix} a & b \\ c & d \end{smallmatrix} \right) \right) := \chi_1(ad - bc)\chi_2\chi_1^{-1}(d) \prod_{\sigma \notin J} \sigma(d)^{r_\sigma}$ if $c/d \in O_F$ (resp. $\mathbf{1}_{x + pO_F}\left(\left(\begin{smallmatrix} a & b \\ c & d \end{smallmatrix} \right) \right) := \chi_1(ad - bc)\chi_2\chi_1^{-1}(d) \prod_{\sigma \notin J} \sigma(d)^{r_\sigma}$ if $c/d \in x + pO_F$) and $\mathbf{1}_{O_F}\left(\left(\begin{smallmatrix} a & b \\ c & d \end{smallmatrix} \right) \right) = 0$ (resp. $\mathbf{1}_{x + pO_F}\left(\left(\begin{smallmatrix} a & b \\ c & d \end{smallmatrix} \right) \right) = 0$) otherwise. It is straightforward to check that for $x \in O_F$

$$\begin{pmatrix} 1 & 0 \\ -x & p \end{pmatrix} \mathbf{1}_{O_F} = \chi_2(p) p^{\sum_{\sigma \notin J} r_\sigma} \mathbf{1}_{x + pO_F},$$

which implies

$$\|\mathbf{1}_{x + pO_F}\| = |\chi_2(p^{-1}) p^{-\sum_{\sigma \notin J} r_\sigma}|_F \|\mathbf{1}_{O_F}\|, \tag{14}$$

where $\| \cdot \|$ is any invariant norm on (11) (induced by a unitary Banach space representation). Taking a set $(x_i)_{i \in I}$ of representatives of O_F / pO_F in O_F, we obviously have $\mathbf{1}_{O_F} = \sum_{i \in I} \mathbf{1}_{x_i + pO_F}$ which implies

$$\|\mathbf{1}_{O_F}\| \leq \sup_{i \in I} \|\mathbf{1}_{x_i + pO_F}\| = |\chi_2(p^{-1}) p^{-\sum_{\sigma \notin J} r_\sigma}|_F \|\mathbf{1}_{O_F}\|.$$

Since $\|\mathbf{1}_{O_F}\| \neq 0$, we deduce $|\chi_2(p^{-1}) p^{-\sum_{\sigma \notin J} r_\sigma}|_F \geq 1$ which is just what we want. $\qquad\square$

Corollary 5.2 *Let* $D = D(\alpha, \widetilde{\alpha}, (k_\sigma, a_\sigma, \widetilde{a}_\sigma)_{\sigma \in S})$ *be a rank 2 filtered φ-module as in §3 (not necessarily weakly admissible) and let* $\Pi(D)$ *be the locally $\mathbb{Q}_p$-analytic representation of* $\text{GL}_2(F)$ *associated to D in §4. If* $\text{soc}'_{\text{GL}_2(F)} \Pi(D)$ *(see (10)) is contained in a unitary Banach space representation of* $\text{GL}_2(F)$ *(for instance if* $\Pi(D)$ *itself is), then D is weakly admissible.*

Proof The central character of $\text{soc}'_{\text{GL}_2(F)} \Pi(D)$ is also the one of $\pi_D(\emptyset, S)$ and the assumption implies that it sends p to an element of valuation zero inside $E^\times$, which immediately gives (4). Assume first $(\alpha \widetilde{\alpha}^{-1})^f \neq q^{\pm 1}$. The locally $\mathbb{Q}_p$-analytic representations $\pi_D(Z_D, Z_D)$ and $\widetilde{\pi}_D(\widetilde{Z}_D, \widetilde{Z}_D)$ both appear as subrepresentations of $\text{soc}'_{\text{GL}_2(F)} \Pi(D)$, hence are contained in a unitary Banach space representation if $\mathrm{soc}'_{\mathrm{GL}_2(F)}\Pi(D)$ is. Applying (13) to $\pi_D(Z_D, Z_D)$ yields

$$[F : \mathbb{Q}_p] - \mathrm{val}_F(\widetilde{\alpha}) - |Z_D| + \sum_{\sigma \in S \setminus Z_D} (k_\sigma - 2) \geq 0,$$

which can be rewritten as

$$\mathrm{val}_F(\widetilde{\alpha}) \leq \sum_{\sigma \in S \setminus Z_D} (k_\sigma - 1)$$

which, combined with (4), is equivalent to

$$\sum_{\sigma \in Z_D} (k_\sigma - 1) \leq \mathrm{val}_F(\alpha). \tag{15}$$

Applying (13) to $\widetilde{\pi}_D(\widetilde{Z}_D, \widetilde{Z}_D)$ yields

$$[F : \mathbb{Q}_p] - \mathrm{val}_F(\alpha) - |\widetilde{Z}_D| + \sum_{\sigma \in S \setminus \widetilde{Z}_D} (k_\sigma - 2) \geq 0,$$

which can be rewritten as

$$\mathrm{val}_F(\alpha) \leq \sum_{\sigma \in S \setminus \widetilde{Z}_D} (k_\sigma - 1). \tag{16}$$

We see that (15) and (16) are just (5), and by Lemma 3.4 this finishes the proof in the case $(\alpha\widetilde{\alpha}^{-1})^f \neq q^{\pm 1}$. Assume now $(\alpha\widetilde{\alpha}^{-1})^f = q$. As $\pi_D(Z_D, Z_D)$ is a subrepresentation of $\mathrm{soc}'_{\mathrm{GL}_2(F)}\Pi(D)$, the first part of the above proof gives (15). If $\widetilde{Z}_D \neq \emptyset$ then $\widetilde{\pi}_D(\widetilde{Z}_D, \widetilde{Z}_D)$ is still a subrepresentation of $\mathrm{soc}'_{\mathrm{GL}_2(F)}\Pi(D)$ and the second part of the above proof gives (16) and hence the result by Lemma 3.4. Assume $\widetilde{Z}_D = \emptyset$. The equality $\mathrm{val}_F(\alpha\widetilde{\alpha}^{-1}) = [F : \mathbb{Q}_p]$ combined with (4) gives:

$$\mathrm{val}_F(\alpha) = \frac{1}{2}\sum_{\sigma \in S} k_\sigma \quad \text{and} \quad \mathrm{val}_F(\widetilde{\alpha}) = \frac{1}{2}\sum_{\sigma \in S}(k_\sigma - 2).$$

We thus have (16) since $\widetilde{Z}_D = \emptyset$ and $k_\sigma \geq 2$ for all σ and we are done by Lemma 3.4. The case $(\alpha\widetilde{\alpha}^{-1})^f = q^{-1}$ is symmetric by exchanging α and $\widetilde{\alpha}$. $\quad\square$

Remark 5.3 One can expect that the converse statement of Corollary 5.2 holds, namely that if D is weakly admissible, then there always exists an invariant norm (or lattice) on $\mathrm{soc}'_{\mathrm{GL}_2(F)}\Pi(D)$. This holds for instance when $F = \mathbb{Q}_p$ but is non-trivial and ultimately rests on the construction of $\Pi(D)$ via (φ, Γ)-modules ([7], [1]).

6 Amice-Vélu and Vishik revisited

We state and (re)prove a slight generalization of a well-known result of Amice-Vélu and Vishik.

Let $U \subseteq O_F$ an open subset, $J \subseteq S$ and $r_\sigma \in \mathbb{Z}_{\geq 0}$ for $\sigma \in S \setminus J$. Denote by $\mathcal{F}(U, J, (r_\sigma)_{\sigma \in S \setminus J})$ the E-vector space of functions $f : U \to E$ such that there exists an open (disjoint) cover $(a_i + \varpi_F^{n_i} O_F)_{i \in I}$ of U such that, for each i, one has an expansion

$$f(z)|_{a_i + \varpi_F^{n_i} O_F} = \sum_{\substack{\underline{m} = (m_\sigma)_{\sigma \in S} \in \mathbb{Z}_{\geq 0}^{[F:\mathbb{Q}_p]} \\ m_\sigma \leq r_\sigma \text{ if } \sigma \notin J}} a_{\underline{m}} \prod_{\sigma \in S} \sigma(z - a_i)^{m_\sigma} \qquad (17)$$

with $|a_{\underline{m}}|_F q^{-n_i(\sum_{\sigma \in S} m_\sigma)} \to 0$ when $\sum_{\sigma \in S} m_\sigma \to +\infty$ ($a_{\underline{m}} \in E$). Recall that $\mathcal{F}(U, J, (r_\sigma)_{\sigma \in S \setminus J})$ is an inductive limit of Banach spaces with injective and compact transition maps ([17, §16]), namely the Banach spaces of functions as in (17) with norm

$$\sup_{\underline{m}} \left(|a_{\underline{m}}|_F \, q^{-n_i(\sum_{\sigma \in S} m_\sigma)} \right).$$

Note that $\mathcal{F}(U, J, (r_\sigma)_{\sigma \in S \setminus J}) \subseteq \mathcal{F}(U, J', (r_\sigma)_{\sigma \in S \setminus J'})$ for any $J \subseteq J'$.

The technical but key lemma that follows is essentially due to Amice-Vélu and Vishik.

Lemma 6.1 *Let B be a p-adic Banach space over E and ι be an E-linear map $\mathcal{F}(U, J, (r_\sigma)_{\sigma \in S \setminus J}) \to B$. Let $\| \cdot \|$ be a norm on B defining its topology and assume that there exist $C \in \mathbb{R}_{>0}$ and $c \in \mathbb{R}_{\geq 0}$ such that, for any $a \in O_F$, $n \in \mathbb{Z}_{\geq 0}$ with $a + \varpi_F^n O_F \subseteq U$ and any $(m_\sigma)_{\sigma \in S} \in \mathbb{Z}_{\geq 0}^{[F:\mathbb{Q}_p]}$ with $m_\sigma \leq r_\sigma$ if $\sigma \notin J$, one has:*

$$\left\| \iota \left(\mathbf{1}_{a + \varpi_F^n O_F}(z) \prod_{\sigma \in S} \sigma(z - a)^{m_\sigma} \right) \right\| \leq C q^{-n(\sum_{\sigma \in S} m_\sigma - c)}, \qquad (18)$$

where $\mathbf{1}_{a + \varpi_F^n O_F}$ is the characteristic function of $a + \varpi_F^n O_F$. Let

$$J' := J \amalg \{\tau \in S \setminus J, c < r_\tau + 1\}.$$

Then ι uniquely extends to an E-linear map $\iota' : \mathcal{F}(U, J', (r_\sigma)_{\sigma \in S \setminus J'}) \to B$ such that the diagram

$$
\begin{array}{ccc}
\mathcal{F}(U, J, (r_\sigma)_{\sigma \in S \setminus J}) & \xrightarrow{\ \iota\ } & B \\
\big\downarrow & \nearrow_{\iota'} & \\
\mathcal{F}(U, J', (r_\sigma)_{\sigma \in S \setminus J'}) & &
\end{array}
$$

commutes and such that (18) holds for all $(m_\sigma)_{\sigma \in S} \in \mathbb{Z}_{\geq 0}^{[F:\mathbb{Q}_p]}$ *with* $m_\sigma \leq r_\sigma$ *if* $\sigma \notin J'$ *(possibly up to increasing C). Moreover,* ι *and* ι' *are continuous.*

Proof First, the map ι is automatically continuous. Indeed, for any $a \in O_F$ and $n \in \mathbb{Z}_{\geq 0}$ as in the statement, the inequality (18) gives that ι is continuous upon restriction to the Banach subspace of analytic functions on $a + \varpi_F^n O_F$ as in (17). Since the topology on $\mathcal{F}(U, J, (r_\sigma)_{\sigma \in S \setminus J})$ is the locally convex topology with respect to these Banach subspaces, this implies ι is continuous. Let $\tau \in J' \setminus J$. By induction it is enough to prove the statement replacing J' by $J \amalg \{\tau\}$. By E-linearity, it is then enough to prove that ι uniquely extends to each function of the form

$$1_{a+\varpi_F^n O_F}(z)\tau(z-a)^{m_\tau} \prod_{\sigma \in S \setminus \tau} \sigma(z-a)^{m_\sigma} \tag{19}$$

with $m_\sigma \leq r_\sigma$ if $\sigma \notin J \amalg \{\tau\}$ and $m_\tau \geq r_\tau + 1$ so that (18) still holds (maybe up to modifying C). Let f be a function as in (19) and let

$$D^- := \{\underline{d} = (d_\sigma)_{\sigma \in S} \in \mathbb{Z}_{\geq 0}^{[F:\mathbb{Q}_p]}, d_\sigma \leq m_\sigma \text{ and } d_\tau \leq r_\tau\},$$
$$D^+ := \{\underline{d} = (d_\sigma)_{\sigma \in S} \in \mathbb{Z}_{\geq 0}^{[F:\mathbb{Q}_p]}, d_\sigma \leq m_\sigma \text{ and } r_\tau + 1 \leq d_\tau\}.$$

Since for any function h on U,

$$1_{a+\varpi_F^n O_F}(z)h(z-a) = \sum_{a' \in a + \varpi_F^n[k_F]} 1_{a'+\varpi_F^{n+1}O_F}(z)h((z-a') + (a'-a)),$$

an easy computation shows we can rewrite f as $f_n^+ + f_n^-$, where

$$f_n^+ := \sum_{a' \in a + \varpi_F^n[k_F]} 1_{a'+\varpi_F^{n+1}O_F}(z)\left(\sum_{\underline{d} \in D^+} \left(a_{\underline{d}}^+ \prod_{\sigma \in S} \sigma(a-a')^{m_\sigma - d_\sigma} \prod_{\sigma \in S} \sigma(z-a')^{d_\sigma}\right)\right),$$

$$f_n^- := \sum_{a' \in a + \varpi_F^n[k_F]} 1_{a'+\varpi_F^{n+1}O_F}(z)\left(\sum_{\underline{d} \in D^-} \left(a_{\underline{d}}^- \prod_{\sigma \in S} \sigma(a-a')^{m_\sigma - d_\sigma} \prod_{\sigma \in S} \sigma(z-a')^{d_\sigma}\right)\right)$$

for some $a_{\underline{d}}^+, a_{\underline{d}}^- \in O_E$. Since

$$\left|a_{\underline{d}}^- \prod_{\sigma \in S} \sigma(a-a')^{m_\sigma - d_\sigma}\right|_F \leq q^{-n(\sum_{\sigma \in S} m_\sigma - d_\sigma)},$$

and since one has by (18) and the definition of D^- for all $(d_\sigma)_{\sigma \in S} \in D^-$

$$\left\|\iota\left(1_{a'+\varpi_F^{n+1}O_F}(z) \prod_{\sigma \in S} \sigma(z-a')^{d_\sigma}\right)\right\| \leq Cq^{-(n+1)(\sum_{\sigma \in S} d_\sigma - c)},$$

we see that

$$\|\iota(f_n^-)\| \leq Cq^c q^{-n(\sum_{\sigma \in S} m_\sigma - c)}.$$

One can start again and write f_n^+ as $f_{n+1}^+ + f_{n+1}^-$ where f_{n+1}^+, f_{n+1}^- are finite linear combinations over O_E of functions $\mathbf{1}_{a''+\varpi_F^{n+2}O_F}(z) \prod_{\sigma \in S} \sigma(z - a'')^{d_\sigma}$ with $r_\tau + 1 \le d_\tau \le m_\tau$ for f_{n+1}^+ and where $\|\iota(f_{n+1}^-)\| \le Cq^c q^{-(n+1)(\sum_{\sigma \in S} m_\sigma - c)}$ by the same proof as before. Iterating this process, we see that for any integer $M \ge n$ the function f in (19) can be written

$$f = f_M^+ + \sum_{i=n}^{M} f_i^-, \tag{20}$$

where $\iota(f_i^-)$ is defined and satisfies $\|\iota(f_i^-)\| \le Cq^c q^{-i(\sum_{\sigma \in S} m_\sigma - c)}$ and where f_M^+ is a finite linear combination over O_E of functions $\mathbf{1}_{a''+\varpi_F^M O_F}(z) \prod_{\sigma \in S} \sigma(z - a'')^{d_\sigma}$ with $d_\sigma \le m_\sigma$ and $r_\tau + 1 \le d_\tau$. As $c < r_\tau + 1 \le \sum_{\sigma \in S} m_\sigma$ (recall $r_\tau + 1 \le m_\tau$), we see that $\iota(f_i^-) \to 0$ in B when $i \to +\infty$. If ι extends to $\mathcal{F}(U, J', (r_\sigma)_{\sigma \in S \setminus J'})$ in such a way that (18) is satisfied (up to modifying C), we see that we have in particular $\|\iota(f_M^+)\| \le Cq^{-M(r_\tau+1-c)}$ and hence $\iota(f_M^+) \to 0$ when $M \to +\infty$. So we must have $\iota(f) = \sum_{i=n}^{+\infty} \iota(f_i^-)$. Conversely, setting $\iota(f) := \sum_{i=n}^{+\infty} \iota(f_i^-)$ implies

$$\|\iota(f)\| \le \sup_{i \ge n} \|\iota(f_i^-)\| = Cq^c \sup_{i \ge n} q^{-i(\sum_{\sigma \in S} m_\sigma - c)} = Cq^c q^{-n(\sum_{\sigma \in S} m_\sigma - c)}$$

and hence (18) is still satisfied replacing C by Cq^c. The continuity of ι' is checked as for ι. $\qquad\square$

7 Weak admissibility and $\mathrm{GL}_2(F)$-unitarity II

Using Lemma 6.1, we show that if a continuous unitary Banach space representation of $\mathrm{GL}_2(F)$ over E contains $\mathrm{soc}'_{\mathrm{GL}_2(F)} \Pi(D)$ (see (10)), then it automatically contains a larger locally $\mathbb{Q}_p$-analytic representation $\Pi(D)^{\mathrm{Amice}}$ which is included in $\Pi(D)$.

We start with the following theorem.

Theorem 7.1 *Fix $J \subseteq S$, $r_\sigma \in \mathbb{Z}_{\ge 0}$ ($\sigma \in S \setminus J$) and $\chi_1, \chi_2 \colon F^\times \to E^\times$ two locally J-analytic multiplicative characters. Define*

$$J' := J \amalg \{\tau \in S \setminus J, - \mathrm{val}_{\mathbb{Q}_p}(\chi_1(p)) < r_\tau + 1\}. \tag{21}$$

Then any continuous E-linear equivariant injection

$$\left(\otimes_{\sigma \notin J} (\mathrm{Sym}^{r_\sigma} E^2)^\sigma \right) \otimes_E \left(\mathrm{Ind}_{B(F)}^{\mathrm{GL}_2(F)} \chi_1 \otimes \chi_2 \right)^{J\text{-}an} \hookrightarrow B, \tag{22}$$

where B is a unitary Banach space representation of $\mathrm{GL}_2(F)$ over E, canoni-

cally extends to an E-linear continuous equivariant injection

$$\left(\otimes_{\sigma\notin J'}(\mathrm{Sym}^{r_\sigma}E^2)^\sigma\right)\otimes_E\left(\mathrm{Ind}_{B(F)}^{\mathrm{GL}_2(F)}\chi_1\otimes\chi_2\prod_{\sigma\in J'\setminus J}\sigma^{r_\sigma}\right)^{J'\text{-}an}\hookrightarrow B.\quad (23)$$

Proof By embedding

$$\left(\otimes_{\sigma\notin J}(\mathrm{Sym}^{r_\sigma}E^2)^\sigma\right)\otimes_E\left(\mathrm{Ind}_{B(F)}^{\mathrm{GL}_2(F)}\chi_1\otimes\chi_2\right)^{J\text{-}an}\qquad (24)$$

inside $\left(\mathrm{Ind}_{B(F)}^{\mathrm{GL}_2(F)}\chi_1\otimes\chi_2\prod_{\sigma\notin J}\sigma(z)^{r_\sigma}\right)^{\mathbb{Q}_p\text{-}an}$, any element f of (24) can be seen as a pair of functions $(f_1\colon O_F\to E,\ f_2\colon\varpi_F O_F\to E)$ by setting

$$f_1(z):=f\left(\begin{pmatrix}0&1\\-1&z\end{pmatrix}\right)\quad\text{and}\quad f_2(z):=\chi_2\chi_1^{-1}(z)\left(\prod_{\sigma\notin J}\sigma(z)^{r_\sigma}\right)f_1\left(\frac{1}{z}\right)\quad (25)$$

where f_2 is the only continuous function on $\varpi_F O_F$ agreeing with the right-hand side of (25) on $\varpi_F O_F\setminus\{0\}$. The map $f\mapsto f_1\oplus f_2$ yields an isomorphism between (24) and $\mathcal{F}(O_F,J,(r_\sigma)_{\sigma\in S\setminus J})\oplus\mathcal{F}(\varpi_F O_F,J,(r_\sigma)_{\sigma\in S\setminus J})$ and the action of $g=\left(\begin{smallmatrix}a&b\\c&d\end{smallmatrix}\right)\in\mathrm{GL}_2(F)$ is given by

$$(gf)_1(z)=\chi_1(\det g)\chi_2\chi_1^{-1}(-cz+a)\prod_{\sigma\notin J}\sigma(-cz+a)^{r_\sigma}f_1\left(\frac{dz-b}{-cz+a}\right)\quad (26)$$

if $\frac{dz-b}{-cz+a}\in O_F$, and

$$(gf)_1(z)=\chi_1(\det g)\chi_2\chi_1^{-1}(dz-b)\prod_{\sigma\notin J}\sigma(dz-b)^{r_\sigma}f_2\left(\frac{-cz+a}{dz-b}\right)$$

if $\frac{dz-b}{-cz+a}\in F\setminus O_F$, and symmetric formulas for $(gf)_2(z)$. Let ι be a continuous injection as in (22) and let $\|\cdot\|$ be an invariant norm on B (which exists by assumption). Let $f:=f_1\oplus 0=f_1$ where $f_1(z):=\prod_{\sigma\in S}\sigma(z)^{m_\sigma}$ for $z\in O_F$ and some $(m_\sigma)_{\sigma\in S}\in\mathbb{Z}_{\geq 0}^{[F:\mathbb{Q}_p]}$ such that $m_\sigma\leq r_\sigma$ if $\sigma\notin J$. By continuity of ι, there is $C\in\mathbb{R}_{>0}$ such that $\|\iota(f)\|\leq C$ for all such f. Using $\|\iota(gf)\|=\|\iota(f)\|$ and then applying (26) with $g=\left(\begin{smallmatrix}1&a/\varpi_F^n\\0&1/\varpi_F^n\end{smallmatrix}\right)$ $(a\in O_F,n\in\mathbb{Z}_{\geq 0})$ gives

$$\left\|\iota\left(1_{a+\varpi_F^n O_F}(z)\prod_{\sigma\in S}\sigma(z-a)^{m_\sigma}\right)\right\|=\left|\chi_1(\varpi_F^n)\prod_{\sigma\in S}\sigma(\varpi_F^n)^{m_\sigma}\right|_F\|\iota(f)\|$$

$$\leq q^{-n\left(\sum_{\sigma\in S}m_\sigma+\mathrm{val}_{\mathbb{Q}_p}(\chi_1(p))\right)}C.$$

Lemma 6.1 applied with $c:=-\mathrm{val}_{\mathbb{Q}_p}(\chi_1(p))$ and the norm induced by B (via ι) gives that $\mathcal{F}(O_F,J,(r_\sigma)_{\sigma\in S\setminus J})\oplus 0\hookrightarrow B$ canonically extends to a continuous map $\mathcal{F}(O_F,J',(r_\sigma)_{\sigma\in S\setminus J'})\oplus 0\to B$. Let $f:=0\oplus f_2=f_2$ where $f_2(z):=\prod_{\sigma\in S}\sigma(z)^{m_\sigma}$ with the m_σ as before. Applying (26) (more precisely its symmetric version for $(gf)_2(z)$) with $g:=\left(\begin{smallmatrix}1/\varpi_F^n&0\\a/\varpi_F^n&1\end{smallmatrix}\right)\in\mathrm{GL}_2(F)$ gives by an analogous proof that

$0 \oplus \mathcal{F}(\varpi_F O_F, J, (r_\sigma)_{\sigma \in S \setminus J}) \hookrightarrow B$ canonically extends to a continuous map $0 \oplus \mathcal{F}(\varpi_F O_F, J', (r_\sigma)_{\sigma \in S \setminus J'}) \to B$. Via the isomorphism

$$\left(\otimes_{\sigma \notin J'} (\mathrm{Sym}^{r_\sigma} E^2)^\sigma \right) \otimes_E \left(\mathrm{Ind}_{B(F)}^{\mathrm{GL}_2(F)} \chi_1 \otimes \chi_2 \prod_{\sigma \in J' \setminus J} \sigma^{r_\sigma} \right)^{J'\text{-an}}$$

$$\simeq \mathcal{F}(O_F, J', (r_\sigma)_{\sigma \in S \setminus J'}) \oplus \mathcal{F}(\varpi_F O_F, J', (r_\sigma)_{\sigma \in S \setminus J'}) \quad (27)$$

(defined as previously), we have that ι extends to an E-linear continuous map as in (23) (still denoted ι) except that it remains to prove that it is equivariant and injective. Let's first indicate how equivariance can be checked. The case g scalar being obvious, it is enough by composition to check equivariance for

$$g \in \left\{ \begin{pmatrix} 0 & \varpi_F \\ 1 & 0 \end{pmatrix}, \begin{pmatrix} 1 & \varpi_F \\ 0 & 1 \end{pmatrix}, \begin{pmatrix} 1 & 0 \\ 0 & \lambda \end{pmatrix}, \lambda \in O_F \setminus \{0\} \right\}.$$

By linearity, it is enough to prove

$$\iota(g(f_1 \oplus 0)) = g\iota(f_1 \oplus 0)$$

for $f_1 = \mathbf{1}_{a+\varpi_F^n O_F}(z) \prod_{\sigma \in S} \sigma(z-a)^{m_\sigma} \in \mathcal{F}(O_F, J', (r_\sigma)_{\sigma \in S \setminus J'})$ and g as above, and an analogous statement with $\iota(g(0 \oplus f_2))$. We treat only the first case, leaving the second to the reader. Going back to the proof of Lemma 6.1, one writes $f_1 = f_M^+ + \sum_{i=n}^M f_i^-$ for all $M \geq n$ as in (20). Because ι is equivariant upon restriction to $\mathcal{F}(O_F, J, (r_\sigma)_{\sigma \in S \setminus J})$ and $\| \cdot \|$ is invariant, it is enough to check that $\|\iota(g(f_M^+ \oplus 0))\| \to 0$ when $M \to +\infty$. One has $\begin{pmatrix} 0 & \varpi_F \\ 1 & 0 \end{pmatrix}(f_M^+ \oplus 0) = 0 \oplus g_M^+$ where g_M^+ is up to scalar the function $z \in \varpi_F O_F \mapsto f_M^+(z/\varpi_F)$. The assertion then follows from the bounds (18) applied to J' and g_M^+ (see the end of the proof of Lemma 6.1). In the other cases, one has up to scalars $\begin{pmatrix} 1 & \varpi_F \\ 0 & 1 \end{pmatrix}(f_M^+ \oplus 0) = f_M^+(\cdot - \varpi_F) \oplus 0$ and $\begin{pmatrix} 1 & 0 \\ 0 & \lambda \end{pmatrix}(f_M^+ \oplus 0) = f_M^+(\lambda \cdot) \oplus 0$, and the assertion follows again easily from (18). Finally, injectivity follows from continuity and equivariance. Indeed, if (23) is not injective, then its non-zero kernel must contain a non-zero topologically irreducible subrepresentation. But by [21, §1.3] (of which Theorem 4.1 is a special case) any such subrepresentation is also a subrepresentation of (24). However, ι in (22) being injective, this is impossible. $\qquad \square$

Let D be a rank 2 filtered φ-module with Hodge–Tate weights $(0, k_\sigma - 1)_{\sigma \in S}$ as in §3 (not necessarily weakly admissible). For $J \subseteq S$ define

$$Z_D(J) := J \amalg \left\{ \tau \in S \setminus J, \mathrm{val}_F(\alpha) \geq k_\tau - 1 + \sum_{\sigma \in J}(k_\sigma - 1) \right\},$$

$$\widetilde{Z}_D(J) := J \amalg \left\{ \tau \in S \setminus J, \mathrm{val}_F(\widetilde{\alpha}) \geq k_\tau - 1 + \sum_{\sigma \in J}(k_\sigma - 1) \right\}.$$

We set

$$\Pi(D)^{\mathrm{Amice}} := \left(\pi_D(\emptyset, S \setminus Z_D(\emptyset)) / F_D \oplus_{\pi(D)} \widetilde{\pi}_D(\emptyset, S \setminus \widetilde{Z}_D(\emptyset)) / \widetilde{F}_D \right) \quad (28)$$

$$\bigoplus \oplus_{\emptyset \subsetneq J \subseteq Z_D} \pi_D(J, J \amalg (S \setminus Z_D(J)))$$

$$\bigoplus \oplus_{\emptyset \subsetneq J \subseteq \widetilde{Z}_D} \widetilde{\pi}_D(J, J \amalg (S \setminus \widetilde{Z}_D(J))).$$

We have $\mathrm{soc}'_{\mathrm{GL}_2(F)} \Pi(D) \subsetneq \Pi(D)^{\mathrm{Amice}}$.

Corollary 7.2 *Let D be a rank 2 filtered φ-module as above.*

(i) If a unitary Banach space representation of $\mathrm{GL}_2(F)$ over E continuously contains $\mathrm{soc}'_{\mathrm{GL}_2(F)} \Pi(D)$ then it continuously contains $\Pi(D)^{\mathrm{Amice}}$ and D is weakly admissible.

(ii) If D is weakly admissible, then $\Pi(D)^{\mathrm{Amice}} \subseteq \Pi(D)$.

Proof (i) The last statement is Corollary 5.2. Let $J \subseteq S$, $J \neq \emptyset$. Rewriting (21) as $J' = J \amalg S \setminus Z(J)$, where $Z(J) := J \amalg \{\tau \in S \setminus J, -\mathrm{val}_{\mathbb{Q}_p}(\chi_1(p)) \geq r_\tau + 1\}$, Theorem 7.1 implies that a unitary Banach space representation B of $\mathrm{GL}_2(F)$ (continuously) contains $\pi_D(J, J)$ (resp. $\widetilde{\pi}_D(J, J)$) if and only if it (continuously) contains $\pi_D(J, J \amalg (S \setminus Z_D(J)))$ (resp. $\widetilde{\pi}_D(J, J \amalg (S \setminus \widetilde{Z}_D(J)))$). Likewise, if B contains $\pi_D(\emptyset, \emptyset) / F_D$ (resp. $\widetilde{\pi}_D(\emptyset, \emptyset) / \widetilde{F}_D$), the same proof as for Theorem 7.1 gives an equivariant continuous map $\pi_D(\emptyset, S \setminus Z_D(\emptyset)) \to B$ with kernel containing F_D (resp. with tildes). If this kernel were strictly bigger than F_D, then it would also contain $\pi_D(\emptyset, \emptyset)$, which is impossible. We conclude the proof of (i) by using the fact that all embeddings $\pi_D(J, J) \hookrightarrow \pi_D(J, J \amalg (S \setminus Z_D(J)))$, etc., are essential (Theorem 4.1), that is, any subrepresentation of the right-hand side intersects non-trivially the left-hand side.

(ii) If $J \subseteq Z_D$ (resp. $J \subseteq \widetilde{Z}_D$), the first inequality in (5) (resp. in (6)) implies $Z_D \subseteq Z_D(J)$ (resp. $\widetilde{Z}_D \subseteq \widetilde{Z}_D(J)$) and thus we have

$$J \amalg S \setminus Z_D(J) \subseteq J \amalg S \setminus Z_D \quad (\text{resp. } J \amalg S \setminus \widetilde{Z}_D(J) \subseteq J \amalg S \setminus \widetilde{Z}_D).$$

Comparing (9) and (28), we see that $\Pi(D)^{\mathrm{Amice}} \subseteq \Pi(D)$. $\square$

We shall say that a continuous 2-dimensional E-linear representation of $\mathrm{Gal}(\overline{\mathbb{Q}_p}/F)$ is ordinary if its semi-simplification has restriction to inertia isomorphic to $\varepsilon^* \oplus 1$ for some integer $*$. It would have been nice to have in many cases $\Pi(D)^{\mathrm{Amice}} = \Pi(D)$; however, this turns out to be quite rare.

Proposition 7.3 *We have $\Pi(D)^{\mathrm{Amice}} = \Pi(D)$ if and only if either $V_{\mathrm{cris}}(D)$ is split ordinary or $V_{\mathrm{cris}}(D)$ is irreducible and $F = \mathbb{Q}_p$.*

Proof Clearly the statement holds if and only if $Z_D(J) = Z_D$ for all $J \subsetneq Z_D$ and $\widetilde{Z}_D(J) = \widetilde{Z}_D$ for all $J \subsetneq \widetilde{Z}_D$. But $Z_D(J) = Z_D$ if and only if

$$\mathrm{val}_F(\alpha) < k_\tau - 1 + \sum_{\sigma \in J}(k_\sigma - 1)$$

for all $\tau \in S \setminus Z_D$, so $Z_D(J) = Z_D$ for all $J \subsetneq Z_D$ if and only if $Z_D(\emptyset) = Z_D$ if and only if

$$\tau \notin Z_D \Rightarrow \mathrm{val}_F(\alpha) < k_\tau - 1 \qquad (29)$$

and likewise with tildes everywhere. Assume first that there are $\tau, \widetilde{\tau} \in S$ with $\tau \neq \widetilde{\tau}$ such that $\tau \notin Z_D$ and $\widetilde{\tau} \notin \widetilde{Z}_D$. By (4), the inequality (29) and its tilde analogue imply $\sum_{\sigma \in S}(k_\sigma - 1) < k_\tau - 1 + k_{\widetilde{\tau}} - 1$ which is impossible. Since $Z_D \cap \widetilde{Z}_D = \emptyset$, we thus have either $Z_D = S$ and $\widetilde{Z}_D = \emptyset$, or $Z_D = \emptyset$ and $\widetilde{Z}_D = S$, or $Z_D = \widetilde{Z}_D = \emptyset$ and $|S| = 1$. The first two cases correspond to $V_{\mathrm{cris}}(D)$ being split ordinary and the last to $F = \mathbb{Q}_p$ and $V_{\mathrm{cris}}(D)$ being indecomposable. Finally, it is straightforward to check that (29) and its tilde analogue hold in the first two cases, and hold in the last if and only if $V_{\mathrm{cris}}(D)$ is moreover irreducible. $\square$

8 Local–global considerations

We briefly put the previous considerations within a global setting.

Let L be a totally real finite extension of $\mathbb{Q}$ with ring of integers O_L. Assume for simplicity that there is a unique prime ideal $\mathfrak{p}$ in O_L above p and let $L_\mathfrak{p}$ denote the completion of L at $\mathfrak{p}$ and $L_{\mathfrak{p},0}$ its maximal absolutely unramified subfield. Denote by $\mathbb{A}_{L,f}^p$ the finite adèles of L outside p. To any quaternion algebra D over L which splits at only one of the infinite places and which splits at $\mathfrak{p}$ and to any compact open subgroup $K_f^p \subset (D \otimes_L \mathbb{A}_{L,f}^p)^\times$, one can associate a tower of algebraic Shimura curves $(S(K_f^p K_{f,p}))_{K_{f,p}}$ over L where $K_{f,p}$ runs over the compact open subgroups of $(D \otimes_L L_\mathfrak{p})^\times \simeq \mathrm{GL}_2(L_\mathfrak{p})$ (see e.g. [4]). Consider

$$\widehat{H}^1(K_f^p) := \left(\varprojlim_n \varinjlim_{K_{f,p}} H^1_{\text{ét}}(S(K_f^p K_{f,p}) \times_L \overline{\mathbb{Q}}, O_E/p^n O_E) \right) \otimes_{O_E} E,$$

which is a p-adic Banach space over E (an open unit ball being the O_E-module $\varprojlim \varinjlim H^1_{\text{ét}}(S(K_f^p K_{f,p}) \times_L \overline{\mathbb{Q}}, O_E/p^n O_E)$) endowed with a linear continuous unitary action of $\mathrm{GL}_2(L_\mathfrak{p}) \times \mathrm{Gal}(\overline{\mathbb{Q}}/L)$ ([5]). We denote by $\widehat{H}^1(K_f^p)^{\mathbb{Q}_p\text{-an}}$ its locally $\mathbb{Q}_p$-analytic vectors for the action of $\mathrm{GL}_2(L_\mathfrak{p})$ ([19]).

Let g be a cuspidal Hilbert eigenform of level prime to p, E a finite extension of $\mathbb{Q}_p$ containing the Galois closure of $L_\mathfrak{p}$ and the Hecke eigenvalues associated to g, and $k_\sigma \geq 2$ for $\sigma \in S := \mathrm{Hom}(L_\mathfrak{p}, E)$ the various weights of g. We assume all the k_σ are congruent modulo 2. We denote by

$$\rho_g : \mathrm{Gal}(\overline{\mathbb{Q}}/L) \to \mathrm{GL}_2(E)$$

the continuous totally odd Galois representation associated to g ([22]). We normalize ρ_g so that the traces of arithmetic Frobeniuses at unramified places are the Hecke eigenvalues. We let

$$(D_g, \varphi, \mathrm{Fil}^{\cdot} D_{g,L_\mathfrak{p}}) := \left((B_{\mathrm{cris}} \otimes_{\mathbb{Q}_p} \rho_g)^{\mathrm{Gal}(\overline{\mathbb{Q}_p}/L_\mathfrak{p})}, \varphi \otimes \mathrm{Id}, (\mathrm{Fil}^{\cdot} B_{\mathrm{dR}} \otimes_{\mathbb{Q}_p} \rho_g)^{\mathrm{Gal}(\overline{\mathbb{Q}_p}/L_\mathfrak{p})} \right).$$

Choose $\eta \colon \mathrm{Gal}(\overline{\mathbb{Q}_p}/L_\mathfrak{p}) \to E^\times$ a crystalline character such that $\rho_g|_{\mathrm{Gal}(\overline{\mathbb{Q}_p}/L_\mathfrak{p})} \otimes \eta$ has Hodge–Tate weights $(0, k_\sigma - 1)_{\sigma \in S}$ (such a character always exists) and define the filtered module $D_{g,\eta}$ as D_g but replace $\rho_g|_{\mathrm{Gal}(\overline{\mathbb{Q}_p}/L_\mathfrak{p})}$ by $\rho_g|_{\mathrm{Gal}(\overline{\mathbb{Q}_p}/L_\mathfrak{p})} \otimes \eta$. If the eigenvalues of $\varphi^{[L_{\mathfrak{p},0}:\mathbb{Q}_p]}$ on $D_{g,\eta}$ (or equivalently on D_g) are distinct, then the locally $\mathbb{Q}_p$-analytic representation $\Pi(D_{g,\eta})$ is well defined (§4). We set

$$\Pi(D_g) := \Pi(D_{g,\eta}) \otimes \eta^{-1} \circ \det.$$

The locally $\mathbb{Q}_p$-analytic representation $\Pi(D_g)$ of $\mathrm{GL}_2(L_\mathfrak{p})$ is easily checked to be independent of the choice of the crystalline character η as above (note that the ratio of two such η is an unramified character of $\mathrm{Gal}(\overline{\mathbb{Q}_p}/L_\mathfrak{p})$).

Conjecture 8.1 Assume that the eigenvalues of $\varphi^{[L_{\mathfrak{p},0}:\mathbb{Q}_p]}$ on D_g are distinct. If $\mathrm{Hom}_{\mathrm{Gal}(\overline{\mathbb{Q}}/L)}(\rho_g^\vee, \widehat{H}^1(K_f^p)) \neq 0$ then there is an integer $n > 0$ (depending on g and K_f^p) such that

$$\Pi(D_g)^n \subseteq \mathrm{Hom}_{\mathrm{Gal}(\overline{\mathbb{Q}}/L)}(\rho_g^\vee, \widehat{H}^1(K_f^p)^{\mathbb{Q}_p\text{-an}}) \tag{30}$$

and such that any $\mathrm{GL}_2(L_\mathfrak{p})$-subrepresentation of $\mathrm{Hom}_{\mathrm{Gal}(\overline{\mathbb{Q}}/L)}(\rho_g^\vee, \widehat{H}^1(K_f^p)^{\mathbb{Q}_p\text{-an}})$ intersects $\Pi(D_g)^n$ non-trivially.

This conjecture is known so far only for $L = \mathbb{Q}$ ([2], [12]). If one knows that $(\mathrm{soc}'_{\mathrm{GL}_2(F)} \Pi(D_{g,\eta}) \otimes \eta^{-1} \circ \det)^n$ embeds into the right-hand side of (30), then Corollary 7.2(i) (and its proof) give that $(\Pi(D_{g,\eta})^{\mathrm{Amice}} \otimes \eta^{-1} \circ \det)^n$ also embeds. But $(\Pi(D_{g,\eta})^{\mathrm{Amice}} \otimes \eta^{-1} \circ \det)^n$ is the "maximum that p-adic analysis will give you". Going to $(\Pi(D_{g,\eta}) \otimes \eta^{-1} \circ \det)^n = \Pi(D_g)^n$ will presumably require some non-trivial arithmetic geometry.

Finally, if $L_\mathfrak{p} \neq \mathbb{Q}_p$, I never expect (30) to be a topological isomorphism for any n as $\mathrm{Hom}_{\mathrm{Gal}(\overline{\mathbb{Q}}/L)}(\rho_g^\vee, \widehat{H}^1(K_f^p)^{\mathbb{Q}_p\text{-an}})$ should determine $\rho_g|_{\mathrm{Gal}(\overline{\mathbb{Q}_p}/L_\mathfrak{p})}$ (which is not the case of $\Pi(D_g)^n$).

9 The case where the Galois representation is reducible

We examine more closely the structure of $\Pi(D)$ when $V_{\mathrm{cris}}(D)$ is reducible and relate it to considerations of [3].

Let us first make a detour via the modulo p theory. Let $\overline{V} \simeq \begin{pmatrix} \overline{\chi}_2 \omega & * \\ 0 & \overline{\chi}_1 \end{pmatrix}$ be a continuous 2-dimensional representation of $\mathrm{Gal}(\overline{\mathbb{Q}_p}/F)$ over k_E where ω is the reduction modulo p of ε and where $\overline{\chi}_1, \overline{\chi}_2 \colon F^\times \to k_E^\times$ are smooth characters. The results of [3, §19] (in the case F is unramified) suggest that, when $\overline{V}$ is non-split, the corresponding "good" representation(s) π of $\mathrm{GL}_2(F)$ over k_E (e.g. the representation(s) $\mathrm{Hom}_{\mathrm{Gal}(\overline{\mathbb{Q}}/L)}(\overline{V}_g, \varinjlim H^1_{\mathrm{\acute{e}t}}(S(K_f^p K_{f,p}) \times_L \overline{\mathbb{Q}}, k_E))$ when $\overline{V}$ globalizes to a representation $\overline{V}_g$ of $\mathrm{Gal}(\overline{\mathbb{Q}}/L)$, see §8) should (generically) have the form

$$\pi = \pi_0 \text{------} \pi_1 \text{------} \cdots \text{------} \pi_{[F:\mathbb{Q}_p]-1} \text{------} \pi_{[F:\mathbb{Q}_p]} \, ,$$

where $\pi_0 = \mathrm{Ind}_{B(F)}^{\mathrm{GL}_2(F)} \overline{\chi}_1 \otimes \overline{\chi}_2$, $\pi_{[F:\mathbb{Q}_p]} = \mathrm{Ind}_{B(F)}^{\mathrm{GL}_2(F)} \overline{\chi}_2 \omega \otimes \overline{\chi}_1 \omega^{-1}$ (smooth parabolic inductions) and where the π_j for $1 \leq j \leq [F : \mathbb{Q}_p] - 1$ are irreducible and are not subquotients of parabolic inductions (extensions are taken in the category of smooth representations of $\mathrm{GL}_2(F)$ over k_E). When $\overline{V}$ is split, one should have $\pi = \oplus_{j=0}^{[F:\mathbb{Q}_p]} \pi_j$. The representations π_j, $1 \leq j \leq [F : \mathbb{Q}_p] - 1$ are still quite mysterious, although one knows a few things about them (such as their $\mathrm{GL}_2(O_F)$-socle, see [3]).

Now let D be a rank 2 filtered φ-module with Hodge–Tate weights $(0, k_\sigma - 1)_{\sigma \in S}$ as in section 3 (not necessarily weakly admissible). For $j \in \{0, \ldots, [F : \mathbb{Q}_p]\}$ let us first define two series $(\Pi(D)_j)_j$ and $(\widetilde{\Pi}(D)_j)_j$ of locally $\mathbb{Q}_p$-analytic representations of $\mathrm{GL}_2(F)$:

$$\Pi(D)_j := \bigoplus_{\substack{J \subseteq Z_D \\ |J| = |Z_D| - j}} \pi_D(J, J \amalg S \setminus Z_D) \qquad \text{if } 0 \leq j \leq |Z_D| - 1,$$

$$\Pi(D)_j := \pi_D(\emptyset, S \setminus Z_D)/F_D \oplus_{\pi(D)} \widetilde{\pi}_D(\emptyset, Z_D)/\widetilde{F}_D \quad \text{if } j = |Z_D|,$$

$$\Pi(D)_j := \bigoplus_{\substack{J \subseteq S \setminus Z_D \\ |J| = j - |Z_D|}} \widetilde{\pi}_D(J, J \amalg Z_D) \qquad \text{if } |Z_D| + 1 \leq j \leq [F : \mathbb{Q}_p],$$

and likewise for $\widetilde{\Pi}(D)_j$ replacing Z_D by $\widetilde{Z}_D$, $\pi_D(J, J \amalg S \setminus Z_D)$ by $\widetilde{\pi}_D(J, J \amalg S \setminus \widetilde{Z}_D)$ and $\widetilde{\pi}_D(J, J \amalg Z_D)$ by $\pi_D(J, J \amalg \widetilde{Z}_D)$. Since $Z_D \subseteq S \setminus \widetilde{Z}_D$ and $\widetilde{Z}_D \subseteq S \setminus Z_D$, the $\Pi(D)_j$ and $\widetilde{\Pi}(D)_j$ are easily checked to be subquotients of $\Pi(D)$ (we leave this to the reader). Note that the above two series coincide (up to numbering) if and only if $Z_D \amalg \widetilde{Z}_D = S$ in which case one has $\Pi(D)_j = \widetilde{\Pi}(D)_{[F:\mathbb{Q}_p]-j}$.

If D is weakly admissible, recall from §3 that $V_{\mathrm{cris}}(D)$ is reducible if and only if either $\mathrm{val}_F(\alpha) = \sum_{\sigma \in Z_D}(k_\sigma - 1)$ or $\mathrm{val}_F(\widetilde{\alpha}) = \sum_{\sigma \in \widetilde{Z}_D}(k_\sigma - 1)$, and that $V_{\mathrm{cris}}(D)$ is reducible split if and only if both equalities hold if and only if $Z_D \amalg \widetilde{Z}_D = S$.

We consider below extensions in the abelian category of admissible locally $\mathbb{Q}_p$-analytic representations of $\mathrm{GL}_2(F)$ over E ([19]).

Theorem 9.1 *Let D be a weakly admissible rank 2 filtered φ-module as in §3.*

(i) *$V_{\mathrm{cris}}(D)$ is indecomposable if and only if*

$$\Pi(D) \simeq \oplus_{j=0}^{|Z_D|-1}\Pi(D)_j$$

$$\bigoplus \quad \Pi(D)_{|Z_D|} \text{———} \Pi(D)_{|Z_D|+1} \text{———} \cdots \text{———} \Pi(D)_{[F:\mathbb{Q}_p]}$$

if and only if

$$\Pi(D) \simeq \oplus_{j=0}^{|\widetilde{Z}_D|-1}\widetilde{\Pi}(D)_j$$

$$\bigoplus \quad \widetilde{\Pi}(D)_{|\widetilde{Z}_D|} \text{———} \widetilde{\Pi}(D)_{|\widetilde{Z}_D|+1} \text{———} \cdots \text{———} \widetilde{\Pi}(D)_{[F:\mathbb{Q}_p]} \, .$$

(ii) *$V_{\mathrm{cris}}(D)$ is reducible split if and only if $\Pi(D) \simeq \oplus_{j=0}^{[F:\mathbb{Q}_p]}\Pi(D)_j$ if and only if $\Pi(D) \simeq \oplus_{j=0}^{[F:\mathbb{Q}_p]}\widetilde{\Pi}(D)_j$.*

(iii) *Let $\chi_1, \chi_2 : F^\times \to O_E^\times$ be two locally $\mathbb{Q}_p$-analytic integral characters. The following are equivalent:*

- *$V_{\mathrm{cris}}(D)$ is reducible and isomorphic to $\begin{pmatrix} \chi_2\varepsilon & * \\ 0 & \chi_1 \end{pmatrix}$.*
- *$\Pi(D)$ contains $(\mathrm{Ind}_{B(F)}^{\mathrm{GL}_2(F)}\chi_1 \otimes \chi_2)^{\mathbb{Q}_p-\mathrm{an}}$ if $\chi_1 \neq \chi_2$ or $(\mathrm{Ind}_{B(F)}^{\mathrm{GL}_2(F)}\chi_1 \otimes \chi_1)^{\mathbb{Q}_p-\mathrm{an}}/\chi_1 \circ \det$ if $\chi_1 = \chi_2$.*

(iv) *Assume $V_{\mathrm{cris}}(D)$ is reducible and isomorphic to $\begin{pmatrix} \chi_2\varepsilon & * \\ 0 & \chi_1 \end{pmatrix}$ with $\chi_1\chi_2^{-1} \notin \{1, \varepsilon^2\}$.*

- *If $\mathrm{val}_F(\alpha) = \sum_{\sigma \in Z_D}(k_\sigma - 1)$, then $\Pi(D)_0 \simeq (\mathrm{Ind}_{B(F)}^{\mathrm{GL}_2(F)}\chi_1 \otimes \chi_2)^{\mathbb{Q}_p-\mathrm{an}}$ and $\Pi(D)_{[F:\mathbb{Q}_p]} \simeq (\mathrm{Ind}_{B(F)}^{\mathrm{GL}_2(F)}\chi_2\varepsilon \otimes \chi_1\varepsilon^{-1})^{\mathbb{Q}_p-\mathrm{an}}$.*
- *If $\mathrm{val}_F(\widetilde{\alpha}) = \sum_{\sigma \in \widetilde{Z}_D}(k_\sigma - 1)$, then $\widetilde{\Pi}(D)_0 \simeq (\mathrm{Ind}_{B(F)}^{\mathrm{GL}_2(F)}\chi_1 \otimes \chi_2)^{\mathbb{Q}_p-\mathrm{an}}$ and $\widetilde{\Pi}(D)_{[F:\mathbb{Q}_p]} \simeq (\mathrm{Ind}_{B(F)}^{\mathrm{GL}_2(F)}\chi_2\varepsilon \otimes \chi_1\varepsilon^{-1})^{\mathbb{Q}_p-\mathrm{an}}$.*

Proof (ii) is straightforward using $Z_D = S \backslash \widetilde{Z}_D$ and (i) is a consequence of some easy combinatorics using $Z_D \subsetneq S \backslash \widetilde{Z}_D$ and Theorem 4.1 that we leave to the reader.

We prove one implication in (iii). Assume first that $\Pi(D)$ contains $(\mathrm{Ind}_{B(F)}^{\mathrm{GL}_2(F)}\chi_1 \otimes \chi_2)^{\mathbb{Q}_p-\mathrm{an}}$ for some integral characters $\chi_1 \neq \chi_2$. From (9), we see that this parabolic induction must be either $\pi_D(Z_D, S)$ or $\widetilde{\pi}_D(\widetilde{Z}_D, S)$. Going back to (8), this implies either

$$\chi_1 = \mathrm{unr}_F(\alpha^{-1}) \prod_{\sigma \in Z_D} \sigma^{k_\sigma - 1} \text{ and } \chi_2 = \mathrm{unr}_F(\widetilde{\alpha}^{-1}) | \cdot |_F^{-1} \prod_{\sigma \in Z_D} \sigma^{-1} \prod_{\sigma \notin Z_D} \sigma^{k_\sigma - 2} \quad (31)$$

or

$$\chi_1 = \mathrm{unr}_F(\widetilde{\alpha}^{-1}) \prod_{\sigma \in \widetilde{Z}_D} \sigma^{k_\sigma - 1} \text{ and } \chi_2 = \mathrm{unr}_F(\alpha^{-1}) | \cdot |_F^{-1} \prod_{\sigma \in \widetilde{Z}_D} \sigma^{-1} \prod_{\sigma \notin \widetilde{Z}_D} \sigma^{k_\sigma - 2}.$$

The integrality of χ_1 implies either $\mathrm{val}_F(\alpha) = \sum_{\sigma \in Z_D}(k_\sigma - 1)$ or $\mathrm{val}_F(\widetilde{\alpha}) = \sum_{\sigma \in \widetilde{Z}_D}(k_\sigma - 1)$. In the first case, we have from the definition (3) of Z_D that $\widetilde{e}_\sigma \in \mathrm{Fil}^0 D_\sigma$ if $\sigma \in Z_D$ and $\widetilde{e}_\sigma \in \mathrm{Fil}^{-(k_\sigma - 1)} D_\sigma$ if $\sigma \notin Z_D$. Hence the crystalline Galois character $V_{\mathrm{cris}}(\prod_{\sigma_0 \in S_0} E\widetilde{e}_{\sigma_0})$ sends p to $\widetilde{\alpha}^{-[F:\mathbb{Q}_p]}$ and has Hodge–Tate weights $((0)_{\sigma \in Z_D}, (k_\sigma - 1)_{\sigma \notin Z_D})$, hence is exactly $\chi_2 \varepsilon$ (using $\varepsilon = | \cdot |_F \prod_{\sigma \in S} \sigma$). Likewise, we have $V_{\mathrm{cris}}(D / \prod_{\sigma_0 \in S_0} E\widetilde{e}_{\sigma_0}) = \chi_1$, and thus $V_{\mathrm{cris}}(D) \simeq \begin{pmatrix} \chi_2 \varepsilon & * \\ 0 & \chi_1 \end{pmatrix}$. The second case is symmetric. Assume now that $\Pi(D)$ contains $\left(\mathrm{Ind}_{B(F)}^{\mathrm{GL}_2(F)} \chi_1 \otimes \chi_1 \right)^{\mathbb{Q}_p\text{-an}} / \chi_1 \circ \det$, which from (9) must be either $\pi_D(\emptyset, S)/F_D$ (with $F_D \neq 0$) or $\widetilde{\pi}_D(\emptyset, S)/\widetilde{F}_D$ (with $\widetilde{F}_D \neq 0$). Thus $\left(\mathrm{Ind}_{B(F)}^{\mathrm{GL}_2(F)} \chi_1 \otimes \chi_1 \right)^{\mathbb{Q}_p\text{-an}}$ is either $\pi_D(\emptyset, S)$ or $\widetilde{\pi}_D(\emptyset, S)$ and the proof is the same as previously.

Now let us prove (iv) and the other implication in (iii). Assume $\mathrm{val}_F(\alpha) = \sum_{\sigma \in Z_D}(k_\sigma - 1)$. By computing $V_{\mathrm{cris}}(\prod_{\sigma_0 \in S_0} E\widetilde{e}_{\sigma_0})$ as before, we see that we can assume that χ_1 and χ_2 are as in (31) (replacing if necessary χ_1 by $\chi_2 \varepsilon$ and χ_2 by $\chi_1 \varepsilon^{-1}$ when $V_{\mathrm{cris}}(D)$ is split). If $Z_D \neq \emptyset$ or $(\alpha \widetilde{\alpha}^{-1})^f \notin q^{\pm 1}$, we have $\Pi(D)_0 = \pi_D(Z_D, S) = \left(\mathrm{Ind}_{B(F)}^{\mathrm{GL}_2(F)} \chi_1 \otimes \chi_2 \right)^{\mathbb{Q}_p\text{-an}}$ and $\Pi(D)$ contains $\Pi(D)_0$ from (i). If $Z_D \neq S$ or $(\alpha \widetilde{\alpha}^{-1})^f \notin q^{\pm 1}$, we have $\Pi(D)_{[F:\mathbb{Q}_p]} = \widetilde{\pi}_D(S \backslash Z_D, S)$, which is easily checked to be $\left(\mathrm{Ind}_{B(F)}^{\mathrm{GL}_2(F)} \chi_2 \varepsilon \otimes \chi_1 \varepsilon^{-1} \right)^{\mathbb{Q}_p\text{-an}}$. This proves the first part of (iv). If $Z_D = \emptyset$ and $(\alpha \widetilde{\alpha}^{-1})^f = q^{\pm 1}$, the above equality for $\mathrm{val}_F(\alpha)$ together with (4) imply $(\alpha \widetilde{\alpha}^{-1})^f = q^{-1}$ and $k_\sigma = 2$ for all $\sigma \in S$, and we see that $\Pi(D)$ contains $\left(\mathrm{Ind}_{B(F)}^{\mathrm{GL}_2(F)} \chi_1 \otimes \chi_1 \right)^{\mathbb{Q}_p\text{-an}} / \chi_1 \circ \det = \pi_D(\emptyset, S)/F_D$. The proof for $\mathrm{val}_F(\widetilde{\alpha}) = \sum_{\sigma \in \widetilde{Z}_D}(k_\sigma - 1)$ is completely symmetric and left to the reader. This finishes the proofs of (iv) and (iii). □

Remark 9.2 There is a variant of Theorem 9.1(iv) when $\chi_1 \chi_2^{-1} \in \{1, \varepsilon^2\}$ that we leave to the reader.

Let us finish the paper with some free speculation. Assume $V_{\mathrm{cris}}(D)$ is reducible as in (iv) above and consider the case, say, $\mathrm{val}_F(\alpha) = \sum_{\sigma \in Z_D}(k_\sigma - 1)$ (the other case being symmetric). Then $\Pi(D)_0$ (resp. $\Pi(D)_{[F:\mathbb{Q}_p]}$) has a unique (admissible) unitary completion $\widehat{\Pi}(D)_0$ (resp. $\widehat{\Pi}(D)_{[F:\mathbb{Q}_p]}$) which is just the continuous parabolic induction of $\chi_1 \otimes \chi_2$ (resp. $\chi_2 \varepsilon \otimes \chi_1 \varepsilon^{-1}$) and which is topologically

irreducible if $(k_\sigma)_{\sigma \in S} \neq (2, \ldots, 2)$ or $(\alpha \widetilde{\alpha}^{-1})^f \neq q$ (resp. or $(\alpha \widetilde{\alpha}^{-1})^f \neq q^{-1}$).
Theorem 9.1 together with the characteristic p considerations at the beginning of this section strongly suggest that there should also exist topologically irreducible admissible unitary completions $\widehat{\Pi}(D)_j$ of $\Pi(D)_j$ for $1 \leq j \leq [F : \mathbb{Q}_p] - 1$, containing $\Pi(D)_j$ and which may – or may not – only depend on D, together with an admissible unitary Banach space representation B of $\mathrm{GL}_2(F)$ of the form:

$$B \simeq \widehat{\Pi}(D)_0 \, \text{——} \, \widehat{\Pi}(D)_1 \, \text{——} \, \cdots \, \text{——} \, \widehat{\Pi}(D)_{[F:\mathbb{Q}_p]-1} \, \text{——} \, \widehat{\Pi}(D)_{[F:\mathbb{Q}_p]}$$

when $V_{\mathrm{cris}}(D)$ is non-split and of the form $B \simeq \oplus_{j=0}^{[F:\mathbb{Q}_p]} \widehat{\Pi}(D)_j$ when $V_{\mathrm{cris}}(D)$ is split, and such that B completely determines D. Here, admissibility for Banach space representations is as in [20].

For simplicity (and because it is speculative) we just focus on the case $[F : \mathbb{Q}_p] = 2$ (and keep the same assumptions as above). We write $S = \{\sigma, \sigma'\}$. Assuming a unitary Banach space representation as $\widehat{\Pi}(D)_1$ exists, one can consider its locally $\mathbb{Q}_p$-analytic vectors $\widehat{\Pi}(D)_1^{\mathbb{Q}_p - \mathrm{an}}$. One has an exact sequence (in the category of admissible locally $\mathbb{Q}_p$-analytic representations of $\mathrm{GL}_2(F)$):

$$0 \longrightarrow \Pi(D)_1 \longrightarrow \widehat{\Pi}(D)_1^{\mathbb{Q}_p - \mathrm{an}} \longrightarrow \Pi(D)_1^? \longrightarrow 0,$$

where

$$\Pi(D)_1 = \widetilde{\pi}_D(\{\sigma\}, \{\sigma\}) \oplus \widetilde{\pi}_D(\{\sigma'\}, \{\sigma'\}) \quad \text{if} \quad Z_D = \emptyset,$$

$$\Pi(D)_1 = \pi_D(\emptyset, S \setminus Z_D) \oplus_{\pi(D)} \widetilde{\pi}_D(\emptyset, Z_D) \quad \text{if} \quad |Z_D| = 1,$$

$$\Pi(D)_1 = \pi_D(\{\sigma\}, \{\sigma\}) \oplus \pi_D(\{\sigma'\}, \{\sigma'\}) \quad \text{if} \quad Z_D = S.$$

Granting the existence of such an "extra-constituent" $\Pi(D)_1^?$, we can speculate that the "complete" locally $\mathbb{Q}_p$-analytic representation(s) $\Pi(D)^?$ of $\mathrm{GL}_2(F)$ associated to D should be a non-split extension:

$$0 \longrightarrow \Pi(D) \longrightarrow \Pi(D)^? \longrightarrow \Pi(D)_1^? \longrightarrow 0$$

(even if $V_{\mathrm{cris}}(D)$ is irreducible) and that the "parameters" giving the isomorphism class of this extension should be related to (and determine) the values of a_σ, $a_{\sigma'}$, $\widetilde{a}_\sigma$ and $\widetilde{a}_{\sigma'}$ (up to modifications as in Lemma 3.2).

References

[1] Berger L., Breuil C., *Sur quelques représentations potentiellement cristallines de* $\mathrm{GL}_2(\mathbb{Q}_p)$, Astérisque 330, 2010, 155–211.

[2] Breuil C., Emerton M., *Représentations p-adiques ordinaires de* $GL_2(\mathbb{Q}_p)$ *et compatibilité local-global*, Astérisque 331, 2010, 255–315.

[3] Breuil C., Paškūnas V., *Towards a modulo p Langlands correspondence for* GL_2, Memoirs of A.M.S., published online 23 May 2011.

[4] Carayol H., *Sur les représentations l-adiques associées aux formes modulaires de Hilbert*, Ann. Scient. E.N.S. 19, 1986, 409–468.

[5] Calegari F., Emerton M., *Completed Cohomology - A Survey*, 239–257 in this volume.

[6] Colmez P., *La série principale unitaire de* $GL_2(\mathbb{Q}_p)$*: vecteurs localement analytiques*, preprint 2010.

[7] Colmez P., *Représentations de* $GL_2(\mathbb{Q}_p)$ *et* (φ, Γ)*-modules*, Astérisque 330, 2010, 281–509.

[8] Colmez P., Fontaine J.-M., *Construction des représentations semi-stables*, Inventiones Math. 140, 2000, 1–43.

[9] Dousmanis G., *Rank two filtered* (φ, N)*-modules with Galois descent data and coefficients*, to appear in Trans. of A.M.S.

[10] Emerton M., *Jacquet modules of locally analytic representations of p-adic reductive groups I. Constructions and first properties*, Ann. Scient. É.N.S. 39, 2006, 775–839.

[11] Emerton M., *Jacquet modules of locally analytic representations of p-adic reductive groups II. The relation to parabolic induction*, to appear in J. Institut Math. Jussieu.

[12] Emerton M., *Local-global compatibility in the p-adic Langlands program for* $GL_{2/\mathbb{Q}}$, preprint 2010.

[13] Fontaine J.-M., *Représentations p-adiques semi-stables*, Astérisque 223, 1994, 113–184.

[14] Frommer H., *The locally analytic principal series of split reductive groups*, preprint 2003.

[15] Liu R., *Locally analytic vectors of some crystabelian representations of* $GL_2(\mathbb{Q}_p)$, Compositio Math., to appear.

[16] Orlik S., Strauch M., *On the irreducibility of locally analytic principal series representations*, Representation Theory 14, 2010, 713–746.

[17] Schneider P., *Nonarchimedan Functional Analysis*, Springer Monographs in Maths, 2002.

[18] Schneider P., Teitelbaum J., *Locally analytic distributions and p-adic representation theory*, J. Amer. Math. Soc. 15, 2002, 443–468.

[19] Schneider P., Teitelbaum J., *Algebras of p-adic distributions and admissible representations*, Inventiones Math. 153, 2003, 145–196.

[20] Schneider P., Teitelbaum J., *Banach space representations and Iwasawa theory*, Israel J. Math. 127, 2002, 359–380.

[21] Schraen B., *Représentations p-adiques de* $GL_2(F)$ *et catégories dérivées*, Israel J. Math. 176, 2010, 307–362.

[22] Taylor R., *On Galois representations associated to Hilbert modular forms*, Inventiones Math. 98, 1989, 265–280.

Completed cohomology – a survey

Frank Calegari

Northwestern University

Matthew Emerton

Northwestern University

This note summarizes the theory of p-adically completed cohomology. This construction was first introduced in paper [8] (although insufficient attention was given there to the integral aspects of the theory), and then further developed in the papers [4] and [12]. The papers [8] and [4] may give the impression that p-adically completed cohomology is some sort of auxiliary construction that can be used to prove theorems (of either a p-adic or classical nature) about automorphic forms. However, we believe that p-adically completed cohomology is in fact an object of fundamental importance, and that it provides the best approximation that we know of to spaces of "p-adic automorphic forms". (In particular, unlike the spaces that go by this name that are sometimes constructed by arithmetico-geometric means in the theory of modular curves, or more generally Shimura varieties, p-adically completed cohomology admits a representation of the p-adic group, and thus allows the introduction of representation-theoretic methods into the study of p-adic properties of automorphic forms.)

A systematic exposition of the theory, and of its (largely conjectural, at this point) applications to the p-adic aspects of the Langlands correspondence between automorphic eigenforms and Galois representations, will be given in the paper [6]. These notes provide a summary of some of the basic points of the theory, as well as one of the main conjectures of [6] (Conjecture 5 below). The anticipated connection of the theory with the p-adic aspects of the Langlands program are discussed in the final Section 8. We don't attempt to formulate any formal conjectures in this section, but simply try to indicate, in very general and somewhat idealized terms, what form we expect this connection to take. Our main goal for doing so here is to explain some the motivations behind making Conjecture 5. It remains to be seen how closely our expectations, and our conjecture, conform to the reality of the situation.

Non-abelian Fundamental Groups and Iwasawa Theory, eds. John Coates, Minhyong Kim, Florian Pop, Mohamed Saïdi and Peter Schneider. Published by Cambridge University Press. ©Cambridge University Press 2012.

1 Definitions

Let G_0 be a pro-finite group, assumed to admit a countable basis of neighbour-hoods of the identity, consisting of normal open subgroups, say

$$\cdots \subset G_r \subset \cdots \subset G_1 \subset G_0.$$

Suppose given a tower of topological spaces

$$\cdots \to X_r \to \cdots \to X_1 \to X_0,$$

each equipped with an action of G_0, such that:

(i) the maps $X_{r+1} \to X_r$ are G_0-equivariant;
(ii) G_r acts trivially on X_r, and realizes X_r as a G_0/G_r-torsor over X_0. (In particular, all the maps in the tower are finite coverings.)

In this context we may define the p-adically completed homology and coho-mology modules attached to the tower $X_\bullet$ (having fixed the prime p), namely:

$$\tilde{H}_\bullet := \varprojlim_r H_\bullet(X_r, \mathbb{Z}_p)$$

and

$$\tilde{H}^\bullet := \varprojlim_s \varinjlim_r H^\bullet(X_r, \mathbb{Z}/p^s\mathbb{Z}).$$

We can also consider Borel–Moore and compactly supported variants:

$$\tilde{H}_\bullet^{BM} := \varprojlim_r H_\bullet^{BM}(X_r, \mathbb{Z}_p)$$

and

$$\tilde{H}_c^\bullet := \varprojlim_s \varinjlim_r H_c^\bullet(X_r, \mathbb{Z}/p^s\mathbb{Z}).$$

There are natural maps $\tilde{H}_\bullet \to \tilde{H}_\bullet^{BM}$ and $\tilde{H}_c^\bullet \to \tilde{H}^\bullet$.

From now on we suppose that each X_r is a manifold which is homotopic to a finite simplicial complex. This ensures us that all the homology spaces (usual or Borel–Moore) and cohomology spaces (usual or compactly supported) of X_r that we have written down are finitely generated over the indicated ring of coefficients, and that they are related by appropriate forms of duality.[1]

We also suppose from now on that G_0 is a p-adic analytic group. Without this hypothesis, it seems impossible to control the inverse limits involved in our constructions. On the other hand, with this hypothesis, one can control

---

[1] The paper [12] provides a sheaf-theoretic description of $\tilde{H}^\bullet$ and $\tilde{H}_c^\bullet$, which allows for more flexibility in the basic set-up of the theory than we permit here. However, we will have no need for this extra generality in the applications that we have in mind.

these limits in a very satisfactory way. Indeed, Lazard [14] has proved that completed group ring $\mathbb{Z}_p[[G_0]]$ is Noetherian, and Schneider and Teitelbaum [16] have explained how this result can be applied to control various analytic difficulties that arise in the study of p-adic Banach-space representations of G_0.

In particular, we have the following result (which strengthens the results stated in [8], in so far as it deals explicitly with homology as well as cohomology, and pays attention to the integral aspects of the constructions, and not just to the objects obtained after tensoring up with $\mathbb{Q}_p$), whose proof will appear in [6].

Theorem 1 *(1) The natural action of $\mathbb{Z}_p[[G_0]]$ on the modules $\tilde{H}_\bullet$ and $\tilde{H}_\bullet^{BM}$ makes each of them a finitely generated left module over $\mathbb{Z}_p[[G_0]]$. Furthermore, the canonical topology on each of these modules (obtained by writing it as a quotient of a finite direct sum of copies of $\mathbb{Z}_p[[G_0]]$) coincides with projective limit topology.*

Note that this implies in particular that each of the torsion submodules $\tilde{H}_\bullet[p^\infty]$ and $\tilde{H}_\bullet^{BM}[p^\infty]$ is finitely generated over $\mathbb{Z}_p[[G_0]]$, and so in particular, is of bounded exponent.
(2) The spaces $\tilde{H}^\bullet$ and $\tilde{H}_c^\bullet$ are p-adically complete and their p-power torsion submodules have bounded exponent.
(3) There are short exact sequences

$$0 \to \mathrm{Hom}_{\mathrm{cont}}(\tilde{H}_{\bullet-1}, \mathbb{Q}_p/\mathbb{Z}_p) \to \tilde{H}^\bullet \to \mathrm{Hom}_{\mathrm{cont}}(\tilde{H}_\bullet, \mathbb{Z}_p) \to 0$$

(here cont means continuous with respect to the canonical topology on the source, and the evident topology – either discrete or p-adic – on the target) and

$$0 \to \mathrm{Hom}(\tilde{H}^{\bullet+1}[p^\infty], \mathbb{Q}_p/\mathbb{Z}_p) \to \tilde{H}_\bullet \to \mathrm{Hom}(\tilde{H}^\bullet, \mathbb{Z}_p) \to 0$$

(since the natural topology on $\tilde{H}^\bullet$ is the p-adic topology, and since all $\mathbb{Z}_p$-linear maps are automatically p-adically continuous, there is no need to explicitly specify any continuity conditions on the maps appearing in this exact sequence), as well as similar short exact sequences relating $\tilde{H}_\bullet^{BM}$ and $\tilde{H}_c^\bullet$. These exact sequences are compatible with the maps $\tilde{H}_\bullet \to \tilde{H}_\bullet^{BM}$, and the maps $\tilde{H}_c^\bullet \to \tilde{H}^\bullet$.
(4) For each $r \geq 0$, there are Hochschild–Serre type spectral sequences

$$E_2^{i,j} := H^i(G_r, \tilde{H}^j) \implies H^{i+j}(X_r, \mathbb{Z}_p)$$

and

$$E_2^{i,j} := H^i(G_r, \tilde{H}_c^j) \implies H_c^{i+j}(X_r, \mathbb{Z}_p),$$

compatible with the maps $\tilde{H}_c^\bullet \to \tilde{H}^\bullet$.

Remark 2   If M is any continuous G_0-module over $\mathbb{Z}_p$, then we may associate a family of local systems $\mathcal{M}_\bullet$ on the tower $X_\bullet$ to M, and there is an analogue of part (4) of the theorem for the cohomology of the local system $\mathcal{M}_r$.

Remark 3   The theorem shows the advantages of working with both homology and cohomology. It is on the homology side that the algebraic aspects of the theory are most transparent, while the cohomology side is better adapted to comparison with the situation at finite levels.

It is useful to make some additional definitions on the cohomology side, as follows:

$$\hat{H}^\bullet := \text{ the } p\text{-adic completion of } \varinjlim_r H^\bullet(X_r, \mathbb{Z}_p)$$

$$= \varprojlim_s (\varinjlim_r H^\bullet(X_r, \mathbb{Z}_p)/p^s H^\bullet(X_r, \mathbb{Z}_p))$$

and

$$T_p H^\bullet := \text{ the } p\text{-adic Tate module of } \varinjlim_r H^\bullet(X_r, \mathbb{Z}_p)$$

$$= \varprojlim_s \varinjlim_r H^\bullet(X_r, \mathbb{Z}_p)[p^s]$$

(the transition maps being given by multiplication by p). Note that the p-adic completion kills the p-divisible part of $\varinjlim_r H^\bullet(X_r, \mathbb{Z}_p)$, while the p-adic Tate module knows only about the torsion subgroup of this divisible part. There is an exact sequence

$$0 \to \hat{H}^\bullet \to \tilde{H}^\bullet \to T_p H^{\bullet+1} \to 0. \tag{1}$$

Since $T_p H^{\bullet+1}$ is p-torsion-free and p-adically complete, we may take the $\mathbb{Z}_p$ dual of this exact sequence to obtain a short exact sequence

$$0 \to \text{Hom}(T_p H^{\bullet+1}, \mathbb{Z}_p) \to \text{Hom}(\tilde{H}^\bullet, \mathbb{Z}_p) \to \text{Hom}(\hat{H}^\bullet, \mathbb{Z}_p) \to 0.$$

Recall from Theorem 1 (3) that $\text{Hom}(\tilde{H}^\bullet, \mathbb{Z}_p)$ is the p-torsion free quotient of $\tilde{H}_\bullet$.

There are similar constructions for compactly supported cohomology, and their formation is compatible with the maps $\tilde{H}_c^\bullet \to \tilde{H}^\bullet$.

2 Non-commutative Iwasawa theory

If M is any finitely generated left $\mathbb{Z}_p[[G_0]]$-module, then we define

$$E^\bullet(M) := \text{Ext}^\bullet(M, \mathbb{Z}_p[[G_0]]).$$

These are again naturally $\mathbb{Z}_p[[G_0]]$-modules. (We compute the Exts via the left $\mathbb{Z}_p[[G_0]]$-module structure on $\mathbb{Z}_p[[G_0]]$, leaving a right $\mathbb{Z}_p[[G_0]]$-module structure on $E^\bullet(M)$. We convert this back to a left-module structure via the canonical anti-involution of $\mathbb{Z}_p[[G_0]]$ induced by $g \mapsto g^{-1}$.) We note that these modules are unchanged (up to a natural isomorphism, and applying the forgetful functor from $\mathbb{Z}_p[[G_0]]$-modules to $\mathbb{Z}_p[[G_0']]$-modules) if we replace G_0 by any open subgroup G_0'.

We define the *codimension* of M (or codim M for short) to be the minimal value of i such that $E^i(M) \neq 0$. If G_0 is commutative, then this agrees with the usual notion of codimension of support of the coherent sheaf associated to M on $\text{Spec}\,\mathbb{Z}_p[[G_0]]$. In general, even if G_0 is non-commutative (which is typically the case), this notion behaves entirely analogously to the usual notion of codimension of support of a sheaf in algebraic geometry [18]. In addition to the codimension of M, it is sometimes convenient to speak of the *dimension* of M (or the *Iwasawa dimension* of M, for emphasis), which we define to be the quantity $1 + \dim G_0 - \text{codim}\,M$. (If we think of codim M as being the codimension of the support of the – in general non-existent – coherent sheaf associated to M on the – in general non-existent – Spec of $\mathbb{Z}_p[[G_0]]$, then the Iwasawa dimension of M would be the dimension of its support.)

We also remark that if we shrink G_0 sufficiently (i.e. replace G_0 by G_r for a large enough value of r), then $\mathbb{Z}_p[[G_0]]$ admits a skew-field of fractions, say $\mathcal{L}$, and hence we can speak of a finitely generated $\mathbb{Z}_p[[G_0]]$-module M being torsion-free (the natural map $M \to \mathcal{L} \otimes M$ is injective) or torsion ($\mathcal{L} \otimes M = 0$), and we can speak of the rank of the (torsion-free part of) M (i.e. the rank of the $\mathcal{L}$-vector space $\mathcal{L} \otimes M$). Having positive rank is equivalent to being of codimension 0, while being torsion (or equivalently, being of rank 0) is equivalent to being of positive codimension.

3 Poincaré duality

We suppose that X_0 (and hence every X_r) is equidimensional of some dimension d. We then have two spectral sequences that express Poincaré duality in the p-adically completed situation (the detailed consructions of which will appear in [6]). Note that both work with homology, rather than cohomology. (The reason for this is that the functors E^i intervene.)

Here are the spectral sequences:

$$E_2^{i,j} := E^i(\tilde{H}_j) \implies \tilde{H}^{BM}_{d-i-j}$$

and

$$E_2^{i,j} := E^i(\tilde{H}_j^{BM}) \implies \tilde{H}_{d-i-j}.$$

They are compatible with the maps $\tilde{H}_\bullet \to \tilde{H}_\bullet^{BM}$.

We point out the amplitude of the δ-functor $E^\bullet$ is given by the dimension of the group G_0. In applications, this can often be quite a bit larger than the dimension d of the spaces X_r, which can lead to interesting tension in these spectral sequences.

4 A simple example of everything so far

An illustrative example is given by taking $G_0 = \mathbb{Z}_p$, $G_r := p^r\mathbb{Z}_p$ for $r > 0$, and $X_r := \mathbb{R}/p^r\mathbb{Z}$ for each $r \geq 0$. We let $G_0/G_r := \mathbb{Z}_p/p^r\mathbb{Z}_p \xrightarrow{\sim} \mathbb{Z}/p^r\mathbb{Z}$ act on X_r in the obvious manner by translations. Thus each X_r is a circle, and each map $X_{r+1} \to X_r$ is a degree p covering map. Note that $\mathbb{Z}_p[[G_0]] = \mathbb{Z}_p[[T]]$.

Clearly $\tilde{H}_0 = \mathbb{Z}_p$, with trivial G_0-action, while $\tilde{H}_1 = 0$ (the inverse limit of a sequence of copies of $\mathbb{Z}_p$ under the multiplication by p map obviously vanishes). Similarly $\tilde{H}^0 = \mathbb{Z}_p$, while $\tilde{H}^1 = 0$. (Since the spaces X_r are compact, Borel–Moore homology agrees with usual homology, and compactly supported cohomology agrees with usual cohomology.)

Since G_r is procyclic, we see that $H^0(G_r, \mathbb{Z}_p) = H^1(G_r, \mathbb{Z}_p) = \mathbb{Z}_p$, and thus we do indeed recover the cohomology of the circle X_r from the completed cohomology via the Hochschild–Serre type spectral sequence.

Similarly, using the resolution

$$0 \to \mathbb{Z}_p[[T]] \xrightarrow{T\cdot} \mathbb{Z}_p[[T]] \to \mathbb{Z}_p \to 0,$$

we compute that $E^0(\mathbb{Z}_p) = 0$, while $E^1(\mathbb{Z}_p) = \mathbb{Z}_p$. This is immediately seen to be consistent with the Poincaré duality spectral sequence.

5 Congruence quotients of symmetric spaces

We now suppose that $\mathbb{G}$ is a connected reductive linear algebraic group over $\mathbb{Q}$. We let $\mathbb{A}$ denote the adèle ring over $\mathbb{Q}$, let $\mathbb{A}^\infty$ denote the ring of finite adèles, and let $\mathbb{A}^{\infty,p}$ denote the ring of prime-to-p finite adèles. We fix a compact open subgroup $K^{\infty,p}$ of $\mathbb{G}(\mathbb{A}^{\infty,p})$ (the tame level), take G_0 to be a sufficiently small compact open subgroup of $\mathbb{G}(\mathbb{Q}_p)$, and let $(G_r)_{r\geq 1}$ be a sequence of normal open subgroups of G_0 that form a neighbourhood basis of the identity in G_0.

We let K_∞° denote the connected component of the identity of a maximal compact subgroup K_∞ of $\mathbb{G}(\mathbb{R})$, and let A_∞° denote the connected component of the identity of the group of real points A_∞ of a maximal $\mathbb{Q}$-split torus in the centre of $\mathbb{G}$.

We define

$$X_r := \mathbb{G}(\mathbb{Q}) \backslash \mathbb{G}(\mathbb{A}) / A_\infty^\circ K_\infty^\circ G_r K^{\infty, p}.$$

The X_r form a tower with an action of G_0 satisfying the axioms of Section 1, and so the preceding theory applies. Furthermore, the usual argument about limits of (co)homology of arithmetic quotients in the adèlic setting shows that $\tilde{H}_\bullet$, $\tilde{H}^\bullet$, etc., all inherit not just an action of G_0, but of the entire p-adic group $\mathbb{G}(\mathbb{Q}_p)$. They also inherit an action of a suitably completed Hecke algebra $\mathbb{T}$ (built up out of spherical Hecke operators at primes $\ell \neq p$ over which $\mathbb{G}$ and the tame level $K^{\infty, p}$ are unramified), and an action of the component group $\pi_0 := (A_\infty K_\infty)/(A_\infty^\circ K_\infty^\circ)$.

The following theorem is the main result of [4]. (The result about $T_p H^{\bullet+1}$ being torsion, even in the middle dimension in the discrete series case, is not stated there, but follows from the proof.)

Theorem 4 *If, in the preceding setting, the group $\mathbb{G}$ is semi-simple, then the modules $\tilde{H}_n$ are torsion $\mathbb{Z}_p[[G_0]]$-modules, unless the group $\mathbb{G}(\mathbb{R})$ admits discrete series and n is the "middle dimension" (i.e. one half of the dimension of the X_r). In this latter case, $\mathrm{Hom}(\hat{H}^n, \mathbb{Z}_p)$ (and hence $\tilde{H}_n$) has positive rank, while $\mathrm{Hom}(T_p H^{n+1}, \mathbb{Z}_p)$ is torsion over $\mathbb{Z}_p[[G_0]]$.*

We believe that this is actually only the beginning of the story with regard to the codimensions of the various $\tilde{H}_\bullet$. In the following section, we will go on to describe how we expect the rest of the story to play out. But first, we will close this section by discussing the boundary long exact sequence.

Each X_r embeds as an open subset of its Borel–Serre compactification $\overline{X}_r$. We let ∂X_r denote $\overline{X}_r \setminus X_r$; it is the boundary of $\overline{X}_r$. As r varies the spaces $\overline{X}_r$ form a tower to which the p-adically completed cohomology machinery applies, and hence we may define completed (co)homology spaces $\tilde{H}_\bullet(\partial)$ and $\tilde{H}^\bullet(\partial)$. There are long exact sequences

$$\cdots \to \tilde{H}_\bullet(\partial) \to \tilde{H}_\bullet \to \tilde{H}_\bullet^{BM} \to \tilde{H}_{\bullet-1}(\partial) \to \cdots \tag{2}$$

and

$$\cdots \to \tilde{H}^{\bullet-1}(\partial) \to \tilde{H}_c^\bullet \to \tilde{H}^\bullet \to \tilde{H}^\bullet(\partial) \to \cdots. \tag{3}$$

The complement ∂X_r is a union of strata $\partial X_{\mathbb{P}, r}$ indexed by the (equivalence classes of) proper parabolic subgroups $\mathbb{P}$ of $\mathbb{G}$. For fixed $\mathbb{P}$ and varying r the strata $\partial X_{\mathbb{P}, r}$ again form a tower to which the p-adically completed cohomology machine applies, allowing us to define $\tilde{H}_\bullet(\partial_{\mathbb{P}})$ and $\tilde{H}^\bullet(\partial_{\mathbb{P}})$. There is a Mayer–Vietoris spectral sequence relating the various $\tilde{H}_\bullet(\partial_{\mathbb{P}})$ (resp. $\tilde{H}^\bullet(\partial_{\mathbb{P}})$) to $\tilde{H}_\bullet(\partial)$ (resp. $\tilde{H}^\bullet(\partial)$).

If we let $\mathbb{P} = \mathbb{MN}$ be a Levi decomposition of the proper parabolic $\mathbb{P}$, then each $\partial_{\mathbb{P},r}$ is a bundle over a union of congruence quotients associated to $\mathbb{M}$, whose fibre is a compact nilmanifold (of dimension equal to $\dim \mathbb{N}$). A generalization to a nilmanifolds of the example of Section 4 shows that these nilmanifolds contribute to completed (co)homology only in degree 0, and so $\tilde{H}_\bullet(\partial_{\mathbb{P}})$ and $\tilde{H}^\bullet(\partial_{\mathbb{P}})$ are supported in the same degrees as the p-adically completed (co)homology associated to $\mathbb{M}$. To describe the precise relationship between the p-adically completed cohomology for $\partial_{\mathbb{P}}$ and for $\mathbb{M}$, it is convenient to take a direct limit over all tame levels. We then find (at least when p is odd) that:

$$\varinjlim_{\text{tame levels}} \tilde{H}^\bullet(\partial_{\mathbb{P}})$$

$$\xrightarrow{\sim} \operatorname{Ind}_{\mathbb{P}(\mathbb{A}_f)}^{\mathbb{G}(\mathbb{A}_f)}\Big(\varinjlim_{\text{tame levels}} \text{invariants under } \ker(\pi_0^{\mathbb{M}} \to \pi_0)$$

$$\text{of the } \tilde{H}^\bullet \text{ associated to } \mathbb{M}\Big), \quad (4)$$

where the induction is p-adically completed at p, and smooth at primes away from p. (We have written $\pi_0^{\mathbb{M}}$ to denote the analogue for the Levi subgroup $\mathbb{M}$ of the component group π_0 which was defined above for $\mathbb{G}$. Since π_0 and $\pi_0^{\mathbb{M}}$ are finite 2-groups, there is the possibility of extra complications when $p = 2$, which we won't attempt to address here.)

We note that much of this general set-up for congruence quotients has been described by Richard Hill in [12, §4].

6 Conjectures on codimensions

We maintain the set-up of the preceding section. We write $G_\infty := \mathbb{G}(\mathbb{R})$, and we begin by defining two quantities associated to $\mathbb{G}$:

$$l_0 := \text{rank of } G_\infty - \text{rank of } A_\infty K_\infty,$$

and

$$q_0 = (\text{ dimension of } G_\infty - \text{ dimension of } A_\infty K_\infty - l_0)/2 = (d - l_0)/2,$$

where d denotes the dimension of the quotients X_r. Note that if $\mathbb{G}$ is semisimple, then these quantities coincide with the quantities denoted by the same symbols in Borel–Wallach [2]. Namely, l_0 denotes the "defect" of G_∞ with regard to possessing discrete series, while q_0 denotes the first "interesting" dimension for the (co)homology of X_r.

We now state our basic conjecture on codimensions.

Conjecture 5 (1) If $n < q_0$, then the codimension of $\tilde{H}_n$ is greater than $l_0 + q_0 - n$.

(2) The codimension of $\tilde{H}_{q_0}$ equals l_0.

(3) If $n < q_0$, then the codimension of $\tilde{H}_n^{BM}$ is greater than $l_0 + q_0 - n$.

(4) The codimension of $\tilde{H}_{q_0}^{BM}$ equals l_0.

(5) $\tilde{H}_{q_0}$ is p-torsion-free.

(6) $\tilde{H}_{q_0}^{BM}$ is p-torsion-free.

(7) $\tilde{H}_n = 0$ if $n > q_0$.

(8) $\tilde{H}_n^{BM} = 0$ if $n > q_0$.

In fact, these conjectures are not independent, as the following theorem makes clear.

Theorem 6 *(1) If parts (1) and (2) (resp. parts (3) and (4)) of Conjecture 5 hold for G, and for all the proper Levi subgroups of G, then parts (3) and (4) (resp. parts (1) and (2)) of the conjecture also holds for G (and also for all its Levi subgroups), as do parts (7) and (8).*

(2) If part (7) (resp. part (8)) of Conjecture 5 holds for $G \times G$, then part (5) (resp. part (6)) of the conjecture holds for G.

Proof Suppose that parts (1) and (2), or parts (3) and (4), of the conjecture hold for all the proper Levi subgroups of G. Applying the theorem inductively, we conclude that parts (1), (2), (3), (4), (5), and (6) of the conjecture hold for these Levi subgroups. Formula (4) then allows us to bound from below the codimensions of the p-adically completed homology spaces associated to the various boundary strata. A comparison of the invariants l_0 and q_0 for G and for its Levi subgroups, along with a consideration of the Mayer–Vietoris spectral sequence that computes the p-adically completed homology of the boundary in terms of the p-adically completed cohomology of its various strata, as well as of the long exact sequence (2), then shows that parts (1) and (2) of the conjecture for G are equivalent to parts (3) and (4) for G. Looking at the Poincaré duality spectral sequence then shows that parts (7) and (8) of the conjecture also hold.

Part (2) of the theorem follows by applying a Kunneth-type theorem to compare the completed cohomology for G to that for $G \times G$. ☐

Remark 7 In particular, the preceding theorem shows that if either parts (1) and (2), or part (3) and (4), of Conjecture 5 hold for every group $\mathbb{G}$, then all parts of the conjecture hold for all groups $\mathbb{G}$.

Remark 8   Richard Hill has also made a conjecture about the vanishing of completed cohomology of arithmetic quotients, namely [12, Conj. 3]. Conjecture 5 implies Hill's conjecture, but is quite a bit stronger in general. The various vanishing results proved in [12, §5], being consistent with [12, Conj. 3], are thus also consistent with our general conjecture.

Remark 9   Eric Urban has made a conjecture somewhat analogous to Conjecture 5 concerning the dimensions of irreducible components of "eigenvarieties" [17, Conj. 5.7.3]. In fact, since "eigenvarieties" can also be constructed from completed cohomology via applying the locally analytic Jacquet module functor of [7] (see [8]), it seems likely that the eventual relationship of Conjecture 5 to Urban's conjecture will be more than one of mere analogy. On the other hand, neither conjecture seems to be an immediate logical consequence of the other, and a precise understanding of their mutual relationship remains to be found.

There are several different heuristics and motivations behind Conjecture 5 (as well as a small amount of actual evidence); some of these heuristics are explained in Section 8. For now, let us remark that it is consistent with Künneth, and it is consistent with the Poincaré duality spectral sequence. Of course, in the case when $\mathbb{G}$ is semi-simple, it is also consistent with Theorem 4, which shows at least that the codimension of $\tilde{H}_n$ is zero (respectively, positive) exactly when it is predicted to be so by the conjecture.

We now discuss some illustrative examples.

Example 10   If $\mathbb{G}$ is a torus, then $q_0 = 0$, and a generalization of the analysis of the example in Section 4 shows that in this case, Conjecture 5 is equivalent to Leopoldt's conjecture. (See [12, Cor. 5].)

Example 11   If $\mathbb{G} = \mathrm{SL}_2$, then $l_0 = 0$ and $q_0 = 1$, and one finds that $\tilde{H}_0 = \mathbb{Z}_p$ has codimension 3, while $\tilde{H}_1$ has codimension zero. (Apply Theorem 4.)

Example 12   If K is a quadratic imaginary field, and $\mathbb{G} = \mathrm{SL}_{2/K}$ (regarded as a group over $\mathbb{Q}$ by restriction of scalars, as usual), then $l_0 = 1$ and $q_0 = 1$, and one finds that $\tilde{H}_0 = \mathbb{Z}_p$, and so has codimension 6, that $\tilde{H}_1$ has codimension 1, and that $\tilde{H}_2 = 0$. (Apply Theorem 4 together with the Poincaré duality spectral sequence. Note that the term $E^6(\tilde{H}_0) = E^6(\mathbb{Z}_p) = \mathbb{Z}_p$ plays a crucial role in showing that $\tilde{H}_1 \neq 0$.)

Example 13   If $\mathbb{G}$ is semi-simple and simply connected and satisfies the congruence subgroup property, then $\tilde{H}_1$ is finite.

Example 14 If $G = \mathrm{Sp}_4$, then $l_0 = 0$ and $q_0 = 3$. One finds that $\tilde{H}_0 = \mathbb{Z}_p$, $\tilde{H}_1$ is finite (since $\mathbb{G}$ satisfies the congruence subgroup property), $\tilde{H}_2$ has codimension at least 1 (by Theorem 4), and $\tilde{H}_3$ has codimension 0 (again by Theorem 4). The Poincaré duality spectral sequence then shows that $\tilde{H}_n = 0$ for $n \geq 4$.

Remark 15 It is not clear in general how one might go about computing the codimension of support of (or any other information about) the completed cohomology. However, suppose one knew that the $\mathbb{Z}_p$-cohomology of each X_r was p-torsion-free (or, more generally, of bounded exponent). The Tate module term in the exact sequence (1) would then vanish, and hence there would be an isomorphism $\hat{H}^{\bullet} \overset{\sim}{\longrightarrow} \tilde{H}^{\bullet}$. Now $\hat{H}^{\bullet}$ is the p-adic completion of the classical cohomology, and this classical cohomology may be described in terms of classical automorphic forms (at least after tensoring with $\mathbb{Q}_p$ over $\mathbb{Z}_p$). Thus it seems not totally inconceivable that one might be able to prove results about (e.g. the codimension of support of) $\hat{H}^{\bullet}$. (This is done for modular curves in [9], although admittedly the arguments there rely on the full strength of the p-adic Langlands program for $\mathrm{GL}_2(\mathbb{Q}_p)$.) Consequently, it seems worth investigating and attempting to establish torsion-freeness results for the cohomology of congruence quotients, say in the Shimura variety context.

7 Mod p analogues

Returning to the notation of Section 1, we observe that we may define mod p analogues of completed homology and cohomology. Namely, we write

$$\tilde{H}_{\bullet, \mathbb{F}_p} := \varprojlim_r H_{\bullet}(X_r, \mathbb{F}_p)$$

and

$$H^{\bullet}_{\mathbb{F}_p} := \varinjlim_r H^{\bullet}(X_r, \mathbb{F}_p).$$

We can also consider Borel–Moore and compactly supported variants:

$$\tilde{H}^{BM}_{\bullet, \mathbb{F}_p} := \varprojlim_r H^{BM}_{\bullet}(X_r, \mathbb{F}_p)$$

and

$$H^{\bullet}_{c, \mathbb{F}_p} := \varinjlim_r H^{\bullet}_c(X_r, \mathbb{F}_p).$$

There are natural maps $\tilde{H}_{\bullet, \mathbb{F}_p} \rightarrow \tilde{H}^{BM}_{\bullet, \mathbb{F}_p}$ and $H^{\bullet}_{c, \mathbb{F}_p} \rightarrow H^{\bullet}_{\mathbb{F}_p}$. Note that in the case of cohomology, there is no completion at all. We then have the following result (whose proof will appear in [6]), which provides the analogue of Theorem 1 for $\mathbb{F}_p$-coefficients, as well as universal coefficient type results relating the completed (co)homology to its mod p variant.

Theorem 16 *(1) There is a natural action of the completed group ring $\mathbb{F}_p[[G_0]]$ on each of $\tilde{H}_{\bullet,\mathbb{F}_p}$ and $\tilde{H}_{\bullet,\mathbb{F}_p}^{BM}$, which makes each of these spaces a finitely generated left module over $\mathbb{F}_p[[G_0]]$. Furthermore, the canonical topology on each of these modules (obtained by writing it as a quotient of a finite direct sum of copies of $\mathbb{F}_p[[G_0]]$) coincides with projective limit topology.*

(2) The spaces $H_{\mathbb{F}_p}^\bullet$ and $H_{c,\mathbb{F}_p}^\bullet$ are admissible smooth representations of G_0 over $\mathbb{F}_p$.

(3) There are natural isomorphisms

$$H_{\mathbb{F}_p}^\bullet \xrightarrow{\sim} \mathrm{Hom}_{\mathrm{cont}}(\tilde{H}_{\bullet,\mathbb{F}_p}, \mathbb{F}_p)$$

(here cont *means continuous with respect to the canonical topology on the source, and the evident topology – either discrete or p-adic – on the target) and*

$$\tilde{H}_{\bullet,\mathbb{F}_p} \xrightarrow{\sim} \mathrm{Hom}(H_{\mathbb{F}_p}^\bullet, \mathbb{F}_p)$$

(since the natural topology on $\tilde{H}^\bullet$ is the p-adic topology, and since all $\mathbb{Z}_p$-linear maps are automatically p-adically continuous, there is no need to explicitly specify any continuity conditions on the maps appearing in this exact sequence), as well as similar isomorphisms relating $\tilde{H}_{\bullet,\mathbb{F}_p}^{BM}$ and $H_{c,\mathbb{F}_p}^\bullet$. These isomorphisms are compatible with the maps $\tilde{H}_{\bullet,\mathbb{F}_p} \to \tilde{H}_{\bullet,\mathbb{F}_p}^{BM}$, and the maps $H_{c,\mathbb{F}_p}^\bullet \to H_{\mathbb{F}_p}^\bullet$.

(4) For each $r \geq 0$, there are Hochschild–Serre type spectral sequences

$$E_2^{i,j} := H^i(G_r, H^j) \implies H^{i+j}(X_r, \mathbb{F}_p)$$

and

$$E_2^{i,j} := H^i(G_r, H_c^j) \implies H_c^{i+j}(X_r, \mathbb{F}_p),$$

compatible with the maps $H_{c,\mathbb{F}_p}^\bullet \to H_{\mathbb{F}_p}^\bullet$.

(5) There are short exact sequences

$$0 \to \tilde{H}_\bullet/p\tilde{H}_\bullet \to \tilde{H}_{\bullet,\mathbb{F}_p} \to \tilde{H}_{\bullet-1}[p] \to 0$$

and

$$0 \to \tilde{H}^\bullet/p\tilde{H}^\bullet \to H_{\mathbb{F}_p}^\bullet \to \tilde{H}^{\bullet+1}[p] \to 0,$$

as well as similar isomorphisms relating $\tilde{H}_{\bullet,\mathbb{F}_p}^{BM}$ to $\tilde{H}_\bullet^{BM}$ and $H_{c,\mathbb{F}_p}^\bullet$ to $\tilde{H}_c^\bullet$.

If M is any finite generated left $\mathbb{F}_p[[G_0]]$-module, we define $E^{\bullet}_{\mathbb{F}_p}(M) :=$ $\text{Ext}^{\bullet}(M, \mathbb{F}_p[[G_0]])$. We can then define the $\mathbb{F}_p$-*codimension* of M (or $\text{codim}_{\mathbb{F}_p} M$, for short) to be the minimal i such that $E^i_{\mathbb{F}_p}(M) \neq 0$. Of course, we may also regard M as a left $\mathbb{Z}_p[[G_0]]$-module, and there are short exact sequences

$$0 \to E^{\bullet}(M) \to E^{\bullet}_{\mathbb{F}_p}(M) \to E^{\bullet+1}(M) \to 0$$

(as follows by applying $\text{Ext}^{\bullet}(M, -)$ to the short exact sequence $0 \to \mathbb{Z}_p[[G_0]] \xrightarrow{p\cdot} \mathbb{Z}_p[[G_0]] \to \mathbb{F}_p[[G_0]] \to 0$), from which one deduces that the $\mathbb{F}_p$-codimension of M is one less than the codimension of M. In particular, we find that the Iwasawa dimension of M is equal to $\dim G_0 - \text{codim}_{\mathbb{F}_p} M$.

If M is a finitely generated $\mathbb{Z}_p[[G_0]]$-module, then M/pM is a finitely generated $\mathbb{F}_p[[G_0]]$-module, and we have the following simple lemma.

Lemma 17 *If M is a finitely generated $\mathbb{Z}_p[[G_0]]$-module which is furthermore p-torsion free, then the codimension of M and the $\mathbb{F}_p$-codimension of M/pM coincide.*

Proof Computing $E^i(M)$ using a projective resolution of M, one immediately finds (using the p-torsion-freeness of M) that

$$E^{\bullet}(M)/pE^{\bullet}(M) \xrightarrow{\sim} E^{\bullet}_{\mathbb{F}_p}(M/pM).$$

The lemma follows from this. $\square$

Returning to the case when M is a finitely generated $\mathbb{F}_p[[G_0]]$-module, we note that $V := \text{Hom}_{\text{cont}}(M, \mathbb{F}_p)$, when equipped with the contragredient action of G_0, is then an admissible smooth representation of G_0 over $\mathbb{F}_p$. (We also recall that, conversely, M can be recovered from V via the double duality isomorphism $M = \text{Hom}(V, \mathbb{F}_p)$.) One nice aspect of the mod p situation is that the Iwasawa dimension of M can be recovered from the behaviour of the corresponding smooth representation V "at finite level", as the following theorem shows. (For the first statement of the following result, see [1]; the proof of the second statement will appear in [6].)

Theorem 18 *Let M be a finitely generated $\mathbb{F}_p[[G_0]]$-module, and write $V :=$ $\text{Hom}_{\text{cont}}(M, \mathbb{F}_p)$, so that V is an admissible smooth G_0-representation over $\mathbb{F}_p$. If the Iwasawa dimension of M is equal to d, then we have that $\dim H^0(G_r, V) \sim$ $\lambda \cdot [G_0 : G_r]^{\frac{d}{\dim G_0}}$, for some $\lambda > 0$, while $\dim H^i(G_r, V) = O([G_0 : G_r]^{\frac{d}{\dim G_0}})$ for all $i > 0$.*

Remark 19 If G is (the group of $\mathbb{Q}_p$-points of) a reductive group over $\mathbb{Q}_p$, if V is an irreducible admissible smooth representation of G over $\mathbb{C}$, and if $\{G_r\}_{r \geq 0}$

is a neighbourhood basis of the identity of G consisting of compact open subgroups, then $\dim H^0(G_r, V) \sim \lambda \cdot [G_0 : G_r]^{\frac{d_{GK}}{\dim G}}$, for some $\lambda > 0$, where d_{GK} is the Gelfand–Kirillov dimension of V. Theorem 18 thus suggests an analogy between Gelfand–Kirillov dimensions of admissible smooth representations in characteristic zero, and Iwasawa dimensions of duals to admissible smooth representations in characteristic p. It remains to be seen how fruitful this analogy is.

Turning now to the context of Sections 5 and 6, we believe that the analogue of Conjecture 5 above, with codimensions replaced by $\mathbb{F}_p$-codimensions, should hold for $\tilde{H}_{\bullet,\mathbb{F}_p}$ and $\tilde{H}^{BM}_{\bullet,\mathbb{F}_p}$. In fact, Conjecture 5 and its mod p analogue are far from independent. The following result gives one example of how they are related.

Proposition 20 *If $\tilde{H}_{q_0}$ is p-torsion-free, then $\tilde{H}_{q_0}$ has codimension equal to l_0 if and only if $\tilde{H}_{q_0,\mathbb{F}_p}$ has $\mathbb{F}_p$-codimension equal to l_0.*

Proof This follows immediately from part (5) of Theorem 16, together with Lemma 17.   □

We also have the following interpretation of (a somewhat weakened form) of the mod p analogue of Conjecture 5 in terms of the rate of growth of $\mathbb{F}_p$-cohomology up the congruence tower X_r.

Proposition 21 *The following are equivalent:*

(1) $\tilde{H}_{i,\mathbb{F}_p}$ has codimension $> l_0$ for $i < q_0$, and $\tilde{H}_{q_0,\mathbb{F}_p}$ has codimension l_0;

(2) $\dim H^i(X_r, \mathbb{F}_p) = O(V(X_r))^{1-\frac{l_0+1}{\dim G}}$ if $i < q_0$, while

$$\dim H^{q_0}(X_r, \mathbb{F}_p) \asymp V(X_r)^{1-\frac{l_0}{\dim G}}.$$

Proof The equivalence of the two conditions follows from Theorem 18 together with part (4) of Theorem 16.   □

Of course, there is an analogous statement relating the codimension of the modules $\tilde{H}^{BM}_{i,\mathbb{F}_p}$ to the rate of growth of compactly supported $\mathbb{F}_p$-cohomology up the tower.

Remark 22 If we assume that not only the equivalent conditions (1) and (2) hold, but that the analogous statements also hold for each Levi factor of G, then we may use a Poincaré duality argument, together with a comparison of $\tilde{H}^{BM}_{\bullet,\mathbb{F}_p}$ and $\tilde{H}_{\bullet,\mathbb{F}_p}$, to deduce that furthermore $\dim H^i(X_r, \mathbb{F}_p) = O(V(X_r))^{1-((l_0+1)/d)}$ if $i > l_0 + q_0$.

8 Heuristics related to the *p*-adic Langlands programme

Throughout this section we fix an isomorphism $\iota : \mathbb{C} \xrightarrow{\sim} \overline{\mathbb{Q}}_p$. If $\mathbb{G}$ is a reductive group over $\mathbb{Q}$, then the Langlands programme [13], when coupled with the conjecture of Fontaine–Mazur [10], posits the existence of a correspondence of the form

$$\{\text{algebraic automorphic Hecke eigenforms on } \mathbb{G}(\mathbb{A})\}$$

$$\longleftrightarrow$$

$$\{\text{continuous representations } G_{\mathbb{Q}} \to \mathbb{G}^L(\overline{\mathbb{Q}}_p) \text{ which are unramified} \tag{5}$$

$$\text{outside finitely many primes and are de Rham at } p\},$$

where $G_{\mathbb{Q}}$ denotes the absolute Galois group of $\mathbb{Q}$, and $\mathbb{G}^L$ denotes the *L*-group of G (thought of as a reductive group over $\mathbb{Q}$). There are many subtleties involved in making a conjecture of this form precise, beginning perhaps with the precise definition of *algebraic* in the context of automorphic forms, and we won't attempt to discuss them here, or to state a precise conjecture, referring instead to the recent paper [3], which gives this issue the careful consideration that it deserves.

We content ourselves by remarking that the matching (such as it exists) between automorphic forms and Galois representations is supposed to be made in the following manner (which depends on our chosen isomorphism $\iota : \mathbb{C} \xrightarrow{\sim} \overline{\mathbb{Q}}_p$): an algebraic automorphic Hecke eigenform f and a *p*-adic Galois representation ρ match if, for some sufficiently large finite set of primes S, containing p together with all the primes dividing either the conductor of f or the conductor of ρ, we have that for each prime $\ell \notin S$, the ℓth Langlands parameter computed by applying the Satake isomorphism (suitably normalized) to the system of Hecke eigenvalues at the prime ℓ attached to f, which is a semi-simple conjugacy class in $\mathbb{G}^L(\mathbb{C})$, coincides (under the isomorphism $\mathbb{G}^L(\mathbb{C}) \xrightarrow{\sim} \mathbb{G}^L(\overline{\mathbb{Q}}_p)$ induced by ι) with the semi-simple part of the conjugacy class in $\mathbb{G}^L(\mathbb{C})$ obtained by applying ρ to the conjugacy class of Frobenius elements at ℓ in $G_{\mathbb{Q}}$. We refer to [3, Conj. 3.2.1] and the subsequent remarks for a more precise statement and discussion.

We note in addition that to an algebraic automorphic eigenform we can associate a collection of integral "Hodge numbers" ([3, Rem. 3.2.3]; these are constructed using the Langlands parameter of the representation of $\mathbb{G}(\mathbb{R})$ generated by f, or alternatively, using the infinitesimal character of f), and that if f and ρ match in the above sense, then the Hodge numbers associated to f should match with the Hodge–Tate weights of the de Rham representation $\rho_{|G_{\mathbb{Q}_p}}$.

Now the property of being algebraic is defined in terms of these "Hodge numbers" being integral [3] (the precise definition of algebraic then depending on exactly how the Hodge numbers are normalized [3]). It is then possible to enlarge the automorphic side of the conjectured correspondence by considering automorphic forms that are non-algebraic, i.e. whose associated "Hodge numbers" are not integral. It is also possible to enlarge the Galois side of the conjectured correspondence by considering automorphic forms that are non-de Rham. These representations have associated Hodge–Sen–Tate weights which are typically non-integral.

Unfortunately, it seems that when we enlarge the two sides of the correspondence in this manner, the enlarged sets are not related to each other. To understand why, note that on the automorphic side we are enlarging the set of allowed Hodge numbers in an archimedean manner, while on the Galois side, we are doing so in a p-adic manner, and there is no reason (that we know of) to think that these two kinds of enlargements should have any relationship to one another.

What we would like, in order to extend the Langlands correspondence so as to allow non-de Rham Galois representations to be included, is to introduce a notion of p-adic automorphic form which generalizes the notion of algebraic automorphic form, whose associated "Hodge numbers" (whatever they might be) are p-adic rather than archimedean. We believe that p-adically completed cohomology provides a partial (and slightly indirect) solution to the problem of defining such a notion of p-adic automorphic form.

We put ourselves in the context of Section 5. As noted there, the completed homology modules $\tilde{H}_n$ admit an action of a completed Hecke algebra $\mathbb{T}$.

Franke's theorem [11] describing the cohomology of congruence quotients in terms of automorphic forms shows that any system of Hecke eigenvalues appearing in some degree of cohomology (with $\mathbb{C}$-coefficients, or more generally with coefficients in some local system associated to an algebraic representation of $\mathbb{G}_{/\mathbb{C}}$) of $\mathbb{G}(\mathbb{Q})\backslash\mathbb{G}(\mathbb{A})/A_\infty K_\infty G_r K^{\infty,p}$, is also a system of Hecke eigenvalues arising from some algebraic automorphic Hecke eigenform (more precisely, a C-algebraic eigenform in the sense of [3]) that is invariant under $K^{\infty,p}$. We call such automorphic Hecke eigenforms *cohomological Hecke eigenforms of tame level $K^{\infty,p}$*.

Using our fixed isomorphism $\iota : \mathbb{C} \to \mathbb{Q}_p$, we may identify cohomology with coefficients in $\mathbb{C}$ (or a local system associated to an algebraic representation of $\mathbb{G}_{/\mathbb{C}}$) with cohomology with coefficients in $\overline{\mathbb{Q}}_p$ (or a local system associated to an algebraic representation of $\mathbb{G}_{/\overline{\mathbb{Q}}_p}$). One can verify that the Hochschild–Serre-type spectral sequence of Theorem 1 (4) (and its generalization for local systems mentioned in Remark 2) is Hecke equivariant, and so we conclude that each cohomological Hecke eigenform of tame level $K^{\infty,p}$ gives rise to a point $\lambda \in \operatorname{Spec} \mathbb{T}(\overline{\mathbb{Q}}_p)$.

We expect that $\operatorname{Spec} \mathbb{T}$ will be topologically finitely generated as a $\mathbb{Z}_p$-algebra. The point of the preceding discussion is that every system of Hecke eigenvalues associated to a cohomological Hecke eigenform of tame level $K^{\infty,p}$ appears as a point of $\operatorname{Spec} \mathbb{T}(\overline{\mathbb{Q}}_p)$. Thus we regard $\operatorname{Spec} \mathbb{T}(\overline{\mathbb{Q}}_p)$ as a space parameterizing "p-adic cohomological systems of Hecke eigenvalues"; points of this space serve as a working model for the notion of (cohomological) p-adic automorphic Hecke eigenvalues.

We then expect that the conjectural correspondence (5) (or, more precisely, its restriction to the cohomological Hecke eigenforms) should extend to identify $\mathbb{T}$ with a space of p-adic Galois representations. Just slightly more precisely, we write the complete ring $\mathbb{T}$ as a product $\mathbb{T} = \prod_{\mathfrak{m}} \mathbb{T}_{\mathfrak{m}}$ of its local factors. Attached to each $\mathfrak{m}$, and to a choice of an embedding $\mathbb{T}/\mathfrak{m} \hookrightarrow \overline{\mathbb{F}}_p$, we expect that there should be an associated Galois representation $\rho_{\mathfrak{m}} : G_{\mathbb{Q}} \to \mathbb{G}^L(\overline{\mathbb{F}}_p)$. (This is quite possibly an oversimplification, as the discussion of [3] shows,[2] or perhaps more accurately, it is an idealization of the actual situation; however, it will serve well enough for our present purpose, which is only to explain our motivations in the most general way.) Associated to $\rho_{\mathfrak{m}}$ is a deformation ring $R_{\rho_{\mathfrak{m}}}$ parameterizing deformations of $\rho_{\mathfrak{m}}$. (Here we should impose ramification conditions away from p expressing the idea that the prime-to-p conductor of the deformations considered is bounded by $K^{\infty,p}$ — this is a relatively minor point. The key point is that we impose no ramification conditions at p. Since we are only discussing heuristics here, we don't try to make our discussion more precise; in particular, we don't attempt to address the precise meaning of $R_{\rho_{\mathfrak{m}}}$ in the case when $\rho_{\mathfrak{m}}$ is "irreducible" (i.e. factors through a parabolic of $\mathbb{G}^L$).) What we then expect is that there will be an isomorphism $\mathbb{T}_{\mathfrak{m}} \xrightarrow{\sim} R_{\rho_{\mathfrak{m}}}$.

Now one can compute the expected dimension of $R_{\rho_{\mathfrak{m}}}$ via the global Euler characteristic formula [15], and what one finds is that expected dimension of $R_{\rho_{\mathfrak{m}}}$ is $1 + \dim \mathbb{B} - l_0$, where $\mathbb{B}$ is a Borel subgroup of $\mathbb{G}$. Thus we expect that each $\mathbb{T}_{\mathfrak{m}}$, and so $\mathbb{T}$ itself, should have Krull dimension $1 + \dim \mathbb{B} - l_0$.

Now consider $\tilde{H}_{q_0}$ as $\mathbb{T}$-module. We in fact hope that it will be faithful (in other words, there are no systems of eigenvalues appearing in $\tilde{H}_n$ for $n \neq q_0$ which don't also appear in $\tilde{H}_{q_0}$), the reason being that q_0 is the first dimension in which tempered representations of G_∞ can contribute cohomology [2], and

[2] To be just slightly less oblique, we are (in the terminology of [3]) considering C-algebraic, rather than L-algebraic, automorphic forms, and the discussion of [3], especially Section 5, shows that it may be necessary to replace G by a z-extension in order to get a correspondence between C-algebraic automorphic eigenforms and p-adic Galois representations.

so is the first "interesting dimension of cohomology". Since it is the first dimension which knows about the interesting automorphic forms, we also expect that it knows about all the interesting Galois representations, and we hope that all the Galois representations are in the Zariski closure of the interesting ones.

Assuming this faithfulness, one can then naively view $\tilde{H}_{q_0}$ as a kind of bundle over $\operatorname{Spec} \mathbb{T}$, and one might then guess the following dimension formula:

$$\text{Iwasawa dim. of } \tilde{H}_{q_0}$$
$$\stackrel{?}{=} \text{Krull dim. of } \operatorname{Spec} \mathbb{T} + \text{Iwasawa dim. of the fibres.} \quad (6)$$

What do we expect the Iwasawa dimension of a fibre to be? The fibre of $\tilde{H}_{q_0}$ over a point $\lambda \in \operatorname{Spec} \mathbb{T}(\overline{\mathbb{Q}}_p)$ associated to a cohomological Hecke eigenform is at least morally dual to the λ-eigenspace in cohomology, and the rough analogy between Iwasawa dimensions and Gelfand–Kirillov dimensions of smooth representations (see Theorem 18 and Remark 19) then suggests that the dimension of this fibre might be equal to the Gelfand–Kirillov dimension of the λ-eigenspace in cohomology, which in turn will be equal to $\dim \mathbb{G}/\mathbb{B}$ in the generic situation. (An irreducible admissible smooth representation of $\mathbb{G}(\mathbb{Q}_p)$ has Gelfand–Kirillov dimension at most equal to $\dim \mathbb{G}/\mathbb{B}$, with equality precisely when it is generic, i.e. when it admits a Whittaker model.) Combining this final heuristic with (6) and our computation of the expected dimension of $\mathbb{T}$, we are left with the following guess:

$$\text{Iwasawa dim. of } \tilde{H}_{q_0} \stackrel{?}{=} 1 + \dim \mathbb{B} - l_0 + \dim \mathbb{G}/\mathbb{B} = 1 + \dim \mathbb{G} - l_0.$$

Equivalently, we guess that the codimension of $\tilde{H}_{q_0}$ is equal to l_0, which is Conjecture 5 (2). The preceding discussion provides one of the main motivations for this conjecture.

Since we expect that the systems of eigenvalues appearing in $\tilde{H}_n$ for $n < q_0$ will be quite special, we expect that the support of $\tilde{H}_n$ in $\operatorname{Spec} \mathbb{T}$ will have positive codimension. This helps to motivate Conjecture 5 (1). As noted in Remark 7, if we believe parts (1) and (2) of Conjecture 5, we are naturally led to believe the entire conjecture.

We close by remarking that a large part of the framework described here has been worked out in detail in the particular case of $\mathbb{G} = \mathrm{GL}_2$ [9].

References

[1] K. Ardakov, K. Brown, *Ring-Theoretic Properties of Iwasawa Algebras: a survey*, Documenta Math., Extra volume in honour of John Coates's 60th birthday (2006), 7–33.

[2] A. Borel, N. Wallach, *Continuous cohomology, discrete subgroups, and represen-tations of reductive groups*, Annals of Mathematics Studies, 94, Princeton University Press, Princeton, N.J.; University of Tokyo Press, Tokyo, 1980. xvii+388 pp.

[3] K. Buzzard, T. Gee, *The conjectural connections between automorphic represen-tations and Galois representations*, preprint (2010).

[4] F. Calegari, M. Emerton, *Bounds for multiplicities of unitary representations of cohomological type in spaces of cusp forms*, Ann. Math. **170** (2009), 1437–1446.

[5] F. Calegari, M. Emerton, *Mod-p cohomology growth in p-adic analytic towers of 3-manifolds*, to appear in Groups, Geometry, and Dynamics.

[6] F. Calegari, M. Emerton, *Completed cohomology of arithmetic groups*, in prepa-ration.

[7] M. Emerton, *Jacquet modules of locally analytic representations of p-adic reduc-tive groups I. Construction and first properties*, Ann. Sci. Math. E.N.S. **39** (2006), 775–839.

[8] M. Emerton, *On the interpolation of systems of eigenvalues attached to automor-phic Hecke eigenforms*, Invent. Math. **164** (2006), 1–84.

[9] M. Emerton, *Local-global compatibility in the p-adic Langlands programme for* $GL_{2/\mathbb{Q}}$, preprint (2010).

[10] J.-M. Fontaine, B. Mazur, *Geometric Galois representations*, Elliptic curves, modular forms, and Fermat's last theorem (J. Coates, S.T. Yau, eds.), Int. Press, Cambridge, MA, 1995, 41–78.

[11] J. Franke, *Harmonic analysis in weighted* L_2*-spaces*, Ann. Sci. École Norm. Sup. (4) 31 (1998), 181–279.

[12] R. Hill, On Emerton's *p*-adic Banach spaces, preprint (2007).

[13] R. P. Langlands, *Automorphic representations, Shimura varieties, and motives. Ein Märchen*, Automorphic forms, representations, and *L*-functions, Proc. Symp. Pure Math. 33 (part 2), 1979, Amer. Math. Soc., 205–246.

[14] M. Lazard, *Groupes analytiques p-adiques*, Publ. Math. IHES **26** (1965).

[15] B. Mazur, *Deforming Galois representations*, Galois groups over $\mathbb{Q}$ (Y. Ihara, K. Ribet, J-P. Serre eds.), Springer Verlag 1989, 385–435.

[16] P. Schneider, J. Teitelbaum, *Banach space representations and Iwasawa theory*, Israel J. Math. **127** (2002), 359–380.

[17] E. Urban, *Eigenvarieties for reductive groups*, to appear in Ann. Math.

[18] O. Venjakob, *On the structure theory of the Iwasawa algebra of a p-adic Lie group*, J. Eur. Math. Soc. 4 (2002), 271–311.

Tensor and homotopy criteria for functional equations of ℓ-adic and classical iterated integrals

Hiroaki Nakamura

Okayama University

Zdzisław Wojtkowiak

Université de Nice-Sophia Antipolis

1 Introduction

The purpose of this paper is to show equivalence of two criteria for functional equations of (complex and ℓ-adic) iterated integrals, one given by D. Zagier in the case of polylogarithms which we generalize to arbitrary iterated integrals and the other given by the second named author. We establish a device for computing a functional equation from a family of morphisms on fundamental groups of varieties, and present some examples showing how our device works commonly both in complex and ℓ-adic cases. Some of our ℓ-adic examples already supply non-trivial arithmetic relations between "ℓ-adic polylogarithmic characters" – functions on the absolute Galois group $\mathrm{Gal}(\overline{\mathbb{Q}}/\mathbb{Q})$ defined by Kummer properties along towers of certain arithmetic sequences – which were introduced in [NW] as generalization of the so-called Soulé characters studied by Ch. Soulé [S1], [S2].

Let $V := \mathbf{P}^1 -$ {several points} be a punctured projective line defined over a subfield K of $\mathbb{C}$. In [W0,W2,W5], the second named author gave conditions to have functional equations of iterated integrals on V in terms of induced morphisms on fundamental groups. In fact, in [W0], he formulated a complex iterated integral as the image of the (universal) unipotent period along a chain from x to z on $V(\mathbb{C})$ by a 1-form on the Lie algebra of the pro-unipotent fundamental group of V. Also in [W5], introduced is an ℓ-adic iterated integral using the action of the absolute Galois group $\mathrm{Gal}(\overline{K}/K)$ on the torsor of paths from x

This work was partially supported by JSPS KAKENHI 21340009.

Non-abelian Fundamental Groups and Iwasawa Theory, eds. John Coates, Minhyong Kim, Florian Pop, Mohamed Saïdi and Peter Schneider. Published by Cambridge University Press. ©Cambridge University Press 2012.

to z (see Section 4 below). Then, the following result has been proved for their functional equations.

Theorem 1.1 ([W0,W2,W5]) *Let K be a subfield of $\mathbb{C}$, and let $\{a_1, \ldots, a_N\}$, $\{b_1, \ldots, b_M\}$ be respectively N- and M-point subsets of K. Consider $X := \mathbf{P}_K^1 - \{a_1, \ldots, a_N, \infty\}$, $Y := \mathbf{P}_K^1 - \{b_1, \ldots, b_M, \infty\}$, and pick any K-rational (possibly tangential) base point v on X. Suppose we have algebraic morphisms $f_i \colon X \to Y$ ($i = 1, \ldots, m$) together with homomorphisms $\psi_i \colon \mathrm{gr}_\Gamma^n \pi_1(Y(\mathbb{C}), f_i(v)) \to \mathbb{Z}$ and constants $c_1, \ldots, c_m \in \mathbb{Z}$ satisfying*

$$\sum_{i=1}^m c_i \, \psi_i \circ \mathrm{gr}_\Gamma^n(f_{i*}) = 0.$$

Here, $f_{i*} \colon \pi_1(X(\mathbb{C}), v) \to \pi_1(Y(\mathbb{C}), f_i(v))$ denotes the induced homomorphism, and gr_Γ^n denotes the nth graded piece with respect to the lower central filtration. Then, we have a functional equation

$$\sum_{i=1}^m c_i \, \mathcal{L}_Y^\psi(f_i(z), f_i(x)) \equiv 0$$

modulo lower degree terms, where $\mathcal{L}_Y^\psi(f_i(z), f_i(x))$ denote, depending on the context, complex or ℓ-adic iterated integrals on Y. (The lower degree terms will be specified later in this paper.)

On the other hand, in [Z], D. Zagier gave conditions for the functional equations of classical polylogarithms in terms of (generalized) Bloch group [Bl1] that is a certain tensor of symmetric and wedge products of multiplicative groups of fields (cf. also [Ga] (1.10)). We generalize Zagier's conditions from [Z] to arbitrary iterated integrals in terms of tensor algebra of the abelianization of π_1.

The aim of this paper is, first to show that the condition on fundamental groups from [W0,W2,W5] and that on tensor algebras are essentially equivalent in a generalized setting of the above iterated integrals. Also we generalize results from [W5] to the case where X is an arbitrary nonsingular variety (not necessarily a punctured projective line) in ℓ-adic case. See Theorem 4.13 and Theorem 4.14 for our main statement of this paper.

Our main tool is a multi-linearized version of the classical Kummer pairing

$$\pi_1(X)^{\mathrm{ab}} \times O_X^\times \longrightarrow \mathbb{Z}$$

for an algebraic variety X, where $\pi_1(X)^{\mathrm{ab}}$ is the abelianized fundamental group of X and $O_X^\times$ is the unit group of the ring of regular functions on X. The pairing in the complex case is given by

$$(\gamma, f(x)) \mapsto \frac{1}{2\pi i} \int_\gamma d \log f(x),$$

and in this paper, both components of γ and $f(x)$ will be multi-linearized to study informations appearing in the higher graded quotients of the fundamental group. We make use of this tool to establish a device computing a functional equation from a family of morphisms satisfying the criteria of Theorem 4.13 and Theorem 4.14.

One new aspect of our device is that it enables us to compute "lower degree terms" of a functional equation of polylogarithms explicitly from given data in both complex and ℓ-adic cases. A difference between complex and ℓ-adic cases appears in that a complex iterated integral is canonically graded while an ℓ-adic iterated integral is not. This causes us, in ℓ-adic case, to need to introduce an extra notion of "(ℓ-adic) error term" whose computation involves a choice of splitting of the lower central filtration in an ℓ-adic fundamental group. Applying our method developed in this paper, we shall deduce in Section 6 the following examples of typical functional equations.

Complex case	ℓ-adic case
$Li_2(z) + Li_2(1 - z) + \log z \log(1 - z) = \frac{\pi^2}{6}$	$\tilde{\chi}_2^z + \tilde{\chi}_2^{1-z} + \rho_z \rho_{1-z} = \frac{1}{24}(\chi^2 - 1)$
$Li_2(z) + Li_2(\frac{z}{z-1}) = -\frac{1}{2}\log^2(1 - z)$	$\tilde{\chi}_2^z + \tilde{\chi}_2^{\frac{z}{z-1}} = \frac{\rho_{1-z}}{2}\left(\chi - \rho_{1-z}\right)$
$Li_m(z) + (-1)^m Li_m(\frac{1}{z})$ $= -\frac{(2\pi i)^m}{m!} B_m(\frac{\log z}{2\pi i})$	$\tilde{\chi}_m^z + (-1)^m \tilde{\chi}_m^{1/z}$ $= -\frac{1}{m}\{B_m(-\rho_z) - B_m \chi^m\}$
$Li_2(\frac{\xi\eta}{(1-\xi)(1-\eta)}) = Li_2(\frac{\xi}{1-\eta}) + Li_2(\frac{\eta}{1-\xi})$ $-Li_2(\xi) - Li_2(\eta) - \log(1 - \xi)\log(1 - \eta)$	$\tilde{\chi}_2^{\frac{\xi\eta}{(1-\xi)(1-\eta)}} = \tilde{\chi}_2^{\frac{\xi}{1-\eta}} + \tilde{\chi}_2^{\frac{\eta}{1-\xi}}$ $-\tilde{\chi}_2^\xi - \tilde{\chi}_2^\eta - \rho_{1-\xi}\rho_{1-\eta}$

Here, $\tilde{\chi}_m^z \colon G_K \to \mathbb{Z}_\ell$ $(m \geq 1)$ (resp. $\chi \colon G_K \to \mathbb{Z}_\ell^\times$; resp. $\rho_z \colon G_K \to \mathbb{Z}_\ell$) denotes the ℓ-adic polylogarithmic character introduced in [NW] (resp. ℓ-adic cyclotomic character; resp. Kummer 1-cocycle along ℓ-power roots of z) for any field $K(\subset \mathbb{C})$ containing z (cf. §5.2), and B_m (resp. $B_m(X)$) is the mth Bernoulli number (resp. polynomial).

The contents of the present paper will be ordered as follows. In Section 2, we review and study basic properties of the multi-Kummer pairing, and in Section 3, we detect a tensor of functions as the multi-Kummer dual of a form on the Lie algebra of the fundamental group. In Section 4, we rephrase conditions of [W0,W2,W4] on a collection of homomorphisms of $\pi_1(X)$ in terms of conditions on tensor and wedge products of functions, and prove our main statements Theorem 4.13 and Theorem 4.14. As a special case, in Section 5, we closely consider the case of polylogarithms. We will present a more refined statement than Theorems 4.13–4.14 specialized to this case. Section 6 is devoted to present several typical examples of functional equations (listed in the above table) exhibiting computation using our device of this paper.

There is also an important family of functional equations of polylogarithms called the *distribution equations*:

$$Li_k(z^n) = n^{k-1}\left(\sum_{i=0}^{n-1} Li_k(\zeta_n^i z)\right) \qquad (\zeta_n = e^{2\pi i/n})$$

which, together with their ℓ-adic analogs, will be treated from our point of view in the forthcoming subsequent paper [NW2].

Acknowledgements: The first named author would like to thank H. Furusho for useful comments on an early version of this paper. The second named author would like to thank very much Max-Planck-Institut für Mathematik in Bonn, where several essential ideas of this paper were worked out during his visits there. Both authors would like to acknowledge O. Gabber's discovery of Heisenberg covers (indicated in [De0]) which gave important inspiration to our previous work [NW] on ℓ-adic polylogarithmic characters.

2 Multi-Kummer characters

Let X be a (nonsingular, absolutely irreducible) algebraic variety defined over a number field K, a subfield of the field of complex numbers $\mathbb{C}$. We denote by $X^{an} = X(\mathbb{C})$ the analytic manifold of the complex points of X, and by $X_{\overline{K}}$ the algebraic variety obtained as the fiber product $X \times_K \overline{K}$, where $\overline{K}$ is the algebraic closure of K in $\mathbb{C}$. Fix a K-rational (tangential) base point v on X. (For a definition of tangential base points, see, e.g., [N0].)

2.1 Complex case We shall write $O(X^{an})$ for the ring of holomorphic functions on X^{an}. If a function $f \in O(X^{an})$ is everywhere non-vanishing on $X(\mathbb{C})$, i.e., f belongs to the unit group $O(X^{an})^\times$, then it gives an analytic morphism of X to the multiplicative group $f: X \to \mathbb{G}_m(\mathbb{C})$. Any topological loop γ based at v on $X(\mathbb{C})$ is mapped by f to a loop $f(\gamma)$ on the punctured plane $\mathbb{G}_m(\mathbb{C}) = \mathbb{C} - \{0\}$ with the winding number

$$\kappa_f(\gamma) := \frac{1}{2\pi i}\int_{f(\gamma)} \frac{dw}{w} = \frac{1}{2\pi i}\int_\gamma d\log f(x)$$

that obviously belongs to $\mathbb{Z}$. From this arises the Kummer pairing (in the complex case)

$$\pi_1(X^{an})^{ab} \times O(X^{an})^\times \longrightarrow \mathbb{Z}.$$

If the function f is of the form $\exp(g)$ for some $g \in O(X^{\mathrm{an}})$, then κ_f kills all loops γ of $\pi_1(X(\mathbb{C}))$. Define

$$O_h^\times(X^{\mathrm{an}}) := O(X^{\mathrm{an}})^\times / \exp(O(X^{\mathrm{an}})).$$

Then, for each positive integer n, we obtain a natural mapping

$$\kappa^{\otimes n} : \bigotimes_{i=1}^n O_h^\times(X^{\mathrm{an}}) \longrightarrow \mathrm{Hom}(\bigotimes_{i=1}^n \pi_1(X(\mathbb{C}))^{\mathrm{ab}}, \mathbb{Z}),$$

which will play a central role below.

Remark. In the sequel, we shall use both notations $A^{\otimes n}$ and $\bigotimes_{i=1}^n A$ to denote the n-times tensor product of A.

2.2 ℓ-adic case By abuse of notation, for an algebraic variety X, we understand $O(X)$ to be the ring of regular (algebraic) functions on X. Recall from [Roq] (cf. also [La1] chapter II), the unit group $O(X)^\times$ modulo $K^\times$ is finitely generated, torsion free abelian group.

Let $f \in O(X_{\overline{K}})^\times$ and pick any γ from the étale fundamental group $\pi_1(X_{\overline{K}}, v)$. Then, we have an algebraic winding number $\kappa_f(\gamma) \in \hat{\mathbb{Z}}$ as follows. Form the fibre product $X' = X \times_{\mathbb{G}_m} \mathbb{G}_m$ induced from the n-power map $\mathbb{G}_m \to \mathbb{G}_m$:

$$
\begin{array}{ccc}
X' & \xrightarrow{\ f_n\ } & \mathbb{G}_m \\
{\scriptstyle p_n(f)}\big\downarrow & & \big\downarrow{\scriptstyle n} \\
X & \xrightarrow{\ f\ } & \mathbb{G}_m,
\end{array}
$$

then $p_n = p_n(f): X' \to X$ is a (not necessarily connected) finite étale cover of X of degree n. The fundamental group $\pi_1(X, v)$ acts, by definition, on the n point set $p_n^{-1}(v)$, and the action of each loop $\gamma \in \pi_1(X, v)$ is represented by a certain residue $\kappa_n \in (\mathbb{Z}/n\mathbb{Z})$ that "rotates" the upper $\mathbb{G}_m$ through the angle $2\pi\kappa_n/n$. Actually, κ_n is determined up to the residue class of $f \in O(X_{\overline{K}})^\times$ modulo $O(X_{\overline{K}})^{\times n}$. Letting n run over all positive integers multiplicatively, the sequence $\{\kappa_n\}$ defines a coherent element $\kappa_f(\gamma) \in \hat{\mathbb{Z}} = \varprojlim_n (\mathbb{Z}/n\mathbb{Z})$ according to any given class of $\widehat{O}^\times(X_{\overline{K}}) := \varprojlim_n O(X_{\overline{K}})^\times / O(X_{\overline{K}})^{\times n}$. Thus, we obtain the Kummer pairing

$$\pi_1(X_{\overline{K}})^{\mathrm{ab}} \times \widehat{O}^\times(X_{\overline{K}}) \longrightarrow \hat{\mathbb{Z}}.$$

If we fix a prime number ℓ and look only at the ℓ-power maps of $\mathbb{G}_m$, then the above pairing terminates at $\mathbb{Z}_\ell$ and $\pi_1(X_{\overline{K}})$ and $\widehat{O}^\times(X_{\overline{K}})$ factor through

$$\pi_1^\ell(X_{\overline{K}}) := \text{the maximal pro-}\ell\text{ quotient of } \pi_1(X_{\overline{K}}),$$

$$\widehat{O}_\ell^\times(X_{\overline{K}}) := \varprojlim_n (O(X_{\overline{K}})^\times / O(X_{\overline{K}})^{\times \ell^n})$$

respectively. We also obtain the natural analog of $\kappa^{\otimes n}$ (written by the same symbol, for simplicity):

$$\kappa^{\otimes n} : \bigotimes_{i=1}^n \widehat{O}_\ell^\times(X_{\overline{K}}) \longrightarrow \text{Hom}(\bigotimes_{i=1}^n \pi_1^\ell(X_{\overline{K}})^{\text{ab}}, \mathbb{Z}_\ell),$$

where the $\otimes$ are understood to be taken over $\mathbb{Z}_\ell$.

Lemma 2.1 *In both complex and ℓ-adic cases, $\kappa^{\otimes n}$ is injective.*

Proof The complex case: taking the cohomology associated to the exact sequence of sheaves $0 \to \mathbb{Z} \to O \xrightarrow{\exp} O^\times \to 0$ on X^{an} (and the universal coefficient theorem for cohomology and the Hurewicz theorem), we have an injection

$$O_h^\times(X^{\text{an}}) \hookrightarrow H^1(X^{\text{an}}, \mathbb{Z}) \xrightarrow{\sim} \text{Hom}(\pi_1(X^{\text{an}})^{\text{ab}}, \mathbb{Z}).$$

This settles the case $n = 1$. For the case $n > 1$, noting that both $O_h^\times(X^{\text{an}})$ and $\text{Hom}(\pi_1(X^{\text{an}})^{\text{ab}}, \mathbb{Z})$ are torsion free, i.e. flat $\mathbb{Z}$-modules, we obtain the injection

$$\bigotimes_{i=1}^n O_h^\times(X^{\text{an}}) \longrightarrow \bigotimes_{i=1}^n \text{Hom}(\pi_1(X^{\text{an}})^{\text{ab}}, \mathbb{Z}) = \text{Hom}(\bigotimes_{i=1}^n \pi_1(X^{\text{an}})^{\text{ab}}, \mathbb{Z})$$

(cf. [B-1] chapter 2 §4 (23) for the latter equality). The ℓ-adic case: since $O(X_{\overline{K}})^{\times \ell^n}$ contains $\overline{K}^\times$, $\widehat{O}_\ell^\times(X_{\overline{K}})$ is a torsion-free $\mathbb{Z}_\ell$-module. This is injectively mapped into $\text{Hom}(\pi_1^\ell(X_{\overline{K}})^{\text{ab}}, \mathbb{Z}_\ell)$, as ℓ^nth roots of each non-constant function give non-trivial extensions of the function field of X when $n \to \infty$. Thus, the case of $n = 1$ follows. For the case $n > 1$, noting again that these are both flat $\mathbb{Z}_\ell$-modules, we complete the proof in the same way. $\square$

Remark 2.2 It is known that $\pi_1(X(\mathbb{C}))$ is a finitely generated group (see [Ra]). Therefore, the domain and the target modules of $\kappa^{\otimes n}$ are finitely generated torsion free $\mathbb{Z}$- or $\mathbb{Z}_\ell$-modules. Moreover, in the complex case, the cohomology sequence in the above proof implies that the cokernel of $O_h^\times(X^{\text{an}}) \hookrightarrow H^1(X^{\text{an}}, \mathbb{Z})$ is injectively mapped into the complex vector space $H^1(X^{\text{an}}, O)$, hence is a (finitely generated) torsion-free $\mathbb{Z}$-module (in particular, it is a "pure" submodule in the sense of [B-2] chapter 1 §2 ex. 24). Meanwhile, in the ℓ-adic case, torsion possibility in the cokernel of $O_\ell^\times(X_{\overline{K}}) \hookrightarrow \text{Hom}(\pi_1^\ell(X_{\overline{K}})^{\text{ab}}, \mathbb{Z}_\ell)$ is a more subtle question.

Remark 2.3 In our argument below, use of $O_h^\times(X^{an})$ may be replaced by $O^\times(X_{\mathbb{C}}^{alg})/\mathbb{C}^\times$, the multiplicative group of the algebraic unit functions modulo constants, as this group is also injectively mapped in $O_h^\times(X^{an})$. This injectivity follows easily from the fact $\exp(O(X^{an})) \cap O^\times(X_{\mathbb{C}}^{alg}) = \mathbb{C}^\times$. In fact, if an analytic function f on X^{an} has $\exp(f)$ being an algebraic regular function on X, then, all $\exp(f/n)$ $(n \geq 1)$ must be univalent algebraic functions on X, while $O^\times(X_{\mathbb{C}}^{alg})/\mathbb{C}^\times$ has no nontrivial divisible elements by the above mentioned result by Roquette [Roq]. Alternatively, one can use the Kummer sequence in Galois cohomology to show the injectivity of $O^\times(X_{\mathbb{C}}^{alg})/O^\times(X_{\mathbb{C}}^{alg})^{\ell^n}$ into $\mathrm{Hom}(\pi_1^\ell(X_{\overline{K}})^{ab}, \mathbb{Z}/\ell^n\mathbb{Z})$, and then take the projective limit $n \to \infty$.

3 Multi-Kummer duals

We recall that the lower central series of a group π is defined inductively by setting

$$\Gamma^1\pi := \pi, \quad \Gamma^n\pi := (\pi, \Gamma^{n-1}\pi) \quad n > 1.$$

The commutator bracket $(x, y) = xyx^{-1}y^{-1}$ then induces the Lie algebra structure on the graded sum

$$\mathrm{Gr\,Lie\,}\pi = \bigoplus_{n=1}^{\infty} \mathrm{gr}^n\pi = \bigoplus_{n=1}^{\infty} (\Gamma^n\pi/\Gamma^{n+1}\pi)$$

in such a way that, for $\alpha \in \mathrm{gr}^n\pi, \beta \in \mathrm{gr}^m\pi$, the Lie bracket is given by

$$[\alpha, \beta] := (\alpha, \beta) \bmod \Gamma^{n+m+1}\pi.$$

We denote by $\bar{x}$ the image of $x \in \pi$ in the abelianization $\pi^{ab} = \mathrm{gr}^1\pi$.

Definition 3.1 Define a natural map

$$\boldsymbol{a}_n(\pi) \colon \bigotimes_{i=1}^{n} \pi^{ab} \longrightarrow \mathrm{gr}^n\pi$$

by induction on n by setting $\boldsymbol{a}_1(\pi) :=$ identity on π^{ab} and

$$\boldsymbol{a}_n(\pi)(\alpha_1 \otimes \cdots \otimes \alpha_n) := [\alpha_1, \boldsymbol{a}_{n-1}(\pi)(\alpha_2 \otimes \cdots \otimes \alpha_n)]$$

for $n > 1$. In the case when π is a pro-ℓ group, we define $\boldsymbol{a}_n(\pi)$ in exactly the same way after taking the lower central series and the tensor products respectively as topological pro-ℓ groups and as $\mathbb{Z}_\ell$-modules. It is also important to consider the case where π is replaced by its quotient by the (closure of, in the pro-ℓ case) double commutator subgroup π'' of π. It is obvious that the maps $a_n(\pi)$, $a_n(\pi/\pi'')$ are *surjective*, therefore they induce injective homomorphisms

$$\operatorname{Hom}(\operatorname{gr}_\Gamma^n(\pi/\pi''), \mathbb{Z}) \hookrightarrow \operatorname{Hom}(\operatorname{gr}_\Gamma^n \pi, \mathbb{Z}) \hookrightarrow \operatorname{Hom}(\bigotimes_{i=1}^n \pi^{\mathrm{ab}}, \mathbb{Z}).$$

We denote by $a_n^*(\pi)$ (resp. $a_n^*(\pi/\pi'')$) the injection of $\operatorname{Hom}(\operatorname{gr}_\Gamma^n \pi, \mathbb{Z})$ (resp. $\operatorname{Hom}(\operatorname{gr}_\Gamma^n(\pi/\pi''), \mathbb{Z})$) into $\operatorname{Hom}(\bigotimes_{i=1}^n \pi^{\mathrm{ab}}, \mathbb{Z})$ induced by $a_n(\pi)$ (resp. $a_n(\pi/\pi'')$).

Now, let us consider permutations of components of $\bigotimes_{i=1}^n \pi^{\mathrm{ab}}$. We introduce special permutation actions σ, τ by

$$\begin{cases} \sigma(\eta_1 \otimes \cdots \otimes \eta_{n-3} \otimes a \otimes b \otimes c) = \eta_1 \otimes \cdots \otimes \eta_{n-3} \otimes b \otimes c \otimes a, \\ \tau(\eta_1 \otimes \cdots \otimes \eta_{n-2} \otimes a \otimes b) = \eta_1 \otimes \cdots \otimes \eta_{n-2} \otimes b \otimes a, \end{cases}$$

and let any permutation $\rho \in S_{n-2}$ act by

$$\rho(\eta_1 \otimes \cdots \otimes \eta_{n-2} \otimes a \otimes b) = \eta_{\rho(1)} \otimes \cdots \otimes \eta_{\rho(n-2)} \otimes a \otimes b.$$

Proposition 3.2 *If a homomorphism* $\varphi \colon \bigotimes_{i=1}^n \pi^{\mathrm{ab}} \to \mathbb{Z}$ *belongs to the image of* $a_n^*(\pi)$, *then, for* $\eta \in \bigotimes_{i=1}^n \pi^{\mathrm{ab}}$,

(i) $\varphi(\eta) + \varphi(\tau(\eta)) = 0,$

(ii) $\varphi(\eta) + \varphi(\sigma(\eta)) + \varphi(\sigma^2(\eta)) = 0.$

If $\varphi \colon \bigotimes_{i=1}^n \pi^{\mathrm{ab}} \to \mathbb{Z}$ *belongs moreover to the image of* $a_n^*(\pi/\pi'')$, *then*

(iii) $\varphi(\eta) = \varphi(\rho(\eta))$ *for all* $\rho \in S_{n-2}.$

Proof Equations (i), (ii) follow immediately from the Lie identities $[A, B] + [B, A] = 0$, $[A, [B, C]] + [B, [C, A]] + [C, [A, B]] = 0$ respectively. Equation (iii) follows by observing that

$$[A, [B, [C, D]]] = [[A, B], [C, D]] + [B, [A, [C, D]]]$$

and the fact that $[[A, B], [C, D]] = 0$ in $\operatorname{Gr}\operatorname{Lie}(\pi/\pi'')$.   □

Remark 3.3 When π is a free group with free generators $y_1, \ldots, y_N$, the Lie algebra $\operatorname{Gr}(\pi) := \oplus_n \operatorname{gr}_\Gamma^n \pi$ can be regarded as the Lie part of the graded free associative algebra $A(\pi) := \oplus_n (\pi^{\mathrm{ab}})^{\otimes n}$. Therefore, there is a natural embedding $\iota_n \colon \operatorname{gr}_\Gamma^n \pi \hookrightarrow (\pi^{\mathrm{ab}})^{\otimes n}$. It is known that ι_n gives a "$\frac{1}{n}$-splitting" of the above surjection $a_n \colon (\pi^{\mathrm{ab}})^{\otimes n} \to \operatorname{gr}_\Gamma^n \pi$ (cf. [MKS] Theorem 5.17). For a general group π, we do not have an analog of the mapping ι_n, so we choose to argue in the principal use of a_n.

Notation 3.4 For an abelian group A, let the symmetric group S_n of degree n act on $A^{\otimes n}$ by the component permutations. We write $\mathfrak{Sym}^n A$ (resp. $\mathfrak{Alt}^n A$) the usual symmetric (resp. alternate) tensor product defined as the maximal quotient of $A^{\otimes n}$ on which S_n acts trivially (resp. by sign-alterations). On the other hand, we regard the wedge tensor product $\bigwedge^n A$ (resp. the symmetric tensor product $\mathrm{Sym}^n A$) as the submodule of $\bigotimes_{i=1}^{n} A$ consisting of those elements which are sign-alternated (resp. invariant) under the component permutations by S_n. We use the following notations for elements in the latter type of symmetric and alternative tensor products of an abelian group A:

$$a \wedge b := a \otimes b - b \otimes a \quad \in A \wedge A \subset \bigotimes_{i=1}^{2} A,$$

$$a^{\odot n} := \underbrace{a \otimes \cdots \otimes a}_{n} \quad \in \mathrm{Sym}^n A \subset \bigotimes_{i=1}^{n} A.$$

(The last notation is borrowed from e.g. [Bl2], [Ga], and will be used only later in Sections 5 and 6. In fact, since our $\mathrm{Sym}^n A$ is now given as the invariant subspace of $A^{\otimes n}$ (unlike in the quotient symmetric tensor space $\mathfrak{Sym}^n A$ of $A^{\otimes n}$), how to define a "canonical $a_1 \odot \cdots \odot a_n$" in $\mathrm{Sym}^n A$ becomes a nontrivial question, even if it may be restricted to our necessary case $A \cong \mathbb{Z}^r$. We leave the question as a problem for future study when necessity arises.) If A is a $\mathbb{Z}_\ell$-module, we shall understand the tensor $\otimes$ as taken in the category of $\mathbb{Z}_\ell$-modules.

Now, suppose that the above π is given as the fundamental group $\pi_1(X(\mathbb{C}))$ of an algebraic variety X over a field $K \subset \mathbb{C}$ as in Section 2. The above properties (i), (iii) of Proposition 3.2 would lead us to consider the following type of commutative diagram:

$$
\begin{array}{ccc}
\bigotimes_{i=1}^{n} O_h^\times & \xrightarrow[\text{inj.}]{\kappa^{\otimes n}} & \mathrm{Hom}((\pi_1^{\mathrm{ab}})^{\otimes n}, \mathbb{Z}) \\
\uparrow{\scriptstyle incl.} & & \uparrow{\scriptstyle incl.} \\
(O_h^\times)^{\otimes n-2} \otimes \wedge^2 O_h^\times & \xrightarrow[\text{inj.}]{} & \mathrm{Hom}((\pi_1^{\mathrm{ab}})^{\otimes n-2} \otimes \mathfrak{Alt}^2 \pi_1^{\mathrm{ab}}, \mathbb{Z}) \xleftarrow{incl.} \mathrm{Hom}(\mathrm{gr}_\Gamma^n \pi_1, \mathbb{Z}) \\
\uparrow{\scriptstyle incl.} & & \uparrow{\scriptstyle incl.} \qquad\qquad\qquad \uparrow{\scriptstyle incl.} \\
\mathrm{Sym}^{n-2} O_h^\times \otimes \wedge^2 O_h^\times & \xrightarrow[\text{inj.}]{} & \mathrm{Hom}(\mathfrak{Sym}^{n-2} \pi_1^{\mathrm{ab}} \otimes \mathfrak{Alt}^2 \pi_1^{\mathrm{ab}}, \mathbb{Z}) \xleftarrow{incl.} \mathrm{Hom}(\mathrm{gr}^n(\pi_1/\pi_1''), \mathbb{Z})
\end{array}
$$

Here, we should like to relate the homomorphisms in $\mathrm{Hom}(\mathrm{gr}_\Gamma^n \pi_1, \mathbb{Z})$ and $\mathrm{Hom}(\mathrm{gr}^n(\pi_1/\pi_1''), \mathbb{Z})$ with tensors of the unit functions. But since $\kappa^{\otimes n}$ is not necessarily surjective in general, we shall first restrict ourselves to the case where surjectivity of $\kappa^{\otimes n}$ is available: to the special case where π is given as the fundamental group of the complex points of the projective t-line $Y =$

$\mathbf{P}_t^1 - \{b_1, \ldots, b_N, \infty\}$ ($N \geq 2$) with y_i a standard loop running once around the puncture b_i based at a fixed (tangential) base point v on Y. Then, π is a free group, freely generated by the $y_1, \ldots, y_N$. Set $[1, N] = \{1, \ldots, N\}$, and, for any n-tuple $\boldsymbol{k} = (k_1, \ldots, k_n) \in [1, N]^n$, write $\bar{y}_{\boldsymbol{k}} = \bar{y}_{k_1} \otimes \cdots \otimes \bar{y}_{k_n} \in \bigotimes_{i=1}^n \pi^{\mathrm{ab}}$. Observe that the group of units on Y is given by

$$(O(Y^{an})^\times \supset) \, O(Y_{\overline{K}})^\times = \overline{K}\left[t, \prod_{i=1}^N (t - b_i)^{-1}\right]^\times = \overline{K}^\times \prod_{i=1}^N (t - b_i)^{\mathbb{Z}}.$$

Proposition 3.5   *Given the above notation for* $Y = \mathbf{P}_t^1 - \{b_1, \ldots, b_N, \infty\}$ *($N \geq 2$), for any* $\varphi \colon \bigotimes_{i=1}^n \pi_1(Y(\mathbb{C}), v)^{\mathrm{ab}} \to \mathbb{Z}$, *we have*

$$\varphi = \kappa^{\otimes n}\left(\sum_{\boldsymbol{k} \in [1,N]^n} \varphi(\bar{y}_{\boldsymbol{k}})(t - b_{k_1}) \otimes \cdots \otimes (t - b_{k_n})\right).$$

Also, its obvious ℓ-adic analog holds, where $\pi_1(Y(\mathbb{C}), v)$, $\mathbb{Z}$ *are replaced by* $\pi_1^\ell(Y_{\overline{K}}, v)$, $\mathbb{Z}_\ell$ *respectively. In particular,* $\kappa^{\otimes n}$ *gives an isomorphism:*

$$O_h^\times(Y^{an})^{\otimes n} \xrightarrow{\sim} \mathrm{Hom}(\pi_1^{\mathrm{ab}}(Y(\mathbb{C}))^{\otimes n}, \mathbb{Z})$$

$$\left(resp. \; O_\ell^\times(Y_{\overline{K}})^{\otimes n} \xrightarrow{\sim} \mathrm{Hom}(\pi_1^{\mathrm{ab}}(Y_{\overline{K}})^{\otimes n}, \mathbb{Z}_\ell)\right).$$

Proof   As the right-hand side equals

$$\sum_{\boldsymbol{k} \in [1,N]^n} \varphi(\bar{y}_{\boldsymbol{k}}) \, \kappa_{t-b_{k_1}} \otimes \cdots \otimes \kappa_{t-b_{k_n}},$$

the multi-linearity of φ reduces the proof to showing

$$\kappa_{t-b_i}(\bar{y}_j) = \delta_{ij} \quad (1 \leq i, j \leq N; \; \delta = \text{Kronecker's delta}).$$

But this is immediate from the definition of the loops y_j. The ℓ-adic case follows in the same way as above.   □

For $Y = \mathbf{P}_t^1 - \{b_1, \ldots, b_N, \infty\}$ ($N \geq 2$), Proposition 3.2 may be rephrased in terms of the coefficients of the above proposition as follows. We let the permutations $\sigma, \tau \in S_n$ and $\rho \in S_{n-2}$ act on $\boldsymbol{k} = (k_1, \ldots, k_n) \in [1, N]^n$ by $\tau(\boldsymbol{k}) = (k_1, \ldots, k_{n-2}, k_n, k_{n-1})$, $\sigma(\boldsymbol{k}) = (k_1, \ldots, k_{n-3}, k_{n-1}, k_n, k_{n-2})$ and $\rho(\boldsymbol{k}) = (k_{\rho(1)}, \ldots, k_{\rho(n-2)}, k_{n-1}, k_n)$.

Lemma 3.6   *If a homomorphism* $\varphi \colon \bigotimes_{i=1}^n \pi^{\mathrm{ab}} \to \mathbb{Z}$ *belongs to the image of* $a_n^*(\pi)$, *then,*

(i) $$\varphi(\bar{y}_{\boldsymbol{k}}) + \varphi(\bar{y}_{\tau(\boldsymbol{k})}) = 0,$$

(ii) $$\varphi(\bar{y}_{\boldsymbol{k}}) + \varphi(\bar{y}_{\sigma(\boldsymbol{k})}) + \varphi(\bar{y}_{\sigma^2(\boldsymbol{k})}) = 0.$$

If $\varphi\colon \bigotimes_{i=1}^{n} \pi^{\mathrm{ab}} \to \mathbb{Z}$ *belongs moreover to the image of* $a_n^*(\pi/\pi'')$, *then,*

(iii) $\varphi(\bar{y}_k) = \varphi(\bar{y}_{\rho(k)})$ *for all* $\rho \in S_{n-2}$.

Corollary 3.7   *Using the same notation as in Proposition 3.5, let* π *denote* $\pi_1(Y(\mathbb{C}), v)$ *in the complex case, and suppose that* φ *lies in the image of* $a_n^*(\pi)$, *i.e.,* $\varphi \in \mathrm{Hom}(\mathrm{gr}_\Gamma^n \pi, \mathbb{Z})$. *Then,*

$$\varphi = \kappa^{\otimes n}\Bigg(\sum_k \sum_{\substack{l_1 < l_2 \\ l_1, l_2 \in [1,N]}} \varphi(\bar{y}_{(k,l_1,l_2)})(t - b_{k_1}) \otimes \cdots \otimes (t - b_{k_{n-2}}) \otimes ((t - b_{l_1}) \wedge (t - b_{l_2})) \Bigg),$$

where $k = (k_1, \ldots, k_{n-1})$ *runs over all tuples belonging to* $[1, N]^{n-2}$. *If moreover* φ *lies in the image of* $a_n^*(\pi/\pi'')$, *i.e.,* $\varphi \in \mathrm{Hom}(\mathrm{gr}_\Gamma^n(\pi/\pi''), \mathbb{Z})$, *then the coefficients* $\varphi(\bar{y}_{(k,l_1,l_2)})$ *are invariant under the action of* S_{n-2} *on the first* $n - 2$ *tensor components, so that the above inside of* $\kappa^{\otimes n}$ *lies in* $\mathrm{Sym}^{n-2} O_h^\times(Y^{\mathrm{an}}) \otimes \wedge^2 O_h^\times(Y^{\mathrm{an}})$. *Exactly parallel statements also hold in the* ℓ*-adic case after substituting* $\pi_1^\ell(Y_{\overline{K}}, v)$, $\mathbb{Z}_\ell$ *and* $O_\ell^\times(Y_{\overline{K}})$ *in the obvious way.*

Proof   This follows as a simple combination of Proposition 3.2(i) and Proposition 3.5.   □

Definition 3.8   We shall call the inside of the right-hand side of the above corollary the *Kummer dual* of φ and write $\widehat{\kappa_{\otimes n}}(\varphi)(t)$. Regarded as an element of $(\bigotimes_{i=1}^{n} O_h^\times(Y^{\mathrm{an}})) \otimes \wedge^2 O_h^\times(Y^{\mathrm{an}})$ in the complex case, and of $(\bigotimes_{i=1}^{n} \widehat{O_\ell^\times}(Y_{\overline{K}})) \otimes \wedge^2 \widehat{O_\ell^\times}(Y_{\overline{K}})$ in the ℓ-adic case, the Kummer dual $\widehat{\kappa_{\otimes n}}(\varphi)(t)$ of φ is uniquely determined by the equality

$$\varphi = \kappa^{\otimes n}(\widehat{\kappa_{\otimes n}}(\varphi))$$

by virtue of the injectivity of $\kappa^{\otimes n}$ shown in Lemma 2.1. Moreover, given a k-morphism of an algebraic variety X with base point v to the punctured projective t-line Y:

$$f\colon X \longrightarrow Y = \mathbf{P}_t^1 - \{b_1, \ldots, b_M, \infty\}$$

and $\varphi \in \mathrm{Hom}(\mathrm{gr}_\Gamma^n(\pi_1(Y(\mathbb{C}), f(v))), \mathbb{Z})$ (or $\in \mathrm{Hom}(\mathrm{gr}_\Gamma^n(\pi_1(Y_{\overline{K}}, f(v))), \mathbb{Z}_\ell)$ in the ℓ-adic case), we shall denote by $\widehat{\kappa_{\otimes n}}(\varphi)(f)$ the pulled-back image of the Kummer dual $\widehat{\kappa_{\otimes n}}(\varphi)(t)$ induced by the mapping $g \mapsto g \circ f$ of the unit functions on Y to those on X.

4 Iterated integrals and their functional equations

We first review the definition of complex and ℓ-adic iterated integrals. For the reader's convenience, we make the paper as complete as possible even if this means that we repeat some arguments from earlier papers of the second named author. Let X be an algebraic variety defined over $K \subset \mathbb{C}$. We assume, for simplicity, that X is nonsingular and absolutely irreducible over K.

4.1 Complex iterated integrals In this subsection, we set $K = \mathbb{C}$ and pick complex points $v, z \in X(\mathbb{C})$ and a path p from v to z on $X(\mathbb{C})$. Given a collection of holomorphic 1-forms $w_1, \ldots, w_n$ on the smooth analytic manifold $X(\mathbb{C})$, we can form an iterated integral $\int_p w_1 \cdots w_n$.

Take a smooth compactification X^* of X with $D := X^* - X$ a normal crossing divisor, and let $\Omega^i (= \Omega^i_{\log}(X)) := \Omega^i(X \log D)(X^*)$ be the space of (global sections of) meromorphic i-forms on X^*, holomorphic on X, with logarithmic singularities along D. It is known that the spaces $\Omega^i_{\log}(X)$ are determined independently of the choice of the compactification X^* of X as above (cf. [Ii] chapter 11).

Let V_i be the dual space of Ω^i and $K^\perp \subset V_1 \wedge V_1$ be the orthogonal space to the kernel of the cup product $\Omega^1 \wedge \Omega^1 \to \Omega^2$.

Definition 4.1 Let $\mathrm{Lie}(V_1) = \bigoplus_{n=1}^\infty \mathrm{Lie}(V_1)_n$ be the free Lie algebra generated by V_1 equipped with natural gradation by $\mathrm{Lie}(V_1)_n$ – the part of homogenous Lie polynomials of degree n. Let $L(V_1, K^\perp)$ be the quotient of $\mathrm{Lie}(V_1)$ modulo the ideal generated by $K^\perp \subset \mathrm{Lie}(V_1)_2 = V_1 \wedge V_1$. We denote by L_X the completion of $L(V_1, K^\perp)$ with respect to the lower central series, and by U_X (resp. $\pi(X)$) the complete Hopf algebra given as the universal enveloping algebra of L_X (resp. the group of the group-like elements of U_X).

Note that $L(V_1, K^\perp)$ has a natural gradation $L(V_1, K^\perp) = \bigoplus_{n=1}^\infty L(V_1, K^\perp)_n$ inherited from that of $\mathrm{Lie}(V_1)$. There is a natural bijection between L_X and $\pi(X)$ given by exp and log that also preserves the lower central filtrations of L_X and of $\pi(X)$ mutually. Thus, we have a *canonical* identification (cf. also [De2] §12):

$$(4.1) \qquad \pi(X)/\Gamma^{N+1}\pi(X) \underset{\exp}{\overset{\log}{\rightleftarrows}} L_X/\Gamma^{N+1}L_X \cong \bigoplus_{n=1}^N L(V_1, K^\perp)_n.$$

It is known that the 1-form $\omega_X \in V_1 \otimes \Omega^1 = \mathrm{Hom}(\Omega^1, \Omega^1)$ corresponding to the identity element gives an integrable connection on the trivial principal bundle $X \times U_X \to X$, i.e., it satisfies $d\omega_X = \omega_X \wedge \omega_X = 0$ (cf. [H] §4, [W2] §1).

Given a path γ from v to z, the associated horizontal section starting from $(v, 1) \in X \times U_X$ over γ terminates at a point $(z, \Lambda_\gamma(z, v))$ that is uniquely determined as long as $\gamma \colon v \rightsquigarrow z$ changes in the same homotopy class. From this, one can define the parallel transport mapping

$$(4.2) \qquad \theta_{v,z,X} \colon \pi_1(X, v, z) \longrightarrow \pi(X) \quad (\gamma \mapsto \Lambda_\gamma(z, v)).$$

This construction is compatible with composition of paths, i.e., for paths $\alpha: v \rightsquigarrow y, \beta: y \rightsquigarrow z$, we have

$$(4.3) \qquad \Lambda_\alpha(y, v)\Lambda_\beta(z, y) = \Lambda_{\alpha\beta}(z, v).$$

It is shown by K.-T. Chen that $\Lambda_\gamma(z, v)$ can be expressed in terms of iterated integrals as

$$(4.4) \qquad \Lambda_\gamma(z, v) = 1 + \int_\gamma \omega_X + \int_\gamma \omega_X \omega_X + \int_\gamma \omega_X \omega_X \omega_X + \cdots.$$

The above constructions of L_X, $\pi(X)$, U_X together with $\Lambda_\gamma(z, v)$ are functorial with respect to morphisms, i.e., for any morphism $f: X \rightarrow Y$, the naturally induced homomorphism $f_*: L_X \rightarrow L_Y$ (that can be extended to the homomorphism $f_*: U_X \rightarrow U_Y$ which also gives $f_*: \pi(X) \rightarrow \pi(Y)$) maps each component of the gradation of L_X into the corresponding component of L_Y, and keeps

$$(4.5) \qquad f_*(\Lambda_\gamma(z, v)) = \Lambda_{f(\gamma)}(f(z), f(v))$$

(see e.g. corollary 1.7 of [W3]). In fact, the pull-back of differentials by a morphism $f: X \rightarrow Y$ keeps the differentials with log poles, i.e., it holds that $f^* \Omega^1_{\log}(Y) \subset \Omega^1_{\log}(X)$ (cf. [Ii] §11.1c). This, together with the fact $f_* \omega_X = f^* \omega_Y$ (where, f_*, f^* act on $V_1 \otimes \Omega^1$ respectively by $f_* \otimes \mathrm{id}$, $\mathrm{id} \otimes f^*$) implies (4.5).

When $v = z$, the monodromy map extends to a surjective homomorphism of complete Hopf algebras

$$(4.6) \qquad \theta_{v,X}: \mathbb{C}\pi_1(X, v)^\wedge \longrightarrow U_X.$$

Notably, R. Hain [H] determined the kernel of $\theta_{v,X}$ to be the ideal $I := F^0 \cap J + F^{-1} \cap J^2 + F^{-2} \cap J^3 + \cdots$, where J is the augmentation ideal and F^i are Hodge filtrations on the Malcev fundamental group algebra $\mathbb{C}\pi_1(X, v)^\wedge$.

Remark 4.2 Note that $\Omega^1_{\log}(X) \subset H^1_{DR}(X/\mathbb{C})$ ([De1]). In the above construction, we need not assume the equality of this inclusion. In particular, we allow the case $H^1(X^*, O_{X^*}) \neq \{0\}$.

Example 4.3 Let X be an elliptic curve $E := \{y^2 = x^3 + ax + b\} \cup \{\infty\}$ minus a set of several points $D = \{p_1, \ldots, p_N\} \cup \{\infty\}$. By Abel's theorem (cf. [Rob] p. 134), every holomorphic 1-form ω with poles included in D is uniquely written in the form

$$\omega = \underbrace{\left(df + \alpha\frac{xdx}{y}\right)}_{\omega_2} + \underbrace{\left(\beta\frac{dx}{y}\right)}_{\omega_1} + \underbrace{\left(\sum_{i=1}^N \gamma_i \frac{y + y(p_i)}{x - x(p_i)}\frac{dx}{y}\right)}_{\omega_3},$$

with $f \in O(E \setminus D)^\times$, $\alpha, \beta, \gamma_i \in \mathbb{C}$. If we write $\omega_2, \omega_1, \omega_3$ respectively for the above three terms $(\ldots)$, then ω_1 is a differential of the 1st kind, $\omega_2 + \omega_1$ is of the 2nd kind and $\omega_1 + \omega_3$ is of the 3rd kind. (Residues of the differential $\frac{y + y(p_i)}{x - x(p_i)} \frac{dx}{y}$ are ± 2.) The space $\Omega^1_{\log}(E \setminus D)$ consists of the differential forms ω with no ω_2, i.e., only of the 1st and 3rd differentials, that is a finite dimensional vector space. (If we allow the part ω_2 to be alive, then we would need to deal with an infinite dimensional space of forms.) Meanwhile the map $\Omega^1(X \log D)(X^*) \to H^1_{DR}(X/\mathbb{C})$ is injective, but generally has a non-trivial cokernel (correspond-ing to the class of anti-holomorphic differentials). For any given morphism $f \colon E \setminus D \to \mathbf{P}^1_t \setminus \{a_1, \ldots, a_M, \infty\}$, the pullback by f sends each $\frac{dt}{t - a_i}$ into $\Omega^1_{\log}(E \setminus D)$. This observation exhibits a point that enables us to carry out the above construction in a functorial way with respect to morphisms, and leads us to generalize our previous result (cf. Theorem 1.1) in the setting of an arbitrary variety X accompanied with morphisms to a punctured line.

Now, suppose $X = \mathbf{P}^1 - \{a_1, \ldots, a_N, \infty\}$ $(N \geq 2)$ and that we are given a form $\psi \colon \mathrm{gr}^n_\Gamma \pi_1(X, v) \to \mathbb{Z}$. In this special case, Hain's kernel is trivial and the monodromy representation $\theta_{v,X}$ gives a canonical isomorphism between $\mathbb{C}\pi_1(X, v)^\wedge$ and U_X. Moreover, since $K^\perp = 0$, one can identify

$$(4.7) \qquad \mathrm{gr}^n \theta_{v,X} \colon \mathrm{gr}^n_\Gamma \pi_1(X, v) \otimes \mathbb{C} \xrightarrow{\sim} \mathrm{Lie}(V_1)_n.$$

Then, the linear form $\psi_\mathbb{C} = \psi \otimes \mathbb{C}$ on $\mathrm{gr}^n_\Gamma \pi_1(X, v) \otimes \mathbb{C}$ composed with (the split-ting of $L_X / \Gamma^{N+1} L_X$) and log of (4.1) can be regarded as a polynomial function on $\pi(X) / \Gamma^{n+1} \pi(X)$. Note also that one can follow the same story even when v is a tangential base point on X (See [W3] §3).

Definition 4.4   Assume $X = \mathbf{P}^1 - \{a_1, \ldots, a_N, \infty\}$ $(N \geq 2)$. We may think of $\psi_\mathbb{C} \circ \log$ as a polynomial function on $\pi(X) / \Gamma^{n+1} \pi(X)$ as above. We shall call

$$\mathcal{L}^\psi_\mathbb{C}(z, v; \gamma) := \psi_\mathbb{C} \left(\log \Lambda_\gamma(z, v)^{-1} \bmod \Gamma^{n+1} \right)$$

the *complex iterated integral* associated to the form $\psi \in \mathrm{Hom}(\mathrm{gr}^n_\Gamma \pi_1(X, v), \mathbb{Z})$ and the path $\gamma \colon v \rightsquigarrow z$ on X.

4.2 ℓ-adic iterated integrals ([W4])

Let v, z be K-rational (possibly tangen-tial) points on X, and p be an étale path from v to z.

There is an isomorphism of pro-ℓ spaces between the pro-ℓ fundamental group $\pi := \pi^\ell_1(X_{\overline{K}}, v)$ and the pro-ℓ torsor of paths $\pi^\ell_1(X_{\overline{K}}, v, z)$ from v to z given by

$$(4.8) \qquad \pi = \pi^\ell_1(X_{\overline{K}}, v) \xrightarrow{\sim} \pi^\ell_1(X_{\overline{K}}, v, z) \quad (\gamma \mapsto p\gamma^{-1}).$$

Through this identification, the natural Galois action on $\pi^\ell_1(X_{\overline{K}}, v, z)$ induces a $\mathbb{Q}_\ell$-linear Galois representation $G_K \to \mathrm{GL}(\hat{U}(\pi))$ $(\sigma \mapsto \sigma_p)$ in the universal enveloping algebra $\hat{U}(\pi)$ of the (complete) ℓ-adic Lie algebra $L(\pi)$ of π. This new action generally depends on the choice of path $p: v \rightsquigarrow y$, and is determined by the formula

$$(4.9) \qquad \sigma_p(S) := \mathfrak{f}_\sigma^p \cdot \sigma(S) \qquad (S \in \pi_1^\ell(X_{\overline{K}}, v), \ \sigma \in G_K),$$

where $\mathfrak{f}_\sigma^p := p \cdot \sigma(p)^{-1}$. To be distinguished from the above action σ_p, we shall also write σ_v to designate the standard Galois action at v: $G_K \to \operatorname{Aut}(\hat{U}(\pi))$ $(\sigma \mapsto \sigma_v)$.

Definition 4.5 Let $\{\Gamma_i\}_{i=1,2,\ldots}$ denote the lower central filtrations of $L(\pi)$, and set

$$G_i(v) := \{\sigma \in G_K \mid \sigma_v : \text{trivial on } L(\pi)/\Gamma^{i+1}L(\pi)\};$$
$$H_i(z, v) := \{\sigma \in G_i(v) \mid \sigma_p : \text{trivial on } L(\pi)/\Gamma^i L(\pi)\}.$$

It follows from [W4], lemma 1.0.5, that $H_i(z, v)$ does not depend on the choice of p. Note that, for $i = 1$, $G_1(v) = H_1(z, v)$.

Now, recall that we have a canonical group homomorphism of π into $\pi(\mathbb{Q}_\ell)$ – the ($\mathbb{Q}_\ell$-valued points) of the pro-algebraic group formed by the group-like elements of $U(\pi)$. There is also a bijective correspondence between $\pi(\mathbb{Q}_\ell)$ and $L(\pi)$ given by log and exp. In this paper, for any element $\mathfrak{f} \in \pi$, we shall write simply $\log(\mathfrak{f}) \in L(\pi)$ to denote the log of the image of $\mathfrak{f}$ in $\pi(\mathbb{Q}_\ell)$. On the other hand, the universal enveloping algebra $U(\pi)$ has the augmentation ideal I and we have $I/I^2 \xrightarrow{\sim} \operatorname{gr}_\Gamma^1(\pi) \otimes \mathbb{Q}_\ell$. A $\mathbb{Q}_\ell$-linear automorphism ε of $\hat{U}(\pi)$ is called unipotent if it acts trivially on I/I^2. We write $\operatorname{Log}(\varepsilon)$ for the $\mathbb{Q}_\ell$-linear endomorphism obtained as the logarithm of a $\mathbb{Q}_\ell$-linear unipotent automorphism ε of $\hat{U}(\pi)$. The following basic facts were proved in [W4].

Proposition 4.6 *Suppose $\sigma \in G_1(v)$. Then:*

(i) *The actions of σ_v and σ_p on $\hat{U}(\pi)$ are unipotent.*

(ii) *$(\operatorname{Log}\sigma_p)(1) \in \hat{U}(\pi)$ is a Lie element, i.e., belongs to the Lie part $L(\pi) \subset \hat{U}(\pi)$.*

(iii) *$\operatorname{Log}(\sigma_p) = L_{(\operatorname{Log}\sigma_p)(1)} + \operatorname{Log}(\sigma_v)$, where L_λ means the left multiplication by λ.*

(iv) *If $\sigma \in H_i(z, v)$, then $(\operatorname{Log}\sigma_p)(1)$ belongs to $\Gamma^i L(\pi)$.*

Definition 4.7 For each $\mathbb{Q}_\ell$-valued form ψ on $L(\pi)$, we define the *naive ℓ-adic iterated integral* $\mathcal{L}_{\mathrm{nv}}^\psi(z, v; p)$ to be the function on G_K given by

$$\mathcal{L}_{\mathrm{nv}}^\psi(z, v; p)(\sigma) := \psi(\log(\mathfrak{f}_\sigma^p)^{-1}) \qquad (\sigma \in G_K),$$

where $\mathfrak{f}_\sigma^p = p \cdot \sigma(p)^{-1}$. The *big ℓ-adic iterated integral* $\mathcal{L}^\psi(z, v; p)$ is defined to be the function on $G_1(v)$ by

$$\mathcal{L}^\psi(z, v; p)(\sigma) := \psi((\mathrm{Log}\sigma_p)(1)) \quad (\sigma \in G_1(v)).$$

If ψ is a(n induced) form on $\mathrm{gr}_\Gamma^i \pi \otimes \mathbb{Q}_\ell \cong \Gamma^i L(\pi)/\Gamma^{i+1} L(\pi)$, then we define the *ℓ-adic iterated integral* $\mathcal{L}^\psi(z, v)$ to be the restriction of $\mathcal{L}^\psi(z, v; p)$ on $H_i(z, v)$, that is,

$$\mathcal{L}^\psi(z, v)(\sigma) := \psi((\mathrm{Log}\sigma_p)(1)) \quad (\sigma \in H_i(z, v)).$$

Proposition 4.8 $\mathcal{L}^\psi(z, v)$ *does not depend on the choice of the path p.* □

In the special case of $X = \mathbf{P}_K^1 - \{a_1, \ldots, a_N, \infty\}$ ($N \geq 2$), we may proceed to some more construction. Fix a generator system $\vec{x} = (x_1, \ldots, x_N)$ of the topological fundamental group $\pi_1(X(\mathbb{C}), v)$ so that x_i ($1 \leq i \leq N$) is a loop based at v running around the puncture a_i once in the anti-clockwise way. Since $\pi = \pi_1^\ell(X_{\overline{K}}, v)$ is canonically isomorphic to the pro-ℓ completion of $\pi_1(X(\mathbb{C}), v)$, we may regard $\vec{x}$ as the topological generator system of π. Then, $X_i = \log(x_i)$ freely generate the complete ℓ-adic Lie algebra $L(\pi)$, i.e., every element of $L(\pi)$ has an expansion as a formal Lie series in $X_1, \ldots, X_N$. In particular, we may define the homogeneous degree n part $L(\pi)_{n,\vec{x}} \subset L(\pi)$ and the decomposition

$$(4.10) \qquad\qquad L(\pi) = \prod_{n=1}^\infty L(\pi)_{n,\vec{x}}$$

depending on the choice of $\vec{x}$. Given a $\mathbb{Z}$-valued form $\psi : \mathrm{gr}^n \pi_1(X(\mathbb{C}), v) \to \mathbb{Z}$ and a generator system $\vec{x} = (x_1, \ldots, x_N)$ of $\pi_1(X(\mathbb{C}), v)$, we obtain a $\mathbb{Q}_\ell$-valued form $\psi_{\vec{x}}$ (or just written ψ) on $L(\pi)$ as the composition of

$$(4.11) \qquad \psi_{\vec{x}} : L(\pi) \xrightarrow{\text{proj.}} L(\pi)_{n,\vec{x}} \xrightarrow{\sim} \mathrm{gr}^n \pi_1(X, v) \otimes \mathbb{Q}_\ell \xrightarrow{\psi \otimes \mathrm{id}} \mathbb{Q}_\ell,$$

where the middle isomorphism is the one induced by mapping $X_i \mapsto \bar{x}_i \in \mathrm{gr}_\Gamma^1(\pi_1(X, v))$. For this isomorphism $\psi_{\vec{x}}$, we may apply Definition 4.7 above to define the *ℓ-adic naive iterated integral* $\mathcal{L}_\mathrm{nv}^{\psi_{\vec{x}}}(z, v; p)$, the *ℓ-adic big iterated integral* $\mathcal{L}_\mathrm{nv}^{\psi_{\vec{x}}}(z, v; p)$ and *ℓ-adic iterated integral* $\mathcal{L}^{\psi_{\vec{x}}}(z, v)$.

The graded quotients $G_n(v)/G_{n+1}(v)$ and $H_n(z, v)/H_{n+1}(z, v)$ have natural G_K-module structures by conjugation. This action is shown to be factored through $G_K/G_1(v)$. Sometimes useful is, denoting by $\mathcal{T}$ a finite collection of pairs (z_j, v_j) of K-rational (tangential) points on X, to consider the intersection

$$(4.12) \qquad\qquad H_n(\mathcal{T}) := \bigcap_{(z_j, v_j) \in \mathcal{T}} H_n(z_j, v_j).$$

In the case of $X = \mathbf{P}_K^1 - \{a_1, \ldots, a_N, \infty\}$, we see that $G_1(v) = G_{K(\mu_{\ell^\infty})}$ and that

$G_K/G_1(v) = \mathrm{Gal}(K(\mu_{\ell^\infty})/K)$ acts on each graded quotient $H_n(\mathcal{T})/H_{n+1}(\mathcal{T})$ via multiplication by the nth power of $\chi\colon G_K \to \mathbb{Z}_\ell^\times$, the cyclotomic character, i.e. makes it isomorphic to a finite sum of the Tate twist $\mathbb{Z}_\ell(n)$. The standard weight argument assures

Proposition 4.9   *Notations being as above for $X = \mathbf{P}_K^1 - \{a_1, \ldots, a_N, \infty\}$ ($N \geq 2$), the natural homomorphisms*

$$H_n(\mathcal{T})/H_{n+1}(\mathcal{T}) \to H_n(\mathcal{S})/H_{n+1}(\mathcal{S}) \to G_n(v)/G_{n+1}(v)$$

($n = 1, 2, \ldots$) are almost surjective (i.e., have finite cokernels) for any subset $\mathcal{S} \subset \mathcal{T}$. $\square$

Let us now take $\mathcal{T}$ to be a collection of (z, v) and $(\vec{a}_i, v)$ with $\vec{a}_i$ being any K-rational tangential base point at a_i for $1 \leq i \leq N$.

Proposition 4.10   *Notations being as above, the ℓ-adic iterated integral $\mathcal{L}^{\psi_{\vec{x}}}(z, v)$ on $H_n(\mathcal{T})$ is independent of the choice of the generator system $\vec{x}$ (i.e., depending only on the order of missing points on $\mathbf{P}_K^1$).*

According to this proposition, we may without ambiguity write $\mathcal{L}^\psi(z, v)$, abbreviating the reference to $\vec{x}$ for the ℓ-adic iterated integral on $H_n(\mathcal{T})$. For details of the above propositions, see [W4].

Before closing this subsection, we note the following lemma that will be applied to control a behavior of naive ℓ-adic iterated integrals under changes of the choice of paths.

Lemma 4.11   *Let X/K be any algebraic variety with K-rational base point v, and let σ_v denote the action of $\sigma \in G_K$ on $L(\pi)$ at v. Then, for every $\sigma \in G_K$ and $s \in \Gamma^n \pi_1^\ell(X_{\overline{K}}, v)$, we have*

$$\log \mathfrak{f}_\sigma^{sp} \equiv \log \mathfrak{f}_\sigma^p + (1 - \sigma_v) \cdot (\log s) \ mod \ \Gamma^{n+1} L(\pi).$$

In particular, $\log \mathfrak{f}_\sigma^p$ is invariant modulo $\Gamma^n L(\pi)$ under the change of $p \mapsto sp$ ($s \in \Gamma^n \pi_1^\ell(X_{\overline{K}}, v)$). In the case $X = \mathbf{P}_K^1 - \{a_1, \ldots, a_N, \infty\}$ ($N \geq 2$), it holds that

$$\log \mathfrak{f}_\sigma^{sp} \equiv \log \mathfrak{f}_\sigma^p + (1 - \chi(\sigma)^n)(\log s) \ mod \ \Gamma^{n+1} L(\pi),$$

where $\chi\colon G_K \to \mathbb{Z}_\ell^\times$ is the cyclotomic character.

Proof   We have $\mathfrak{f}_\sigma^{sp} = sp \cdot \sigma(sp)^{-1} = s\mathfrak{f}_\sigma^p s^{-1} \cdot s\sigma(s)^{-1}$. Taking $\log\colon \pi \to L(\pi)$, we obtain

$$\log \mathfrak{f}_\sigma^{sp} = s(\log \mathfrak{f}_\sigma^p)s^{-1} \boxplus \log(s \cdot \sigma(s)^{-1}),$$

where $\boxplus$ denotes the Baker–Campbell–Hausdorff sum: $S \boxplus T = \log(e^S e^T)$. Since s lies in the center of $\pi/\Gamma^{n+1}\pi$, after taking modulo $\Gamma^{n+1} L(\pi)$, we see the above right-hand side is congruent to $\log \hat{\mathfrak{f}}_{\sigma}^{p} + \log(s) - \log(\sigma(s))$. From this follows the first formula. In the special case of the punctured projective line, σ acts on $\operatorname{gr}_{\Gamma}^{n}\pi$ by $\chi(\sigma)^{n}$-multiplication. Thus, the proof of the lemma is completed.   □

Applying $\psi_{\vec{x}} \colon L(\pi) \to \mathbb{Q}_{\ell}$ to the above lemma gives a generalization of [DW2] Lemma 2.1 to the naive ℓ-adic iterated integrals.

4.3 Functional equations for iterated integrals Suppose now that we are given a K-morphism to a punctured projective t-line:

$$f \colon X \longrightarrow Y = \mathbf{P}_{t}^{1} - \{b_{1}, \ldots, b_{M}, \infty\}.$$

Then, induced are the homomorphisms

$$\operatorname{gr}^{n}(f_{*}) \colon \operatorname{gr}_{\Gamma}^{n}(\pi_{1}(X, v)) \to \operatorname{gr}_{\Gamma}^{n}(\pi_{1}(Y, f(v)))$$

($n = 1, 2, \ldots$), where π_{1} denotes the topological (resp. pro-ℓ) fundamental group of the complex points of X, Y (resp. of $X_{\overline{K}}$, $Y_{\overline{K}}$) in the complex (resp. ℓ-adic) case. Also, the pull-back of functions $g \mapsto g \circ f$ gives rise to the mappings $O_{h}^{\times}(Y^{\mathrm{an}}) \to O_{h}^{\times}(X^{\mathrm{an}})$ and $\widehat{O}_{\ell}^{\times}(Y_{\overline{K}}) \to \widehat{O}_{\ell}^{\times}(X_{\overline{K}})$ in respective cases.

Lemma 4.12 *We have*

$$\kappa^{\otimes n}(g_{1} \otimes \cdots \otimes g_{n})(f_{*}(\alpha_{1}) \otimes \cdots \otimes f_{*}(\alpha_{n}))$$
$$= \kappa^{\otimes n}(g_{1} \circ f \otimes \cdots \otimes g_{n} \circ f)(\alpha_{1} \otimes \cdots \otimes \alpha_{n})$$

for any $g_{1}, \ldots, g_{n} \in O_{h}^{\times}(Y^{\mathrm{an}})$ *(resp.* $\in \widehat{O}_{\ell}^{\times}(Y_{\overline{K}})$*) and for any* $\alpha_{1}, \ldots, \alpha_{n} \in \pi_{1}(X(\mathbb{C}), v)^{\mathrm{ab}}$ *(resp.* $\in \pi_{1}^{\ell}(X_{\overline{K}}, v)^{\mathrm{ab}}$*) in the complex (resp. ℓ-adic) case.*

Proof This follows immediately from the definition of Kummer pairing given in §2.   □

Now we shall state our main theorem. Recall from Definition 3.8 that for any homomorphism $\varphi \colon \operatorname{gr}_{\Gamma}^{n}(\pi_{1}(Y, f(v))) \to \mathbb{Z}$ or $\mathbb{Z}_{\ell}$, the symbol $\widehat{\kappa_{\otimes n}}(\varphi)(f)$ denotes the image of the Kummer dual $\widehat{\kappa_{\otimes n}}(\varphi)(t)$ by the pull-back mapping $g \mapsto g \circ f$.

Theorem 4.13 *Let X be an arbitrary algebraic variety over a subfield K of $\mathbb{C}$, and let $Y := \mathbf{P}_{K}^{1} - \{b_{1}, \ldots, b_{M}, \infty\}$, where $\{b_{1}, \ldots, b_{M}\}$ is a subset of K with cardinality M. Fix any K-rational (possibly tangential) base point v on X. Then, for a collection of algebraic K-morphisms $f_{i} \colon X \to Y$ ($i = 1, \ldots, m$), homomorphisms $\psi_{i} \colon \operatorname{gr}_{\Gamma}^{n}\pi_{1}(Y(\mathbb{C}), f_{i}(v)) \to \mathbb{Z}$ and integers $c_{1}, \ldots, c_{m} \in \mathbb{Z}$, the* *following conditions* (i)$_\mathbb{C}$, (ii)$_\mathbb{C}$ *and* (iii)$_\mathbb{C}$ *are equivalent:*

(i)$_\mathbb{C}$ $\quad\displaystyle\sum_{i=1}^{m} c_i\,\psi_i \circ \mathrm{gr}^n_\Gamma(f_{i*}) = 0 \quad$ in $\mathrm{Hom}\,(\mathrm{gr}^n_\Gamma \pi_1(X(\mathbb{C}), v), \mathbb{Z})$,

(ii)$_\mathbb{C}$ $\quad\displaystyle\sum_{i=1}^{m} c_i\,\widehat{\kappa_{\otimes n}}(\psi_i)(f_i) = 0$

$$in\ \ (\bigotimes_{i=1}^{n-2} O^\times_h(X^{\mathrm{an}})) \otimes (O^\times_h(X^{\mathrm{an}}) \wedge O^\times_h(X^{\mathrm{an}})),$$

(iii)$_\mathbb{C}$ $\quad\displaystyle\sum_{i=1}^{m} c_i\,\mathcal{L}^{\psi_i}_\mathbb{C}(f_i(z), f_i(v); f_i(\gamma)) = 0$

$$for\ each\ path\ \gamma:\ v \rightsquigarrow z\ on\ X(\mathbb{C}).$$

The proof of the above theorem will be given soon after stating the next theorem. In the following ℓ-adic analogs (i)$_\ell$, (ii)$_\ell$ and (iii)$_\ell$ of the above conditions, we have, in general, the equivalence (i)$_\ell$ $\Leftrightarrow$ (ii)$_\ell$ and the implication (i)$_\ell$ (ii)$_\ell$ $\Rightarrow$ (iii)$_\ell$, (iii)$^{\mathrm{nv}}_\ell$. If moreover $\mu_{\ell^\infty} \not\subset K$, then we also have (i)$_\ell$ $\Leftrightarrow$ (ii)$_\ell$ $\Leftrightarrow$ (iii)$^{\mathrm{nv}}_\ell$:

In order to state the condition (iii)$_\ell$, (iii)$^{\mathrm{nv}}_\ell$, we need to introduce a precise notion of "error term" which complement (part of) "lower degree terms" of ℓ-adic iterated integrals.

Definition of ℓ-adic error terms Let X, v, z be as in §4.2. We shall call a $\mathbb{Q}_\ell$-valued function $E(\sigma, p)$ on $(\sigma, p) \in G_K \times \pi^\ell_1(X_{\overline{K}}, v, z)$ an *error term of degree n*, if it satisfies

(E1) $\qquad\qquad E(\sigma, p) = 0$ for $\sigma \in H_n(v, z)$,

(E2) $\qquad\qquad E(\sigma, p) = E(\sigma, sp)$ for $s \in \Gamma^n \pi^\ell_1(X_{\overline{K}}, v)$.

We also introduce some more notations for the above system of a collection of algebraic morphisms $f_i: X \to Y$ ($i = 1, \ldots, m$) with $Y := \mathbf{P}^1_K - \{b_1, \ldots, b_M, \infty\}$. Fix a K-rational (*tangential*) base point w on Y, a generator system $\vec{y} = (y_1, \ldots, y_M)$ of $\pi_1(Y(\mathbb{C}), w)$ such that y_j turns only around the puncture b_j once, and paths $\gamma_i: w \rightsquigarrow f_i(v)$ for $1 \le i \le m$. Let $\vec{y}_i := (\gamma_i^{-1} y_j \gamma_i)_{j=1}^M$ be the generator system of $\pi_1(Y(\mathbb{C}), f_i(v))$ induced from γ_i and $\vec{y}$. We then apply the process of (4.10–4.11) to obtain a $\mathbb{Q}_\ell$-valued form $\psi_{i,\vec{y}_i}$ on the ℓ-adic Lie algebra $L(\pi_{Y, f_i(v)})$ of $\pi^\ell_1(Y_{\overline{K}}, f_i(v))$ for each $1 \le i \le m$.

The following theorem refines [W5], theorem 10.0.7, concerning the functional equations of ℓ-adic iterated integrals. One of our new ingredients of this paper is to formulate those functional equations for arbitrary algebraic varieties X, generalizing results of [W4] and [W5], which were concerned with projective lines minus finite number of points. If we restrict ourselves only to $\sigma \in H_n(v, z) \cap \bigcap_{i=1}^{m} H_n(f_i(v), f_i(z))$, then we can deduce those functional equations from (i)$_\ell$, (ii)$_\ell$ by simply applying the given conditions to definitions of ℓ-adic iterated integrals as in loc.cit. But to treat more general σ, we need to employ some elaborate arguments as presented in the next subsection.

Theorem 4.14 *Let X be an arbitrary algebraic variety over a subfield K of $\mathbb{C}$, and let $Y := \mathbf{P}_K^1 - \{b_1, \ldots, b_M, \infty\}$, where $\{b_1, \ldots, b_M\}$ is a subset of K with cardinality M. Fix any K-rational (possibly tangential) base point v on X. Then, for a collection of algebraic K-morphisms $f_i: X \to Y$ (i = 1, \ldots, m), homomorphisms $\psi_i: \mathrm{gr}_\Gamma^n \pi_1(Y(\mathbb{C}), f_i(v)) \to \mathbb{Z}$ and integers $c_1, \ldots, c_m \in \mathbb{Z}$, the following conditions (i)$_\ell$ and (ii)$_\ell$ are equivalent:*

(i)$_\ell$
$$\sum_{i=1}^{m} c_i \, \psi_i \circ \mathrm{gr}_\Gamma^n(f_{i*}) = 0 \quad \text{in } \mathrm{Hom}\,(\mathrm{gr}_\Gamma^n \pi_1^\ell(X_{\overline{K}}, v), \mathbb{Z}_\ell),$$

(ii)$_\ell$
$$\sum_{i=1}^{m} c_i \, \widehat{\kappa_{\otimes n}}(\psi_i)(f_i) = 0 \quad \text{in } \left(\bigotimes_{i=1}^{n-2} \widehat{O}_\ell^\times(X_{\overline{K}}) \right) \otimes \left(\widehat{O}_\ell^\times(X_{\overline{K}}) \wedge \widehat{O}_\ell^\times(X_{\overline{K}}) \right).$$

Moreover, these conditions imply the following (iii)$_\ell$ and (iii)$_\ell^{\mathrm{nv}}$.

(iii)$_\ell$ *There exists an error term $E: G_{K(\mu_{\ell^\infty})} \times \pi_1^\ell(X_{\overline{K}}, v, z) \to \mathbb{Q}_\ell$ of degree n such that $\sum_{i=1}^{m} c_i \, \mathcal{L}^{\psi_i, \mathcal{I}_i}(f_i(z), f_i(v); f_i(p))(\sigma) = E(\sigma, p)$ ($\sigma \in G_{K(\mu_{\ell^\infty})}$) for each étale path $p: v \rightsquigarrow z \in X(K)$.*

(iii)$_\ell^{\mathrm{nv}}$ *There exists an error term $E_{\mathrm{nv}}: G_K \times \pi_1^\ell(X_{\overline{K}}, v, z) \to \mathbb{Q}_\ell$ of degree n such that $\sum_{i=1}^{m} c_i \, \mathcal{L}_{\mathrm{nv}}^{\psi_i, \mathcal{I}_i}(f_i(z), f_i(v); f_i(p))(\sigma) = E_{\mathrm{nv}}(\sigma, p)$ ($\sigma \in G_K$) for each étale path $p: v \rightsquigarrow z \in X(K)$.*

Finally, if $\mu_{\ell^\infty} \not\subset K$, then (iii)$_\ell^{\mathrm{nv}}$ implies (i)$_\ell$, (ii)$_\ell$.

4.4 Proof of Theorems 4.13 and 4.14 The equivalence of (i) and (ii) follows from Lemma 2.1, Lemma 4.12 and Corollary 3.7 for both complex and ℓ-adic case. We prove (i)$_\mathbb{C}$ implies the functional equation (iii)$_\mathbb{C}$ by essentially tracing lines of [W0], theorem E, and [W5], theorem 11.2.1. Suppose (i)$_\mathbb{C}$ holds. By (4.5), we have $\Lambda_{f_i(\gamma)}(f(z), f(v)) = f_{i*}(\Lambda_\gamma(z, v))$ for each $i = 1, \ldots, m$. Then,

$$\sum_{i=1}^{m} c_i \mathcal{L}_\mathbb{C}^{\psi_i}(f_i(z), f_i(v); f_i(\gamma)) = \sum_{i=1}^{m} c_i \psi_{i\mathbb{C}}(f_{i*}(\log \Lambda_\gamma(z, v)^{-1} \bmod \Gamma^{n+1}))$$

$$= \sum_{i=1}^{m} c_i \psi_{i\mathbb{C}} \circ \mathrm{gr}_\Gamma^n(f_{i*})([\log \Lambda_\gamma(z, v)^{-1}]_n) = 0.$$

Here $[\log \Lambda_\gamma(z, v)^{-1}]_n$ denotes the nth homogeneous component of $\log \Lambda_\gamma(z, v)^{-1} \bmod \Gamma^{n+1}$. Note that we used here the property that f_{i*} preserves the gradation in the complex case, as remarked just before (4.5). Thus the functional equation (iii)$_C$ follows.

Conversely, if (iii)$_C$ holds, then we may argue as in the proof of [W0], theorem 10.5.1. The composition law (4.3) and the surjectivity of $\theta_{v,X}$ (4.6) insure the surjectivity of

$$\theta_{v,z,X}: \pi_1(X, v, z) \to \pi(X) \quad (\gamma \mapsto \Lambda_\gamma(z, v)),$$

hence the images of paths $\gamma: v \leadsto z$ in $\pi(X)$ modulo the lower central subgroups form Zariski dense subsets. Noticing the remark before Definition 4.4, we see that $\psi_{iC} \circ f_{i*} \circ \log$ give polynomial functions on $\pi(X)/\Gamma^{n+1}\pi(X)$. From this follows that the functional equations for all γ implies the equation (i)$_C$.

Next, suppose (i)$_\ell$ holds. Then, since f_i are defined over K, we have

$$(4.13) \qquad \log \mathfrak{f}_\sigma^{f_i(p)} = f_i(\log \mathfrak{f}_\sigma^p) \quad (\sigma \in G_K).$$

Let $L(\pi_{X,v})$, $L(\pi_{Y,f_i(v)})$ denote the ℓ-adic (complete) Lie algebras of $\pi_1^\ell(X_{\overline{K}}, v)$, $\pi_1^\ell(Y_{\overline{K}}, f_i(v))$ respectively, and denote the composition of $\psi_{i,\bar{y}_i}$ with $f_{i*}: L(\pi_{X,v}) \to L(\pi_{Y,f_i(v)})$ by

$$(4.14) \qquad \psi_{i,\bar{y}_i} \circ f_i: L(\pi_{X,v}) \to L(\pi_{Y,f_i(v)}) \to \mathbb{Q}_\ell.$$

Then, from the above (4.13) we see

$$(4.15) \qquad \mathcal{L}_{nv}^{\psi_{i,\bar{y}_i}}(f_i(z), f_i(v); f_i(p)) = \mathcal{L}_{nv}^{\psi_{i,\bar{y}_i} \circ f_i}(z, v; p).$$

However, the necessity of error terms occurs from the fact that, in the ℓ-adic case, we do not have a canonical gradation in $L(\pi_{X,v})$, although we have chosen a collection of splittings of $L(\pi_{Y,f_i(v)})$ (via the path system $\{\gamma_i, y_j\}$ as above) which are compatible with each other for $i = 1, \ldots, m$ but not with $L(\pi_{X,v})$. We anyway take and fix one of the (vector space) complements $L_{<n}$ to $\Gamma^n L(\pi_{X,v}) \subset L(\pi_{X,v})$:

$$(4.16) \qquad L(\pi_{X,v}) = L_{<n} \oplus \Gamma^n L(\pi_{X,v}),$$

and write

$$(4.17) \qquad \log(\mathfrak{f}_\sigma^p)^{-1} = [\log(\mathfrak{f}_\sigma^p)^{-1}]_{<n} + [\log(\mathfrak{f}_\sigma^p)^{-1}]_{\geq n} \quad (\sigma \in G_K).$$

Putting this decomposition of $\log \mathfrak{f}_\sigma^p$ into the map $\sum_i c_i \psi_{i,\bar{y}_i} \circ f_i$, and applying the condition (i)$_\ell$, we obtain

$$(4.18) \quad \sum_i c_i \mathcal{L}_{\mathrm{nv}}^{\psi_{i,\tilde{y}_i} \circ f_i}(z, v; p)$$

$$= \sum_i c_i \psi_{i,\tilde{y}_i} \circ f_i([\log(\mathfrak{f}_\sigma^p)^{-1}]_{<n}) + \sum_i c_i \psi_{i,\tilde{y}_i} \circ f_i([\log(\mathfrak{f}_\sigma^p)^{-1}]_{\geq n})$$

$$= \sum_i c_i \psi_{i,\tilde{y}_i} \circ f_i([\log(\mathfrak{f}_\sigma^p)^{-1}]_{<n}) + 0.$$

Then, noticing that $[\log(\mathfrak{f}_\sigma^p)^{-1}]_{<n}$ vanishes for $\sigma \in H_n(v, z)$ and is invariant under the change $p \mapsto sp$ for $s \in \Gamma^n \pi_1^\ell(X_{\overline{K}}, v)$ (Lemma 4.11), we find the right-hand side above satisfies the conditions (E1) and (E2) of an error term of degree n. This proves (i)$_\ell \Rightarrow$ (iii)$_\ell^{\mathrm{nv}}$. To deduce (iii)$_\ell$, we wish to consider the logarithm of σ_p for $\sigma \in G_{K(\mu_{\ell^\infty})}$, but σ_p may generally not be a unipotent operator on the total $\hat{U}\pi_1^\ell(X_{\overline{K}}, v)$. Consider now the diagonal homomorphism into the direct product:

$$(4.19) \quad \Delta\big(:= \Pi_{i=1}^m f_i\big) : \pi_1^\ell(X_{\overline{K}}, v) \longrightarrow \prod_{i=1}^m \pi_1^\ell(Y_{\overline{K}}, f_i(v))$$

$$(S \mapsto (f_1(S), \dots, f_m(S))),$$

which naturally induces homomorphisms of the ℓ-adic Lie algebra and of the universal enveloping algebra into the corresponding products (denoted also Δ):

$$(4.20) \quad L(\pi_{X,v}) \longrightarrow \bigoplus_{i=1}^m L(\pi_{Y,f_i(v)}), \qquad \hat{U}\pi_1^\ell(X_{\overline{K}}, v) \longrightarrow \prod_{i=1}^m \hat{U}\pi_1^\ell(Y_{\overline{K}}, f_i(v)).$$

Let $\overline{L}_X$, $\overline{U}_X$ denote the images of the last two homomorphisms, and denote by $\bar{f}_{i*}$, for simplicity, the both maps $\overline{L}_X \to L(\pi_{Y,f_i(v)})$ and $\overline{U}_X \to \hat{U}\pi_1^\ell(Y_{\overline{K}}, f_i(v))$ factoring f_i for respective $i = 1, \dots, m$. Introduce the filtration $\{\Gamma^k \overline{L}_X\}_{k=1}^\infty$ as the pull-back of the lower central filtration on $\bigoplus_i L(\pi_{Y,f_i(v)})$. Then, for ψ_i extended to the unique $\mathbb{Q}_\ell$-linear form on $\mathrm{gr}_\Gamma^n L(\pi_{Y,f_i(v)})$ for each $1 \leq i \leq m$, the condition (i)$_\ell$ insures

$$(4.21) \quad \sum_{i=1}^m c_i \psi_i \circ \mathrm{gr}_\Gamma^n(\bar{f}_{i*}) = 0 \text{ in } \mathrm{Hom}(\overline{L}_X, \mathbb{Q}_\ell),$$

whereas the induced operations $\bar{\sigma}_v$ and $\bar{\sigma}_p$ on $\overline{U}_X$ are unipotent for $\sigma \in G_{K(\mu_{\ell^\infty})}$ (as their operations on the ith component are unipotent for every i, cf. Proposition 4.6). Now, since f_i are defined over K, we have

$$(4.22) \quad \sigma_{f_i(p)} \circ f_i = \bar{f}_i \circ \bar{\sigma}_p \circ \Delta = f_i \circ \sigma_p \quad (\sigma \in G_K, \ i = 1, \dots, m),$$

hence it holds on $\overline{U}_X$ that

$$(\mathrm{Log}\sigma_{f_i(p)})(1) = \bar{f}_i(\mathrm{Log}\bar\sigma_p(1)) \quad (\sigma \in G_{K(\mu_{\ell^\infty})}, \; i = 1, \ldots, m).$$

Lemma 4.15 $(\mathrm{Log}\bar\sigma_p)(1)$ *belongs to* $\overline{L}_X$.

Proof   In fact, this follows in exactly the same way as the proof of Proposition 4.6(ii) by taking the logarithm of both sides of $\bar\sigma_p = L_{\bar{\mathfrak{f}}} \circ \bar\sigma_v$, where $\bar{\mathfrak{f}}$ denotes the image of $\mathfrak{f}_\sigma^p$ in $\overline{U}_X$.   □

Again, although we have a collection of compatible splittings of $L(\pi_{Y,f_i(v)})$ (via the path system $\{\gamma_i, y_j\}$ as above), it generally does not induce a compatible splitting on the subspace $\overline{L}_X \subset \bigoplus_i L(\pi_{Y,f_i(v)})$. So we take and fix one (vector space) splitting

$$(4.23) \qquad\qquad \overline{L}_X = \overline{L}_{<n} \oplus \Gamma^n \overline{L}_X,$$

and write

$$(4.24) \qquad (\mathrm{Log}\bar\sigma_p)(1) = [\mathrm{Log}\bar\sigma_p(1)]_{<n} + [\mathrm{Log}\bar\sigma_p(1)]_{\geq n} \quad (\sigma \in G_{K(\mu_{\ell^\infty})}).$$

To deduce (iii)$_\ell$ from (i)$_\ell$, it then only remains to repeat the same argument as above with applying the role of $\log \mathfrak{f}_\sigma^p$ to $\mathrm{Log}\bar\sigma_p(1)$. We may leave the rest to the reader.

Finally, assume $\mu_{\ell^\infty} \not\subset K$ and (iii)$_\ell^{\mathrm{nv}}$. Let us fix a path $p : v \rightsquigarrow z$ on $X_{\overline{K}}$. For each $s \in \Gamma^n \pi_1^\ell(X_{\overline{K}}, v)$ and $\sigma \in G_K$, we have from Lemma 4.11:

$$\mathcal{L}_{\mathrm{nv}}^{\psi_{i,\bar{y}_i}}(f_i(z), f_i(v); f_i(sp))(\sigma)$$

$$= \mathcal{L}_{\mathrm{nv}}^{\psi_{i,\bar{y}_i}}(f_i(z), f_i(v); f_i(p))(\sigma) + (\chi(\sigma)^n - 1)\psi_{i,\bar{y}_i}(f_i(\log s))$$

for $i = 1, \ldots, m$. Since the error term $E_{\mathrm{nv}}(\sigma, p)$ of (iii)$_\ell^{\mathrm{nv}}$ is invariant under the change $p \mapsto sp$, putting the above equation into the functional equation yields

$$(\chi(\sigma)^n - 1)\left(\sum_{i=1}^m c_i\, \psi_{i,\bar{y}_i} \circ f_i\right)(\log(s)) = 0.$$

By assumption of $\mu_{\ell^\infty} \not\subset K$, the factor $\chi(\sigma)^n - 1$ $(\sigma \in G_K)$ runs over a non-trivial subset of $\mathbb{Z}_\ell$. From this together with the observation that the images of $\log(s)$ $(s \in \Gamma^n \pi_1^\ell(X_{\overline{K}}, v))$ generate the vector space $\mathrm{gr}_\Gamma^n \pi_1^\ell(X_{\overline{K}}, v) \otimes \mathbb{Q}_\ell$, we conclude the equation (i)$_\ell$.

The proof of Theorems 4.13 and 4.14 is thus completed. The above proof of Theorem 4.14 also implies the following

Corollary 4.16   *Notations being as in Theorem 4.14 and its proof, suppose condition* (iii)$_\ell^{\mathrm{nv}}$ *holds. Pick a vector space splitting* $L(\pi_{X,v}) = L_{<n} \oplus \Gamma^n L(\pi_{X,v})$

and decompose $\log(\mathfrak{f}_\sigma^p)^{-1} = [\log(\mathfrak{f}_\sigma^p)^{-1}]_{<n} + [\log(\mathfrak{f}_\sigma^p)^{-1}]_{\geq n}$ *according to it. Then the error term of the functional equation is given by*

$$E_{\mathrm{nv}}(\sigma, p) = \sum_i c_i \psi_{i,\bar{y}_i} \circ f_i([\log(\mathfrak{f}_\sigma^p)^{-1}]_{<n}) \quad (\sigma \in G_K, \ p: v \rightsquigarrow z).$$

A similar formula also holds for condition $(iii)_\ell$ *with simple replacements of* $L(\pi_{X,v})$, $\log(\mathfrak{f}_\sigma^p)^{-1}$, G_K *by* $\overline{L}_X$, $(\mathrm{Log}\,\bar{\sigma}_p)(1)$, $G_{K(\mu_{\ell^\infty})}$ *respectively.*

The formula above gives us a way to calculate the error term $E(\sigma, p)$ in a concrete way after picking a splitting of $L(\pi)$. Note that the error term itself is independent of the choice of splitting $L(\pi_{X,v}) = L_{<n} \oplus \Gamma^n L(\pi_{X,v})$, since the left-hand sides of the functional equations $(iii)_\ell$ and $(iii)_{\mathrm{nv}}$ are independent of this choice.

In Theorems 4.13 and 4.14, we restricted the coefficients $c_1, \ldots, c_m$ only to integers. One may ask if a functional equation with more general coefficients exists among complex or ℓ-adic iterated integrals. In fact, it turns out that there are essentially no new such functional equations as shown next.

Definition 4.17 Taking notation as in Theorems 4.13 and 4.14, we say that a family of complex iterated integrals $\mathcal{L}_{\mathbb{C}}^{\psi_i}(f_i(z), f_i(v); f_i(\gamma))$ (resp. ℓ-adic iterated integrals $\mathcal{L}^{\psi_{i,\bar{y}_i}}(f_i(z), f_i(v); f_i(p))$, resp. $\mathcal{L}_{\mathrm{nv}}^{\psi_{i,\bar{y}_i}}(f_i(z), f_i(v); f_i(p))$) $(i = 1, \ldots, m)$ *has a linear functional equation*, if a non-trivial linear combination of them with coefficients in some field F of characteristic 0 becomes zero (resp. becomes an error term) for all paths $\gamma: v \rightsquigarrow z$ (resp. for all étale paths $p: v \rightsquigarrow z$) on X with v fixed, z and γ (resp. p) vary.

Let us here consider the following four conditions:

(a) The family $\{\mathcal{L}_{\mathbb{C}}^{\psi_i}(f_i(z), f_i(v); f_i(\gamma))\}_i$ has a linear functional equation.

(b) The family $\{\mathcal{L}^{\psi_{i,\bar{y}_i}}(f_i(z), f_i(v); f_i(p))\}_i$ has a linear functional equation for all ℓ.

(c) The family $\{\mathcal{L}_{\mathrm{nv}}^{\psi_{i,\bar{y}_i}}(f_i(z), f_i(v); f_i(p))\}_i$ has a linear functional equation for all ℓ.

(d) The family $\{\mathcal{L}_{\mathrm{nv}}^{\psi_{i,\bar{y}_i}}(f_i(z), f_i(v); f_i(p))\}_i$ has a linear functional equation for one ℓ.

Proposition 4.18 *(i) Condition (a) implies all the other conditions (b), (c), (d) and, in each case, the linear combination can be replaced by a (non-trivial) combination with coefficients in $\mathbb{Z}$.*

(ii) If $\mu_{\ell^\infty} \not\subset K$, then conditions (a), (c), (d) are equivalent and, in each case, the linear combination can be replaced by a (nontrivial) combination with coefficients in $\mathbb{Z}$.

Proof (i) Suppose condition (a) holds, i.e., that $(iii)_{\mathbb{C}}$ holds with coefficients $c_1, \ldots, c_m \in F$. Then tracing the same argument as in the proof of Theorem 4.13, one can easily see that

$$\sum_{i=1}^{m} c_i \psi_i \circ \mathrm{gr}_{\Gamma}^n(f_{i*}) = 0$$

in $\mathrm{Hom}(\mathrm{gr}_{\Gamma}^n \pi_1(X(\mathbb{C}), v), F)$. But since $\psi_i \circ \mathrm{gr}_{\Gamma}^n(f_{i*})$ are defined over $\mathbb{Z}$, it follows that the coefficients of this linear equation can be replaced by rational numbers, hence by rational integers. This remark has already been pointed out in [W0], corollary 10.6.7. Then we obtain $(i)_\ell$ for these integer coefficients, which implies (b), (c), (d) with the same coefficients.

(ii) Suppose that condition (d) holds, i.e., that $(iii)_{\ell}^{\mathrm{nv}}$ holds with coefficients $c_1, \ldots, c_m \in F$. Then, again, tracing the argument in the proof of Theorem 4.14, one sees

$$\sum_{i=1}^{m} c_i \psi_i \circ \mathrm{gr}_{\Gamma}^n(f_{i*}) = 0$$

in $\mathrm{Hom}(\mathrm{gr}_{\Gamma}^n \pi_1^\ell(X_{\overline{K}}, v), F\mathbb{Q}_\ell)$. But recalling that $\mathrm{gr}_{\Gamma}^n \pi_1^\ell(X_{\overline{K}}, v)$ is isomorphic to $\mathrm{gr}_{\Gamma}^n \pi_1(X(\mathbb{C}), v) \otimes \mathbb{Z}_\ell$, we may regard $\psi_i \circ \mathrm{gr}_{\Gamma}^n(f_{i*})$ as objects coming from $\mathrm{Hom}(\mathrm{gr}_{\Gamma}^n \pi_1(X(\mathbb{C}), v), \mathbb{Z})$. Therefore, by the same reasoning as in (i), we obtain a linear equation $(i)_{\mathbb{C}}$ with rational integers. This and part (i) conclude the proof. □

Remark 4.19 The condition "(iii) $\sum_{i=1}^{N} n_i \, b_Y(e_i)([f_i]) = 0$" in [W2], theorem 5.1, should be replaced by the condition:

$$\text{"(iii)} \quad \sum_{i=1}^{N} n_i \, \widehat{\kappa_{\otimes n}}(e_i^*)(f_i) = 0 \text{ in } \left(\bigotimes_{i=1}^{n-2} O_h^{\times}(X^{\mathrm{an}}) \right) \otimes \left(O_h^{\times}(X^{\mathrm{an}}) \wedge O_h^{\times}(X^{\mathrm{an}}) \right) \text{".}$$

In the proof of [W2], theorem 5.1, we state that, passing with $\mathrm{Lie}(H(X))$ – free Lie algebra on a vector space $H(X)$ – to a dual object, we get $\mathrm{Lie}(H(X)^*) = \mathrm{Lie}(A^1(X))$ – free Lie algebra on $H(X)^* = A^1(X)$. This is obviously not correct, as a dual of a free Lie algebra on $H(X)$ is not naturally isomorphic to a free Lie algebra on a dual $H(X)^*$. The same remark concerns theorem 10.8.2 in [W0]; condition (i) should be replaced by Zagier's condition.

5 Case of polylogarithms

Now we shall apply results of the previous section in the polylogarithmic case.

5.1 Review of classical polylogarithms Let us set

$$P_0 := \mathbf{P}_K^1 - \{0, 1, \infty\}$$

defined over a field $K \subset \mathbb{C}$. With notation as in 4.1, the space $\Omega^1 = \Omega^1_{\log}(X)$, for $X = P_0(\mathbb{C})$, $D = \{0, 1, \infty\}$ and $X^* = \mathbf{P}^1$, is a two-dimensional vector space generated by the differential form $\omega_0 = \frac{dz}{z}$ and $\omega_1 = \frac{dz}{z-1}$. We take a basis (X, Y) of $V_1 = (\Omega^1)^*$ dual to the basis (ω_0, ω_1) of Ω^1. In this case, $K^\perp = 0$ and $L(V_1, K^\perp)$ is just a free Lie algebra generated by the X and Y.

On the other hand, the topological fundamental group $\pi_1(P_0(\mathbb{C}), \overrightarrow{01})$ is a free group, freely generated by the standard loops x, y running around the punctures $0, 1$ once anticlockwise respectively, so that the Lie algebra Gr Lie $\pi_1(P_0(\mathbb{C}), \overrightarrow{01})$ is freely generated by the images $\bar{x}$ and $\bar{y}$. The natural isomorphism $\mathrm{gr}^1_\Gamma \pi_1(P_0, \overrightarrow{01}) \otimes \mathbb{C} \xrightarrow{\sim} L(V_1, K^\perp)_1 = V_1$ of (4.7) gives then the identification:

$$\frac{\bar{x}}{2\pi i} = X, \qquad \frac{\bar{y}}{2\pi i} = Y.$$

Let us fix a Hall basis of the free Lie algebra Gr Lie $\pi_1(P_0(\mathbb{C}), \overrightarrow{01})$ corresponding to the ordering $\bar{x}, \bar{y}$ of generators. Then, the special elements $e_1, e_2, \ldots$ defined by

$$e_1 := \bar{y}, \quad e_n := [\bar{x}, e_{n-1}] = (\mathrm{ad}\,\bar{x})^{n-1}(\bar{y}) \quad \text{for } n > 1$$

belong to the Hall basis. Write

$$\varphi_n \colon \mathrm{gr}^n_\Gamma \pi_1(P_0(\mathbb{C}), \overrightarrow{01}) \longrightarrow \mathbb{Z}$$

for the linear form dual to e_n with respect the above Hall basis. Since φ_n kills the basis elements other than e_n, especially those of double commutator type, it belongs to $\mathrm{Hom}(\mathrm{gr}_\Gamma(\pi_1/\pi_1''), \mathbb{Z})$. We find its Kummer dual is then given by

$$\widehat{K_{\otimes n}}(\varphi_n) = z^{\otimes n-2} \otimes (z \wedge (z-1)) \in (\mathrm{Sym}^{n-2}O_h^\times) \otimes (\wedge^2 O_h^\times).$$

Definition 5.1 We define the *complex polylogarithm function* $\mathrm{li}_n(z, \gamma)$ as the complex iterated integral associated to the above φ_n and $\gamma \colon \overrightarrow{01} \rightsquigarrow z$ (cf. Definition 4.4):

$$\mathrm{li}_n(z, \gamma) := \mathcal{L}_\mathbb{C}^{\varphi_n}(z, \overrightarrow{01}; \gamma) \; (= \varphi_{n,\mathbb{C}}(\log \Lambda_\gamma(z, \overrightarrow{01})^{-1} \bmod \Gamma^{n+1})).$$

We also recall that *classical polylogarithm functions* $Li_n(z, \gamma)$ $(n = 1, 2, \ldots)$ for a path γ from $\overrightarrow{01}$ to z on $P_0(\mathbb{C})$ are defined as the iterated integrals

$$Li_n(z, \gamma) := \int_\gamma (-w_1) \cdot \underbrace{w_0 \cdots w_0}_{n-1}.$$

Note that $Li_1(z, \gamma) = -\log(1 - z)$.

The following proposition complements lemma 10.6.5 of [W0].

Proposition 5.2   *With notation as above, the complex polylogarithm function is given in terms of the classical polylogarithm functions by the formula:*

$$\mathrm{li}_n(z, \gamma) = \frac{(-1)^{n+1}}{(2\pi i)^n} \sum_{k=0}^{n-1} \frac{B_k}{k!} (\log z)^k Li_{n-k}(z, \gamma) \qquad (n \geq 1).$$

Here, $\log z$ takes the principal branch along γ, and $\{B_n\}_{n=0}^\infty$ is the sequence of Bernoulli numbers defined by $\sum_{n=0}^\infty \frac{B_n}{n!} T^n = \frac{T}{e^T - 1}$.

Proof (cf. [W1], lemma 3.4). We calculate $\Lambda(z) := \Lambda_\gamma(z, \overrightarrow{01})$ for the 1-form $\omega_{P_0} := \frac{dz}{z} X + \frac{dz}{z-1} Y$. This is a group-like element in the non-commutative power series ring $\mathbb{C}\langle\langle X, Y \rangle\rangle$ and its coefficients of the terms X^i, YX^i ($i = 0, 1, 2, \ldots$) look like

$$\Lambda(z) = 1 + \sum_{i=1}^\infty \frac{(\log z)^i}{i!} X^i - \sum_{i=0}^\infty Li_{i+1}(z) YX^i + \cdots \text{ other terms.}$$

(See [F], §3.1, for shapes of coefficients of other terms such as XY, Y^2, XYX.) Noting that the above "X^i-part" can be written as $e^{(\log z)X} - 1$ and taking the logarithm $\log \Lambda(z) = (\Lambda(z) - 1) - \frac{1}{2}(\Lambda(z) - 1)^2 + \frac{1}{3}(\Lambda(z) - 1)^3 - + \cdots$, we find that the coefficient of YX^{n-1} of $\log \Lambda(z)$ comes from the expansion of the product of the series

$$-Li_1(z)Y - Li_2(z)YX - Li_3(z)YX^2 - \cdots$$

with

$$1 - \frac{1}{2}(e^{(\log z)X} - 1)^2 + \frac{1}{3}(e^{(\log z)X} - 1)^3 - + \cdots = \frac{(\log z)X}{e^{(\log z)X} - 1} = \sum_{n=0}^\infty \frac{B_n}{n!}(\log z)^n X^n.$$

On the other hand, if we express the Lie element $\log \Lambda(z)$ as the Lie series with respect to the above fixed Hall basis, the term YX^{n-1} arises only as $(-1)^{n-1}$-multiple of the term appearing in the expansion of $e_n = \mathrm{ad}(\bar{x})^{n-1}(\bar{y}) = (2\pi i)^n \mathrm{ad}(X)^{n-1}(Y)$. The formula follows from this observation and our definition $\mathrm{li}_n(z, \gamma) = \varphi_{n,\mathbb{C}}(\log \Lambda(z)^{-1})$. □

Remark 5.3   As seen from the above proof, we have the equation

$$\log \Lambda_\gamma(z, \overrightarrow{01})^{-1} = -\frac{\log z}{2\pi i}\bar{x} - \frac{\log(1-z)}{2\pi i}\bar{y} + \sum_{m \geq 2} \mathrm{li}_m(z, \gamma) e_m + \text{ other terms.}$$

With regard to this equation, we shall write

$$\mathrm{li}_0(z, \gamma) = -\frac{\log z}{2\pi i}, \qquad \mathrm{li}_1(z, \gamma) = -\frac{\log(1-z)}{2\pi i}.$$

(The latter also follows from Definition 5.1 and Proposition 5.2.)

5.2 ℓ-adic polylogarithms ([W5]) Notations being as in the previous subsection, let K be a subfield of $\mathbb{C}$, $z \in P_0(K)$ and pick an étale path $p: \overrightarrow{01} \rightsquigarrow z$. Regard the pro-$\ell$ fundamental group $\pi_1^\ell(P_0/\bar{K}, \overrightarrow{01})$ as the pro-ℓ completion of $\pi_1(P_0(\mathbb{C}), \overrightarrow{01})$. Let $L(\pi_{\overrightarrow{01}})$ denote the associated complete ℓ-adic Lie algebra consisting of all the formal Lie series in $X := \log(x)$ and $Y := \log(y)$; the fixed generator system $\vec{x} = (x, y)$ of $\pi_1(P_0(\mathbb{C}), \overrightarrow{01})$ as in §5.1 defines the natural extension $\varphi_{n,\vec{x}}: L(\pi_{\overrightarrow{01}}) \to \mathbb{Q}_\ell$ of φ_n as in §4 (4.10–4.11).

Definition 5.4 ([W5] §11) Write I_Y for the ideal of $L(\pi_{\overrightarrow{01}})$ generated by the Lie monomials involving Y twice or more. The ℓ-adic polylogarithm function $\ell i_n(z, p, \vec{x}): G_K \to \mathbb{Q}_\ell$ ($n = 1, 2, ...$) for a path p from $\overrightarrow{01}$ to z on $P_0(K)$ is defined as the naive ℓ-adic iterated integral associated to $\varphi_{n,\vec{x}}$:

$$\ell i_n(z, p, \vec{x})(\sigma) := \mathcal{L}_{\text{inv}}^{\varphi_{n,\vec{x}}}(z, \overrightarrow{01}; p) \ (= \varphi_{n,\vec{x}}(\log(\mathfrak{f}_\sigma^p)^{-1})) \qquad (\sigma \in G_K).$$

More directly, the ℓ-adic polylogarithms can be defined as coefficients of the Lie formal expansion of $\log(\mathfrak{f}_\sigma^p)^{-1}$ in $X = \log x$, $Y = \log y$, i.e.,

$$(5.1) \quad \log \mathfrak{f}_\sigma^p(x, y)^{-1} \equiv \rho_z(\sigma) X + \rho_{1-z}(\sigma) Y$$
$$+ \sum_{m \geq 1} \ell i_{m+1}(z, p, \vec{x})(\sigma) \, \text{ad}(X)^m(Y) \bmod I_Y$$

for $\sigma \in G_K$, where I_Y denotes the ideal generated by the terms with two or more Y, and ρ_z (resp. ρ_{1-z}) is the Kummer cocycle along a carefully chosen system of power roots of z (resp. $1 - z$). With regard to this formula, we shall also define

$$\ell i_0(z, p, \vec{x})(\sigma) = \rho_z(\sigma), \qquad \ell i_1(z, p, \vec{x})(\sigma) = \rho_{1-z}(\sigma).$$

In [NW], we related the above $\ell i_n(z, p, \vec{x})$ with the *ℓ-adic polylogarithmic character* $\tilde{\chi}_m^z: G_K \to \mathbb{Z}_l$ ($m \geq 1$) defined by the Kummer properties for $n \geq 1$:

$$(5.2) \quad \zeta_{\ell^n}^{\tilde{\chi}_m^z(\sigma)} = \sigma \left(\prod_{a=0}^{\ell^n-1} (1 - \zeta_{\ell^n}^{\chi(\sigma)^{-1}a} z^{1/\ell^n})^{\frac{a^{m-1}}{\ell^m}} \right) \Big/ \prod_{a=0}^{\ell^n-1} (1 - \zeta_{\ell^n}^{a+\rho_z(\sigma)} z^{1/\ell^n})^{\frac{a^{m-1}}{\ell^m}},$$

where $(1 - \zeta_{\ell^n}^\alpha z^{1/\ell^n})^{\frac{\beta}{\ell^m}}$ means the βth power of a carefully chosen ℓ^nth root of $(1 - \zeta_{\ell^n}^\alpha z^{1/\ell^n})$ as in loc. cit. depending on $p: \overrightarrow{01} \rightsquigarrow z$. Note, in particular, that $\tilde{\chi}_1^z = \rho_{1-z}$. The formula of [NW], p. 293 corollary, gives an expression for $\ell i_m(z, p, \vec{x})$ in terms of $\tilde{\chi}_m^z$ exactly as in a similar way to Proposition 5.2 of the complex case:

$$(5.3) \quad \ell i_m(z, p, \vec{x})(\sigma) = (-1)^{m+1} \sum_{k=0}^{m-1} \frac{B_k}{k!} (-\rho_z(\sigma))^k \frac{\tilde{\chi}_{m-k}^z(\sigma)}{(m - k - 1)!} \qquad (m \geq 1).$$

(Here we point out a misprint $\sum_{k=0}^m$ in the formula of loc. cit. that should have read $\sum_{k=0}^{m-1}$ as above.) In particular, for $\sigma \in G_{K(z^{1/\ell^\infty})}$, we have

$$(5.4) \qquad \ell i_m(z, p, \vec{x})(\sigma) = (-1)^{m-1} \frac{\tilde{\chi}_m^z(\sigma)}{(m-1)!} \qquad (m \geq 1).$$

Remark 5.5   The ℓ-adic polylogarithms $\ell i_m(z, p, \vec{x})$ introduced in this paper are same as $\ell_m^z(\sigma)$ defined in [NW]. The functions $\ell i_n(z, p, \vec{x})$ give homomorphisms on $G_{K(\mu_{\ell^\infty}, z^{1/\ell^\infty})}$, while, by [NW], prop. 1, $H_2(\vec{01}, z) = G_{K(\mu_{\ell^\infty}, z^{1/\ell^\infty}, (1-z)^{1/\ell^\infty})}$. On $G_1(\vec{01}) = G_{K(\mu_{\ell^\infty})}$, one can easily see their relation to the big ℓ-adic iterated integral:

$$-\ell i_n(z, p, \vec{x})(\sigma) = \mathcal{L}^{\varphi_n, \vec{x}}(z, \vec{01}; p)(\sigma) \qquad (\sigma \in G_1(\vec{01}) = G_{K(\mu_{\ell^\infty})}).$$

(Cf. [W5] cor. 11.0.16, or proof of [NW], lemma 2.)

Remark 5.6   The Galois representation of the pro-ℓ fundamental group of $\mathbf{P}_{\mathbb{Q}}^1 - \{0, 1, \infty\}$ has been studied intensively by Y. Ihara, P. Deligne and other authors (see, e.g., [Ih1], [Ih2], [De2], [HM], [MS]). In particular, the filtration $\{G_n(\vec{01})\}_n$ of $G_{\mathbb{Q}}$ is deeply relevant to the arithmetic of (higher) cyclotomic fields. This would suggest that what is needed is a more consistent study of "z-version" filtrations of $G_{\mathbb{Q}}$ also depending on various quotients of the π_1-torsor. One such important aspect is an ℓ-adic analog of Zagier's conjecture ([W6], [DW2]). Motivic interpretation of this conjecture by Beilinson–Deligne [BD] has greatly inspired creative ideas in our work. The case $\mathbf{P}^1 - \{0, \mu_n, \infty\}$ has already been studied in depth by A. B. Goncharov [Gon] and the second named author ([DW1],[DW2], [W6],[W7],[W8]).

5.3 Functional equations for polylogarithms   Given a K-morphism $f: X \to P_0$ of an arbitrary algebraic variety X to P_0 and a K-rational (tangential) base point v on X, choose a path δ from $\vec{01}$ to $f(v)$ so that we obtain an isomorphism

$$\iota_\delta: \pi_1(P_0, f(v)) \xrightarrow{\sim} \pi_1(P_0, \vec{01})$$

by $x \mapsto \delta x \delta^{-1}$. The induced map $\mathrm{gr}_\Gamma^n(\iota_\delta)$ of $\mathrm{gr}_\Gamma^n(\pi_1(P_0, f(v))) \xrightarrow{\sim} \mathrm{gr}_\Gamma^n(\pi_1(P_0, \vec{01}))$ is *independent of the choice of* δ, as inner automorphisms act trivially on the graded quotients via the (lower) central filtration. From Corollary 3.7 and the above definition of φ_n it follows that the composition $\varphi_n(f) := \varphi_n \circ \mathrm{gr}_\Gamma^n(\iota_\delta)$ yields the identity

$$(5.5) \qquad \varphi_n(f) \circ \mathrm{gr}_\Gamma^n(f_*) = \kappa^{\otimes n}\left(f^{\circ n-2} \otimes (f \wedge (f-1))\right),$$

$$\text{i.e.,} \qquad \widehat{\kappa_{\otimes n}}(\varphi_n(f))(f) = f^{\circ n-2} \otimes (f \wedge (f-1)),$$

where $\widetilde{\kappa_{\otimes n}}(\varphi_n(f))(f)$ designates the pulled-back Kummer dual in the sense of Definition 3.8.

In the ℓ-adic case, we use the natural extensions $\varphi_{n,\vec{x}}: L(\pi_{\overrightarrow{01}}) \to \mathbb{Q}_\ell$ and $\varphi_n(f)_{\vec{x}}: L(\pi_{f(v)}) \to \mathbb{Q}_\ell$ with respect to the fixed generator system $\vec{x} = (x, y)$ (cf. equation (4.11)). They are related by the (naturally induced) isomorphism $\iota_\delta: \hat{U}(\pi_{f(v)}) \xrightarrow{\sim} \hat{U}(\pi_{\overrightarrow{01}})$ by

$$(5.6) \qquad \varphi_n(f)_{\vec{x}}(\lambda) = \varphi_{n,\vec{x}}(\delta\lambda\delta^{-1}) \quad (\lambda \in L(\pi_{f(v)})).$$

Now, given a topological or an étale path q from v to z on X, we have classical and ℓ-adic polylogarithms

$$\mathrm{li}_n(f(z), \delta f(q)), \ \mathrm{li}_n(f(v), \delta); \quad \ell i_n(f(z), \delta f(q), \vec{x}), \ \ell i_n(f(v), \delta, \vec{x}),$$

where $\delta f(q)$ denotes the composition of paths $\delta: \overrightarrow{01} \rightsquigarrow f(v)$ and $f(q): f(v) \rightsquigarrow f(z)$. Later in Proposition 5.10, these will be related to the iterated integrals

$$\mathcal{L}_{\mathbb{C}}^{\varphi_n}(f(z), f(v); f(q)), \quad \mathcal{L}_{\mathrm{nv}}^{\varphi_n(f)_{\vec{x}}}(f(z), f(v); f(q)).$$

that appear in the direct application of the functional equations of Theorems 4.13 and 4.14 to this special case of polylogarithms, that is:

Theorem 5.7 *Let V be an arbitrary algebraic variety over a subfield K of $\mathbb{C}$ with a K-rational (possibly tangential) base point v on V, and let $P_0 = \mathbf{P}_K^1 - \{0, 1, \infty\}$. Then, for algebraic morphisms $f_i: V \to P_0$ $(i = 1, \ldots, m)$ and integers $c_1, \ldots, c_m \in \mathbb{Z}$ together with paths $\delta_i: \overrightarrow{01} \rightsquigarrow f_i(v)$, the following conditions (i)$_\mathbb{C}$, (ii)$_\mathbb{C}$ and (iii)$_\mathbb{C}$ are equivalent:*

(i)$_\mathbb{C}$
$$\sum_{i=1}^m c_i \, \varphi_n(f_i) \circ \mathrm{gr}_\Gamma^n(f_{i*}) = 0 \quad in \ \mathrm{Hom}\,(\mathrm{gr}_\Gamma^n \pi_1(V(\mathbb{C}), v), \mathbb{Z}),$$

(ii)$_\mathbb{C}$
$$\sum_{i=1}^m c_i \, f_i^{\odot n-2} \otimes (f_i \wedge (f_i - 1)) = 0$$

$$in \ \mathrm{Sym}^{n-2} O_h^\times(V^{an}) \otimes (O_h^\times(V^{an}) \wedge O_h^\times(V^{an})),$$

(iii)$_\mathbb{C}$
$$\sum_{i=1}^m c_i \, \mathcal{L}_{\mathbb{C}}^{\varphi_n}(f_i(z), f_i(x); f_i(\gamma)) = 0$$

for each path $\gamma : x \rightsquigarrow z$ on $V(\mathbb{C})$.

For the following ℓ-adic analogs (i)$_\ell$, (ii)$_\ell$ and (iii)$_\ell$ of the conditions above, we have, in general, the equivalence (i)$_\ell$ ⇔ (ii)$_\ell$ and the implication (i)$_\ell$ (ii)$_\ell$ ⇒

(iii)$_\ell$. *If $\mu_{\ell^\infty} \not\subset K$, then we have also* (i)$_\ell$ (ii)$_\ell$ $\Leftrightarrow$ (iii)$_\ell$:

(i)$_\ell$
$$\sum_{i=1}^{m} c_i\, \varphi_n(f_i) \circ \mathrm{gr}_\Gamma^n(f_{i*}) = 0 \quad \text{in } \mathrm{Hom}\,(\mathrm{gr}_\Gamma^n \pi_1^\ell(V_{\overline{K}}, v), \mathbb{Z}_\ell),$$

(ii)$_\ell$
$$\sum_{i=1}^{m} c_i\, f_i^{\odot n-2} \otimes (f_i \wedge (f_i - 1)) = 0$$

$$\text{in } \mathrm{Sym}^{n-2}\widehat{O}_\ell^\times(V_{\overline{K}}) \otimes (\widehat{O}_\ell^\times(V_{\overline{K}}) \wedge \widehat{O}_\ell^\times(V_{\overline{K}})),$$

(iii)$_\ell$ *There exists an error term $E \colon G_K \times \pi_1^\ell(V_{\overline{K}}, v, z) \to \mathbb{Q}_\ell$*

of degree n such that

$$\sum_{i=1}^{m} c_i\, \mathcal{L}_{\mathrm{inv}}^{\varphi_n(f_i)\bar{x}}(f_i(z), f_i(v); f_i(p))(\sigma) = E(\sigma, p) \quad (\sigma \in G_K)$$

for each étale path $p \colon x \rightsquigarrow z$.

Proof   This is only a special case of Theorem 4.13 and Theorem 4.14.   □

We remark that, in conditions (i)$_\mathbb{C}$ and (i)$_\ell$, $\mathrm{Hom}(\mathrm{gr}_\Gamma^n \pi_1, -)$ may be replaced by $\mathrm{Hom}(\mathrm{gr}_\Gamma^n(\pi_1/\pi_1''), -)$ according to the last half statement of Corollary 3.7. For later convenience when computing the error term $E(\sigma, p)$ in ℓ-adic cases, we shall rephrase Corollary 4.16 in this special case using (5.6). Let $\pi_{V,v}$ be the pro-unipotent completion of the pro-ℓ fundamental group $\pi_1^\ell(V_{\overline{K}}, v)$ over $\mathbb{Q}_\ell$, and let $L(\pi_{V,v})$ denote its (complete) Lie algebra equipped with the lower central filtration $L(\pi_{V,v}) = \Gamma^1 L(\pi_{V,v}) \supset \Gamma^2 L(\pi_{V,v}) \supset \cdots$.

Corollary 5.8   *Using notation as in Theorem 5.7, suppose condition (iii)$_\ell$ holds. Pick a vector space splitting $L(\pi_{V,v}) = L_{<n} \oplus \Gamma^n L(\pi_{V,v})$ and decompose $\log(\mathfrak{f}_\sigma^p)^{-1} = [\log(\mathfrak{f}_\sigma^p)^{-1}]_{<n} + [\log(\mathfrak{f}_\sigma^p)^{-1}]_{\geq n}$ according to it. Then the error term of the functional equation is given by*

$$E(\sigma, p) = \sum_i c_i\, \varphi_{n,\bar{x}}(\delta_i \cdot f_i([\log(\mathfrak{f}_\sigma^p)^{-1}]_{<n}) \cdot \delta_i^{-1}) \quad (\sigma \in G_K,\ p \colon v \rightsquigarrow z).$$

Next, we shall consider the problem of expressing the iterated integrals in the functional equations (iii)$_\mathbb{C}$, (iii)$_\ell$ by polylogarithms explicitly. For this purpose, we shall introduce a useful series of polynomials in several variables as follows.

Proposition 5.9 (Polylog-BCH formula)   *Let $\{a_i\}_{i=0}^\infty$, $\{b_i\}_{i=0}^\infty$ be countably many mutually commuting variables, and let X, Y be non-commutative variables. Consider the Lie algebra consisting of the Lie formal series in X and Y with coefficients in the polynomial ring $\mathbb{Q}[a_i, b_i]_{i=0}^\infty$. Then, there exists a unique* *sequence of polynomials*

$$P_n = P_n(\{a_i\}_{i=0}^n, \{b_i\}_{i=0}^n) \qquad (n = 0, 1, 2, \dots)$$

characterized by

$$\left(a_0 X + a_1 Y + \sum_{i=1}^{\infty} a_{i+1} \, \mathrm{ad}(X)^i(Y)\right) \boxplus \left(b_0 X + b_1 Y + \sum_{i=1}^{\infty} b_{i+1} \, \mathrm{ad}(X)^i(Y)\right)$$

$$\equiv P_0 X + P_1 Y + \sum_{i=1}^{\infty} P_i \, \mathrm{ad}(X)^i(Y) \quad mod \; I_Y,$$

where ⊞ denotes the Baker–Campbell–Hausdorff sum: $S \boxplus T = \log(e^S e^T)$, and I_Y is the ideal generated by the Lie monomials having Y twice or more.

Proof We put $S = a_0 X + \sum_{i \geq 0} a_i \, \mathrm{ad}(X)^i(Y)$ and $T = b_0 X + \sum_{i \geq 0} b_i \, \mathrm{ad}(X)^i(Y)$ into the Baker–Campbell–Hausdorff formula:

$$S \boxplus T = S + T + \frac{1}{2}[S, T] + \frac{1}{12}[S[S, T]] - \frac{1}{12}[T[S, T]] - \frac{1}{24}[S[T[S, T]]] + \cdots .$$

Observing $[S, T] \equiv (a_0 b_1 - b_0 a_1)[X, Y] + a_0 \sum_{i \geq 2} b_i \, \mathrm{ad}(X)^i(Y) \pmod{I_Y}$, we have only to treat the terms of the form $[U_1[U_2[\cdots[U_n, [S, T]]\cdots]]]$ where $U_i = S$ or T, because the other type of terms vanish modulo I_Y. Further calculations show that each of these terms with $k + 1$-times of S and $l + 1$-times of T contributes $b_0^l \sum_{j=0}^{\infty} (a_0 b_{j+1} - b_0 a_{j+1}) \, \mathrm{ad}(X)^{k+l+j}(Y)$ regardless of the order of S and T. So these terms having the same numbers of S and T can be counted together; their appearances are summed up as the coefficient of $S^{k+1} T^{l+1}$ of $S \boxplus T$ (in the non-commutative power series ring in S and T) multiplied by $(-1)^l$. The coefficients of these terms are explicitly calculated by K. Goldberg [Gol, th. 3 and (10)] (see also [K]) showing that those coefficients $c(s, t)$ of $S^s T^t$ in $S \boxplus T$ and their generating functions are given by

(G1) $c(s, t) = \dfrac{(-1)^s}{s! \, t!} \displaystyle\sum_{i=1}^{t} \binom{t}{i} B_{s+t-i} \quad (s, t \geq 1),$

(G2) $uv \displaystyle\sum_{k,l=0}^{\infty} c(k+1, l+1) u^k v^l = \dfrac{ue^u(e^v - 1) - ve^v(e^u - 1)}{e^u - e^v} \quad (=: uv \, G(u, v)).$

Putting these together, if we write $G(a_0 t, -b_0 t) = \sum_{i=0}^{\infty} C_i(a_0, b_0) t^i$, then we obtain

$$P_n = a_n + b_n + (a_0 b_1 - b_0 a_1) C_{n-2}(a_0, b_0) + \cdots + (a_0 b_{n-1} - b_0 a_{n-1}) C_0(a_0, b_0)$$

for $n \geq 2$. This completes the proof of our proposition. □

Example 5.10   Taking notation as in the proof above, one can write

$$(5.7) \quad G(a_0 t, -b_0 t) = \sum_{i=0}^{\infty} C_i(a_0, b_0) t^i = \frac{1}{b_0 t} \left(1 - \frac{e^{a_0 t} - 1}{a_0} \cdot \frac{a_0 + b_0}{e^{(a_0 + b_0)t} - 1} \right),$$

which is more useful in computer calculations (The form of the right-hand side given above is derived in [K].) The last equation in the proof above yields a formula for the generating function of the polynomials $\{P_n\}$ of the above proposition as follows:

$$(5.8) \quad \sum_{n=0}^{\infty} P_n t^n = \sum_{n=0}^{\infty} (a_n + b_n) t^n + t \left(a_0 \sum_{n=1}^{\infty} b_n t^n - b_0 \sum_{n=1}^{\infty} a_n t^n \right) \cdot G(a_0 t, -b_0 t).$$

The first several polynomials P_n read

$$P_0 = a_0 + b_0,$$

$$P_1 = a_1 + b_1,$$

$$P_2 = a_2 + b_2 + \frac{1}{2}(a_0 b_1 - b_0 a_1),$$

$$P_3 = a_3 + b_3 + \frac{1}{2}(a_0 b_2 - b_0 a_2) + \frac{1}{12}(a_0^2 b_1 - a_0 a_1 b_0 - a_0 b_0 b_1 + a_1 b_0^2),$$

$$P_4 = a_4 + b_4 + \frac{1}{2}(a_0 b_3 - b_0 a_3) + \frac{1}{12}(a_0^2 b_2 - a_0 a_2 b_0 - a_0 b_0 b_2 + a_2 b_0^2)$$
$$\quad - \frac{1}{24}(a_0^2 b_0 b_1 - a_0 a_1 b_0^2),$$

$$P_5 = a_5 + b_5 + \frac{1}{2}(a_0 b_4 - b_0 a_4) + \frac{1}{12}(a_0^2 b_3 - a_0 a_3 b_0 - a_0 b_0 b_3 + a_3 b_0^2)$$
$$\quad - \frac{1}{24}(a_0^2 b_0 b_2 - a_0 a_2 b_0^2) - \frac{1}{180}(a_0 a_1 b_0^3 - a_0^2 a_1 b_0^2 - a_0^2 b_0^2 b_1 + a_0^3 b_0 b_1)$$
$$\quad - \frac{1}{720}(a_1 b_0^4 - a_0 b_0^3 b_1 - a_0^3 a_1 b_0 + a_0^4 b_1).$$

Concerning symmetric properties of our polynomials P_n, it firstly follows from $(e^S e^T)^{-1} = e^{-T} e^{-S}$ that

$$(5.9) \quad P_n(\{a_i\}_{i=0}^{n}, \{b_i\}_{i=0}^{n}) = -P_n(\{-b_i\}_{i=0}^{n}, \{-a_i\}_{i=0}^{n}).$$

For our later calculations, useful are also the formulas of $G(a_0 t, b_0 t)$ in special cases of a_0 or $b_0 = 0$. These are easily obtained from de l'Hospital's formula as follows:

$$(5.10) \quad G(0, -b_0 t) = -\sum_{n=1}^{\infty} \frac{B_n}{n!} (b_0 t)^{n-1}, \quad G(a_0 t, 0) = 1 + \sum_{n=1}^{\infty} \frac{B_n}{n!} (a_0 t)^{n-1}.$$

Note $B_1 = -\frac{1}{2}$ so that the above series both start from the constant term $\frac{1}{2}$.

Now we arrive at the stage where translations of the functional equations (iii)$_\mathbb{C}$, (iii)$_\ell$ of Theorem 5.6 are available in terms of polylogarithms.

Proposition 5.11 *Given a K-morphism $f\colon V \to P_0$ as in the beginning of this subsection, we can express the complex and ℓ-adic iterated integrals associated to φ_n and the image of a path $q\colon v \rightsquigarrow z$ on V in P_0 in terms of polylogarithms along the paths $\delta\colon \overrightarrow{01} \rightsquigarrow f(v)$ and $\delta f(q)\colon \overrightarrow{01} \rightsquigarrow f(z)$ as follows:*

(i) $\mathcal{L}_\mathbb{C}^{\varphi_n}(f(z), f(v); f(q)) = P_n(\{\mathrm{li}_i(f(z), \delta f(q))\}_{i=0}^n, \{-\mathrm{li}_i(f(v), \delta)\}_{i=0}^n),$

(ii) $\mathcal{L}_{\mathrm{inv}}^{\varphi_n(f)_{\bar{x}}}(f(z), f(v); f(q)) = P_n(\{-\ell i_i(f(v), \delta, \vec{x})\}_{i=0}^n, \{\ell i_i(f(z), \delta f(q), \vec{x})\}_{i=0}^n).$

For the definitions of li_0, li_1, ℓi_0 and ℓi_1, see Remark 5.3 and Definition 5.4.

Proof By the chain rule (4.3), we have

$$\Lambda_{f(q)}(f(z), f(q))^{-1} = \Lambda_{\delta f(q)}(f(z), \overrightarrow{01})^{-1} \cdot \left(\Lambda_\delta(f(v), \overrightarrow{01})^{-1}\right)^{-1}.$$

Hence

$$\log(\Lambda_{f(q)}(f(z), f(q))^{-1})$$
$$= \left(\log(\Lambda_{\delta f(q)}(f(z), \overrightarrow{01})^{-1})\right) \boxplus \left(-\log(\Lambda_\delta(f(v), \overrightarrow{01})^{-1})\right).$$

Expanding both sides above as Lie formal series in $\bar{x}$ and $\bar{y}$ modulo I_Y, we obtain (i). For the ℓ-adic case, since $\mathfrak{f}_\sigma^{\delta f(q)} = \delta \mathfrak{f}_\sigma^{f(q)} \delta^{-1} \mathfrak{f}_\sigma^\delta$, we have

$$\log(\delta(\mathfrak{f}_\sigma^{f(q)})^{-1}\delta^{-1}) = \left(-\log(\mathfrak{f}_\sigma^\delta)^{-1}\right) \boxplus \left(\log(\mathfrak{f}_\sigma^{\delta f(q)})^{-1}\right).$$

By Definition 4.7 and (5.6), the left-hand side of (ii) may be written as

$$\mathcal{L}_{\mathrm{inv}}^{\varphi_n(f)_{\bar{x}}}(f(z), f(v); f(q))(\sigma) = \varphi_n(f)_{\bar{x}}(f(\log(\mathfrak{f}_\sigma^{f(q)})^{-1}))$$
$$= \varphi_{n,\bar{x}}(\delta \log(\mathfrak{f}_\sigma^{f(q)})^{-1}\delta^{-1}).$$

These equations complete the proof of (ii).   □

The following proposition computes the "BCH-conjugation" of polylog Lie series.

Proposition 5.12 *Suppose that A, B are Lie formal power series in X, Y in the form modulo I_Y:*

$$\begin{cases} A \equiv a_0 X + a_1 Y + \sum_{i=1}^\infty a_{i+1}\,\mathrm{ad}(X)^i(Y), \\ B \equiv b_0 X + b_1 Y + \sum_{i=1}^\infty b_{i+1}\,\mathrm{ad}(X)^i(Y). \end{cases}$$

Then, we have

$$\log(e^A e^B e^{-A}) \equiv B + \sum_{n=1}^\infty \sum_{k=1}^\infty (a_0 b_n - b_0 a_n)\frac{a_0^{k-1}}{k!}(\mathrm{ad}\,X)^{n+k-1}(Y) \quad mod\ I_Y.$$

In particular, if $a_0 = 0$, then, $\log(e^A e^B e^{-A}) \equiv B - b_0[X, A]$ mod I_Y.

Proof This follows from the well-known formula $e^A e^B e^{-A} = \sum_{n=0}^{\infty} \frac{(\mathrm{ad}\,A)^n(B)}{n!}$ by simple computation.   □

5.4 Drinfeld associators under S_3
We may consider $\Lambda_\gamma(w, v)$ for any path $\gamma \colon v \rightsquigarrow w$ between tangential basepoints

$$v, w \in \mathfrak{B} := \{\overrightarrow{01}, \overrightarrow{0\infty}, \overrightarrow{\infty 1}, \overrightarrow{10}, \overrightarrow{1\infty}, \overrightarrow{\infty 0}\}.$$

See [De2], [W3] for precise definitions. We shall call $\Lambda_\gamma := \Lambda_\gamma(w, v)$ the Drinfeld associator for γ. The fundamental groupoid $\pi_1(P_0(\mathbb{C}), \mathfrak{B})$ is generated by the standard paths $\langle a, b \rangle$, $[a_b^c]$ ($\{a, b, c\} = \{0, 1, \infty\}$), where $\langle a, b \rangle$ denotes the path from $\overrightarrow{ab}$ to $\overrightarrow{ba}$ along $\mathbf{P}^1(\mathbb{R})$, and $[a_b^c]$ denotes the half anticlockwise rotation from $\overrightarrow{ab}$ to $\overrightarrow{ac}$ (cf. [N2-I]). By the chain rule (4.3) and the pushforward property (4.5) (extended to those Λ_γ between tangential base points), we may compute any Drinfeld associator by compositions of the S_3-transforms of $\Lambda_{\langle 0,1 \rangle}$, $\Lambda_{[0_1^\infty]} = e^{\pi i X}$. Usually, the non-commutative power series $\Lambda(X, Y) := \Lambda_{\langle 0,1 \rangle}$ is called 'the' Drinfeld associator whose coefficients are given by multiple zeta values ([Dr]; see, e.g., [F]). For the sequel of this article, we only recall its polylogarithmic part:

$$\log(\Lambda_{\langle 0,1 \rangle})^{-1} \equiv \sum_{m=2}^{\infty} (-1)^{m+1} \zeta(m)(\mathrm{ad}\,X)^{m-1}(Y) \qquad \mathrm{mod}\ I_Y.$$

It is useful to recall the following action of S_3-automorphisms of P_0 on X, Y to compute the other Λ_γ.

$f(z)$	z	$\frac{z}{z-1}$	$\frac{1}{z}$	$1 - z$	$\frac{1}{1-z}$	$\frac{z-1}{z}$
$f_*(X)$	X	X	$-X - Y$	Y	Y	$-X - Y$
$f_*(Y)$	Y	$-X - Y$	Y	X	$-X - Y$	X

For our later application, let us illustrate the computation of the polylog part of $\log(\Lambda_\gamma)^{-1}$ for $\delta = \langle 0, 1 \rangle [1_0^\infty] \langle 1, \infty \rangle$. By the chain rule and the pushforward property, it follows that

$$\Lambda_\delta = \Lambda_{\langle 0,1 \rangle} e^{\pi i Y} f_*(\Lambda_{\langle 0,1 \rangle})^{-1} = \Lambda(X, Y) e^{\pi i Y} \Lambda(-X - Y, Y)^{-1},$$

where $f(z) = \frac{1}{z}$. Evaluating $\log(\Lambda_\delta)^{-1}$ after the polylog BCH formula (Proposition 5.9), we find from $\zeta(2n) = (2\pi i)^{2n} \frac{-B_{2n}}{2 \cdot (2n)!}$:

$$(5.11) \qquad \log(\Lambda_{(0,1)[1_0^\infty](1,\infty)})^{-1} \equiv \sum_{n=1}^{\infty} (2\pi i)^n \frac{B_n}{n!} (\mathrm{ad}\, X)^{n-1}(Y)$$

$$\equiv \sum_{n=1}^{\infty} \frac{B_n}{n!} (\mathrm{ad}\, \bar{x})^{n-1}(\bar{y}) \quad \mathrm{mod}\ I_Y.$$

As the ℓ-adic correspondent of Drinfeld associators, we may consider the functions $\mathfrak{f}_\sigma^\gamma$ ($\sigma \in G_{\mathbb{Q}}$) for $\gamma \in \pi_1(P_0/\overline{\mathbb{Q}}, \mathfrak{B})$. The basic associator is $\mathfrak{f}_\sigma := \mathfrak{f}_\sigma^{(0,1)}$ whose polylog part is known essentially by Ihara's work (cf. [NW]) as

$$(5.12) \qquad \log(\mathfrak{f}_\sigma)^{-1} \equiv \sum_{m \geq 1} \ell i_{m+1}^{\overrightarrow{10}}(\sigma)(\mathrm{ad}\, X)^m(Y)$$

$$\equiv \sum_{m \geq 1} (-1)^m \frac{\tilde{\chi}_{m+1}^{\overrightarrow{10}}(\sigma)}{m!} (\mathrm{ad}\, X)^m(Y)\ \mathrm{mod}\ I_Y$$

for $\sigma \in G_{\mathbb{Q}}$. The coefficient character $\tilde{\chi}_m^{\overrightarrow{10}}$ is the $(1 - \ell^{m-1})^{-1}$-multiple of the so-called Soulé character χ_m which, over $G_{\mathbb{Q}(\zeta_{\ell^\infty})}$, vanishes for even $m \geq 2$, and is non-trivial for odd $m \geq 3$. Precise formulas for the Soulé characters of even degrees $m = 2k$ ($k = 1, 2, \ldots$) over $G_{\mathbb{Q}}$ have also been calculated by several authors. For $m = 2$, [LS], pp. 582–583, expressed $\tilde{\chi}_2^{\overrightarrow{10}}(\sigma)$ as $\frac{1}{24}(\chi(\sigma)^2 - 1)$ by using the ℓ-adic cyclotomic character $\chi : G_{\mathbb{Q}} \to \mathbb{Z}_\ell^\times$, and [Ih1] theorem (p. 115) remarked a complete formula for all m (without proof). In [W10], the second named author gave a proof for the case of even degrees $m \geq 2$ which has essentially the same nature as the first one below. Note that these formulae should be regarded as the ℓ-adic analogs of the classical formulae $\zeta(2k) = -(2\pi i)^{2k} \frac{B_{2k}}{2 \cdot (2k)!}$ for which a "geometric proof" via the use of associators was presented in [De2], §18.17.

Proposition 5.13   *For $k = 1, 2, \ldots$, we have*

$$\tilde{\chi}_{2k}^{\overrightarrow{10}}(\sigma) = \frac{B_{2k}}{2(2k)}(\chi(\sigma)^{2k} - 1) \qquad (\sigma \in G_{\mathbb{Q}}).$$

First proof   Fix $\sigma \in G_{\mathbb{Q}}$ and consider the power-conjugate form of the σ-action on the standard loop $z := (xy)^{-1}$ around the puncture ∞ in the pro-ℓ fundamental group $\pi_1^\ell(P_0/\overline{\mathbb{Q}}, \overrightarrow{01})$. This is known, for example, from [Ih1] p. 106, [N1] (A10), [N2-I] prop. 2.11, and the σ-actions on both sides of $z = y^{-1} x^{-1}$ yield the equation

$$G_\sigma\, z^{\chi(\sigma)}\, G_\sigma^{-1} = \mathfrak{f}_\sigma(y, x)\, y^{-\chi(\sigma)}\, \mathfrak{f}_\sigma(y, x)^{-1} \cdot x^{-\chi(\sigma)},$$

where $G_\sigma := \mathfrak{f}_\sigma(y, x) y^{\frac{1-\chi(\sigma)}{2}} \mathfrak{f}_\sigma(z, y)$. We shall evaluate the log of both sides of the above equation modulo I_Y. First, by a similar computation to (5.11), it follows that

$$\log G_\sigma \equiv \frac{1-\chi(\sigma)}{2} Y + \sum_{n=1}^{\infty} 2\ell i_{2n}^{\vec{10}}(\sigma)(\operatorname{ad} X)^{2n-1}(Y) \bmod I_Y.$$

Applying this and the well-known formula $\log(e^\alpha e^\beta) \equiv \beta + \sum_{n=0}^{\infty} \frac{B_n}{n!} (\operatorname{ad}\beta)^n(\alpha)$ mod $\deg(\alpha) \geq 2$ to Proposition 5.12, we obtain the following congruence modulo I_Y:

$$\log(G_\sigma \cdot (xy)^{-\chi(\sigma)} \cdot G_\sigma^{-1})$$
$$\equiv \chi(\sigma) \log(y^{-1} x^{-1}) + \chi(\sigma)(\operatorname{ad} X) \log G_\sigma$$
$$\equiv -\chi(\sigma) X + \sum_{n=0}^{\infty} (-1)^{n+1} \chi(\sigma) \frac{B_n}{n!} (\operatorname{ad} X)^n(Y)$$
$$+ \chi(\sigma)(\operatorname{ad} X) \left\{ \frac{1-\chi(\sigma)}{2} Y + 2 \sum_{n=1}^{\infty} \ell i_{2n}^{\vec{10}}(\sigma)(\operatorname{ad} X)^{2n-1}(Y) \right\}.$$

On the other hand, since $\log \mathfrak{f}_\sigma$ and $\log(y^{\chi(\sigma)})$ have no terms of X, by Proposition 5.12, it follows that $\log(\mathfrak{f}_\sigma(y, x) y^{-\chi(\sigma)} \mathfrak{f}_\sigma(y, x)^{-1}) \equiv -\chi(\sigma) Y$, hence the log of LHS is congruent to

$$\log(y^{-\chi(\sigma)} x^{-\chi(\sigma)})$$
$$\equiv -\chi(\sigma) X + \sum_{n=0}^{\infty} (-\chi(\sigma))^{n+1} \frac{B_n}{n!} (\operatorname{ad} X)^n(Y)$$
$$\equiv -\chi(\sigma) X - \chi(\sigma) Y - \frac{\chi(\sigma)^2}{2} [X, Y] + \sum_{n=1}^{\infty} (-\chi(\sigma))^{2n} \frac{B_{2n}}{(2n)!} (\operatorname{ad} X)^{2n}(Y)$$

modulo I_Y. Comparing the coefficients of $(\operatorname{ad} X)^{2n}(Y)$ of the RHS's settles the desired formula.   □

Second proof   We shall make use of the explicit formula (5.2) for $\tilde{\chi}_{2k} := \tilde{\chi}_{2k}^{\vec{10}}$:

(5.13) $$\zeta_{\ell^n}^{\tilde{\chi}_{2k}(\sigma)} = \frac{\sigma \left(\prod_{a=0}^{\ell^n-1} (1 - \zeta_{\ell^n}^{\chi(\sigma)^{-1} a})^{\frac{a^{2k-1}}{\ell^n}} \right)}{\prod_{a=0}^{\ell^n-1} (1 - \zeta_{\ell^n}^{a+\rho_z(\sigma)})^{\frac{a^{2k-1}}{\ell^n}}}.$$

Fix any $\sigma \in G_\mathbb{Q}$ and set $c := \chi(\sigma) \in \mathbb{Z}_\ell^\times$. Pick $\bar{c} \in \mathbb{Z}_{>0}$ such that $c\bar{c} \equiv 1$ mod ℓ^{2n}. Choose any decomposition of the index set

$$S_+ \sqcup S_- \sqcup S_0 = \{1, \ldots, \ell^n - 1\}, \quad S_0 := \begin{cases} \emptyset & (\ell \neq 2), \\ \{\frac{1}{2}\ell^n\} & (\ell = 2), \end{cases}$$

so that $a \in S_+$ if and only if $\ell^n - a \in S_-$. We are going to rewrite both of the numerator and the denominator of RHS of (5.13) by using

(5.14)
$$(1 - \zeta_{\ell^n}^{-a})^{\frac{1}{\ell^n}} = (1 - \zeta_{\ell^n}^{a})^{\frac{1}{\ell^n}} \cdot \zeta_{2\ell^{2n}}^{\ell^n - 2a}.$$

We remark that, for any $a \in S_+$, the quotient of factors corresponding to $\ell^n - a \in S_-$ in the numerator and denominator of (5.13) may be replaced by that of factors corresponding to $-a$. Therefore, we may and do regard the set S_- to be $\{-a \mid a \in S_+\}$. First we shall consider the case $\ell \neq 2$. It is easy to see that the denominator of (5.13) amounts to the product

$$\prod_{a \in S_+} \zeta_{2\ell^{2n}}^{2a^{2k} - a^{2k-1}\ell^n}$$

To apply (5.14) for the numerator of (5.13), we first need to replace the exponent $c^{-1}a$ of ζ_{ℓ^n} by the least residue modulo ℓ^n, i.e., by $\bar{c}a - [\frac{\bar{c}a}{\ell^n}]\ell^n$ (we denote by $[*]$ the largest integer $\leq *$). From this remark, it easily amounts to

$$\prod_{a \in S_+} \zeta_{2\ell^{2n}}^{2a^{2k} - ca^{2k-1}\ell^n - 2c\ell^n a^{2k-1}[\frac{\bar{c}a}{\ell^n}]}.$$

Thus, writing the fractional part as $\{*\} := * - [*]$, we obtain the congruence modulo ℓ^n

$$\tilde{\chi}_{2k}(\sigma) \equiv \sum_{a \in S_+} a^{2k-1} \left(-c \left[\frac{\bar{c}a}{\ell^n} \right] - \frac{c}{2} + \frac{1}{2} \right)$$

$$\equiv \sum_{a \in S_+} a^{2k-1} \left(c \left\{ \frac{\bar{c}a}{\ell^n} \right\} - \left\{ \frac{a}{\ell^n} \right\} + \frac{1-c}{2} \right).$$

Observe that in the above sum S_+ may be replaced by S_- (i.e., giving the same sum). So we may take $\frac{1}{2} \sum_{a \in S}$ instead of $\sum_{a \in S_+}$. Then, applying [La2] p. 39 and then p. 36, we find

$$\tilde{\chi}_{2k}(\sigma) \equiv \frac{1}{2} \sum_{a \in S} \frac{\ell^{n(2k-1)}}{2k} \left[c^{2k} B_{2k} \left(\left\{ \frac{\bar{c}a}{\ell^n} \right\} \right) - B_{2k} \left(\left\{ \frac{a}{\ell^n} \right\} \right) \right]$$

$$\equiv \frac{1}{2} \frac{c^{2k} - 1}{2k} B_{2k}(0)$$

modulo $\frac{\ell^n}{2(2k)D_k} \mathbb{Z}$ (where D_k is the least common multiple of the denominators of coefficients of the polynomial $B_{2k}(X)$). Letting then the projective limit $n \to \infty$, we obtain the desired formula. In the case of $\ell = 2$, we have to take care of the factor coming from the index set S_0. But it turns out to form only a bounded value on S_0 converging to "measure zero", having no essential effect on the final conclusion of the above argument.   □

Remark 5.14   Let δ be a path from $\overrightarrow{01}$ to $z = -1$ on $\mathbf{P}^1(\mathbb{C}) - \{0, 1, \infty\}$ defined as the composition of the anticlockwise half-turn around $z = 0$ and the simple move to $z = -1$ along the reals. By a method similar to that in the second proof of Proposition 5.13, one can show

$$\tilde{\chi}_2^{z=-1}(\sigma) = -\frac{1}{48}(\chi(\sigma)^2 - 1) - \frac{\chi(\sigma)}{2}\rho_2(\sigma) \qquad (\sigma \in G_\mathbb{Q}),$$

where $\tilde{\chi}_2^{z=-1}$ is taken along the above δ. (Changing δ to its complex conjugate $\bar{\delta}$ above changes the sign of $\frac{\chi(\sigma)}{2}\rho_2(\sigma)$ on the right-hand side.) We point out that $\frac{1}{48}(\chi(\sigma)^2 - 1)$ is generally not in $\mathbb{Z}_2$, while $\frac{1}{48}(\chi(\sigma)^2 - 1) \pm \frac{\chi(\sigma)}{2}\rho_2(\sigma)$ does always belong to $\in \mathbb{Z}_2$. This is the ℓ-adic analog of the well-known formula "$Li_2(-1) = -\frac{\pi^2}{12}$" in the complex case (cf. [Le]).

6 Examples

6.1 $Li_2(z) + Li_2(1 - z)$ Let $V = P_0 = \mathbf{P}_z^1 - \{0, 1, \infty\}$ defined over a subfield $K \subset \mathbb{C}$, and consider two morphisms $f_1, f_2 \colon V \to P_0$ defined by $f_1(z) = z$, $f_2(z) = 1 - z$. Then, the images of $v = \overrightarrow{01}$ on V by these morphisms are given by $f_1(v) = \overrightarrow{01}$ and $f_2(v) = \overrightarrow{10}$. Let x, y be the standard loops based at $\overrightarrow{01}$ on $V = P_0$ taken as in §5.1, and $\bar{x}, \bar{y}$ be their images in $\mathrm{gr}^1\pi_1(P_0(\mathbb{C}), \overrightarrow{01})$. The space $\mathrm{gr}^2\pi_1(P_0(\mathbb{C}), \overrightarrow{01})$ is a free $\mathbb{Z}$-module generated by $[\bar{x}, \bar{y}]$, and $\varphi_2 \colon \mathrm{gr}^2\pi_1(P_0(\mathbb{C}), \overrightarrow{01}) \to \mathbb{Z}$ is just taking the coefficient of $[\bar{x}, \bar{y}]$. To connect $\overrightarrow{01}$ to $f_i(v)$ ($i = 1, 2$), we take the path $\delta_1 \colon \overrightarrow{01} \rightsquigarrow f_1(v)$ to be trivial and the path $\delta_2 \colon \overrightarrow{01} \rightsquigarrow f_2(v)$ to be the real segment $[0, 1]$ (i.e., $\delta_2 = \langle 0, 1 \rangle$ in the notation of §5.4). In the sequel, we shall write $\delta := \delta_2$. It is easy to see that $\varphi_2(f_1) \circ \mathrm{gr}_\Gamma^2(f_{1*})([\bar{x}, \bar{y}]) = 1$ and $\varphi_2(f_2) \circ \mathrm{gr}_\Gamma^2(f_{2*})([\bar{x}, \bar{y}]) = -1$, hence that the condition (i)$_\mathbb{C}$ of Theorem 5.7 holds in the above setting, i.e.,

$$(6.1) \qquad \varphi_2(f_1) \circ \mathrm{gr}_\Gamma^2(f_{1*}) + \varphi_2(f_2) \circ \mathrm{gr}_\Gamma^2(f_{2*}) = 0$$

in $\mathrm{Hom}(\mathrm{gr}_\Gamma^2\pi_1(P_0(\mathbb{C}), \overrightarrow{01}), \mathbb{Z})$. This just reflects the simple equation (ii)$_\mathbb{C}$:

$$(6.2) \qquad z \wedge (z - 1) + (1 - z) \wedge (-z) = 0$$

in $\wedge^2 O_h^\times(V^{an})$. The ℓ-adic analogs (i)$_\ell$, (ii)$_\ell$ also hold in the obvious way.

Now, we shall consider the functional equation (iii)$_\mathbb{C}$ in Theorem 5.7. For any path $\gamma \colon v \rightsquigarrow z$, it reads

$$(6.3) \qquad \mathcal{L}_\mathbb{C}^{\varphi_2}(z, \overrightarrow{01}; \gamma) + \mathcal{L}_\mathbb{C}^{\varphi_2}(1 - z, \overrightarrow{10}; f_2(\gamma)) = 0.$$

Let us apply Proposition 5.11(i) to each term of the above. Since $f_1(v) = $

$\overrightarrow{01}$ and δ_1 is trivial, for the first term, the sequence $\{b_i\} = \{0\}$. This implies $\mathcal{L}_{\mathbb{C}}^{\varphi_2}(z, \overrightarrow{01}; \gamma) = \mathrm{li}_2(z, \gamma)$. For the second term, to apply Proposition 5.11, we must calculate $P_2(\{a_i\}_{i=0}^2, \{b_i\}_{i=0}^2)$ for

$$(6.4) \qquad \{a_i\}_{i=0}^2 = \{\mathrm{li}_0(1 - z, \delta f_2(\gamma)), \mathrm{li}_1(1 - z, \delta f_2(\gamma)), \mathrm{li}_2(1 - z, \delta f_2(\gamma))\},$$

$$(6.5) \qquad \{b_i\}_{i=0}^2 = \{0, 0, -\mathrm{li}_2(\overrightarrow{10}, \delta)\}$$

to get $\mathcal{L}_{\mathbb{C}}^{\varphi_2}(1 - z, \overrightarrow{10}; f_2(\gamma)) = \mathrm{li}_2(1 - z, \delta f_2(\gamma)) - \mathrm{li}_2(1, \delta)$. Thus we obtain a functional equation of complex dilogarithms:

$$(6.6) \qquad \mathrm{li}_2(z, \gamma) + \Big(\mathrm{li}_2(1 - z, \delta f_2(\gamma)) - \mathrm{li}_2(1, \delta)\Big) = 0.$$

We may further rewrite it in terms of classical dilogarithms using Proposition 5.2. Noting the Bernoulli numbers $B_0 = 1$, $B_1 = -1/2$, we find that

$$(6.7) \qquad \mathrm{li}_2(z, \gamma) = \frac{1}{4\pi^2}\left(Li_2(z) + \frac{1}{2}\log(1 - z)\log z\right),$$

$$(6.8) \qquad \mathrm{li}_2(1 - z, \delta f_2(\gamma)) = \frac{1}{4\pi^2}\left(Li_2(1 - z) + \frac{1}{2}\log(1 - z)\log z\right),$$

$$(6.9) \qquad \mathrm{li}_2(\overrightarrow{10}, \delta) = \frac{1}{4\pi^2}Li_2(1).$$

Summing up, we obtain the well-known equation (cf. [Le]):

$$(6.10) \qquad Li_2(z) + Li_2(1 - z) + \log z \log(1 - z) = Li_2(1).$$

Note that $Li_2(1) = \zeta(2) = \frac{\pi^2}{6}$.

Next, we shall consider the ℓ-adic analog in this case. Theorem 5.7 (iii)$_\ell$ reads:

$$(6.11) \qquad \mathcal{L}_{\mathrm{nv}}^{\varphi_2(f_1)_{\tilde{x}}}(z, \overrightarrow{01}; \gamma)(\sigma) + \mathcal{L}_{\mathrm{nv}}^{\varphi_2(f_2)_{\tilde{x}}}(1 - z, \overrightarrow{10}; f_2(\gamma))(\sigma) = E(\sigma, \gamma)$$

for $\sigma \in G_K$. Let us first examine the left-hand side above. From (5.3) it follows immediately that the first term is equal to

$$(6.12) \qquad \ell i_2(z, p, \vec{x}) = -\left\{\tilde{\chi}_2^z(\sigma) + \frac{1}{2}\rho_z(\sigma)\rho_{1-z}(\sigma)\right\}.$$

The second term can be calculated after Proposition 5.11 as $P_2(\{a_i\}_{i=0}^2, \{b_i\}_{i=0}^2)$ with

$$a_i = -\ell i_i(\overrightarrow{10}, \delta, \vec{x}), \quad b_i = \ell i_i(1 - z, \delta f_2(\gamma), \vec{x}) \qquad (i = 0, 1, 2).$$

Writing $X = \log x$, $Y = \log y$, we know (cf. (5.1), (5.4))

$$\log(\mathfrak{f}_\sigma^\delta)^{-1} \equiv \sum_{m\geq 1} \ell i_{m+1}^{\overrightarrow{10}}(\sigma)(\operatorname{ad} X)^m(Y)$$

$$\equiv \sum_{m\geq 1}(-1)^m \frac{\tilde{\chi}_{m+1}^{\overrightarrow{10}}(\sigma)}{m!}(\operatorname{ad} X)^m(Y) \quad \operatorname{mod} I_Y.$$

It follows then from Proposition 5.13 that $(a_0, a_1, a_2) = (0, 0, \frac{1}{24}(\chi(\sigma)^2 - 1))$ and hence, the second term on the left-hand side of (6.11) turns out to be

$$\ell i_2(1 - z, \delta f_2(\gamma), \vec{x})(\sigma) - \ell i_2(\overrightarrow{10}, \delta, \vec{x})(\sigma)$$

$$= -\left(\tilde{\chi}_2^{1-z}(\sigma) + \frac{1}{2}\rho_z(\sigma)\rho_{1-z}(\sigma)\right) + \frac{1}{24}(\chi(\sigma)^2 - 1).$$

To estimate $E(\sigma, \gamma)$, the right-hand side of (6.11), we shall make use of Corollary 5.8. Let us take a decomposition of the ℓ-adic Lie algebra

$$(6.13) \qquad L(\pi_{V,v}) = L_{<2} \oplus \Gamma^2 L(\pi_{V,v}), \quad L_{<2} := \mathbb{Q}_\ell \log x + \mathbb{Q}_\ell \log y$$

according to the Lie series expansion with respect to $(\log(x), \log(y))$ in the sense of (4.10). Now, for the generator system $\vec{x} = \vec{y}_1 = (x, y)$, it is easy to see $\varphi_{2,\vec{x}}([\log(\mathfrak{f}_\sigma^\gamma)^{-1}]_{<2}) = 0$, as by definition $\varphi_2(f_1)_{\vec{y}_1}$ quarries out the degree 2 part. For $\vec{y}_2 = \delta^{-1}(x, y)\delta = (f_2(y), f_2(x))$, since $f_2(\log \mathfrak{f}_\sigma^\gamma)$ is just obtained from $\log \mathfrak{f}_\sigma^\gamma$ after replacing $\log(x), \log(y)$ by $\log(f_2(x)), \log(f_2(y))$ respectively, we see also $\varphi_{2,\vec{x}}(\delta[\log(\mathfrak{f}_\sigma^\gamma)^{-1}]_{<2}\delta^{-1}) = 0$. Therefore, in this special case, *the error term vanishes for all $\sigma \in G_K$.* Summing up our above arguments, we obtain the functional equation

$$(6.14) \qquad \tilde{\chi}_2^z(\sigma) + \tilde{\chi}_2^{1-z}(\sigma) + \rho_z(\sigma)\rho_{1-z}(\sigma) = \frac{1}{24}(\chi(\sigma)^2 - 1) \qquad (\sigma \in G_K).$$

Question The above ℓ-adic functional equation (6.14) suggests a possibility of reducing Galois transformations of

$$\prod_{a=0}^{\ell^n - 1}(1 - \zeta_{\ell^n}^a z^{1/\ell^n})^{\frac{a}{\ell^n}}(1 - \zeta_{\ell^n}^a(1 - z)^{1/\ell^n})^{\frac{a}{\ell^n}}$$

to simpler invariants ρ_z, ρ_{1-z} and χ, in a somewhat purely arithmetic way as in the second proof of Proposition 5.13. It seems to the authors a non-trivial question.

6.2 $Li_2(z) + Li_2(z/(z-1))$ We apply a similar argument to the above subsection to $f_1(z) = z$ and $f_2(z) = \frac{z}{z-1}$. In this case, $f_2(v) = \overrightarrow{0\infty}$, so we substitute the half anticlockwise rotation from $\overrightarrow{01}$ to $\overrightarrow{0\infty}$ for $\delta_2 : \overrightarrow{01} \rightsquigarrow f_2(v)$, and set $\delta := \delta_2$ which is $[0_1^\infty]$ in the notation of §5.4. Then, $(f_2(x), f_2(y)) = \delta^{-1}(x, y^{-1}x^{-1})\delta$. For (i)$_\mathbb{C}$, observing that $gr_\gamma^2(\iota_\delta \circ f_2)$ sends $[\bar{x}, \bar{y}]$ to $[\bar{x}, -\bar{x} - \bar{y}]$, we find (6.1) also holds in this case. The condition (ii)$_\mathbb{C}$ can be checked by

$$(6.15) \qquad (z) \wedge (z-1) + \left(\frac{z}{z-1}\right) \wedge \left(\frac{z}{z-1} - 1\right) = 0.$$

The consequent functional equation (iii)$_\mathbb{C}$ reads

$$(6.16) \qquad \mathcal{L}_\mathbb{C}^{\varphi_2}(z, \overrightarrow{01}; \gamma) + \mathcal{L}_\mathbb{C}^{\varphi_2}\left(\frac{z}{z-1}, \overrightarrow{10}; f_2(\gamma)\right) = 0.$$

for any path $\gamma : v \rightsquigarrow z$. Let us apply Proposition 5.11(i). In the same way as in the previous example, $\mathcal{L}_\mathbb{C}^{\varphi_2}(z, \overrightarrow{01}; \gamma) = li_2(z, \gamma)$. For the second term, we calculate $P_2(\{a_i\}_{i=0}^2, \{b_i\}_{i=0}^2) = a_2 + b_2 + \frac{1}{2}(a_0 b_1 - b_0 a_1)$, where

$$(6.17) \quad \{a_i\}_{i=0}^2 = \left\{ li_0\left(\frac{z}{z-1}, \delta f_2(\gamma)\right), li_1\left(\frac{z}{z-1}, \delta f_2(\gamma)\right), li_2\left(\frac{z}{z-1}, \delta f_2(\gamma)\right) \right\},$$

$$(6.18) \quad \{b_i\}_{i=0}^2 = \{-li_0(\overrightarrow{0\infty}, \delta), -li_1(\overrightarrow{0\infty}, \delta), -li_2(\overrightarrow{0\infty}, \delta)\} = \left\{\frac{\pi i}{2\pi i}, 0, 0\right\},$$

to get

$$\mathcal{L}_\mathbb{C}^{\varphi_2}\left(\frac{z}{z-1}, \overrightarrow{10}; f_2(\gamma)\right)$$

$$= li_2\left(\frac{z}{z-1}, \delta f_2(\gamma)\right) - \frac{1}{2}(-li_0(\overrightarrow{0\infty}, \delta)) \cdot li_1\left(\frac{z}{z-1}, \delta f_2(\gamma)\right)$$

$$= li_2\left(\frac{z}{z-1}, \delta f_2(\gamma)\right) - \frac{1}{4} li_1\left(\frac{z}{z-1}, \delta f_2(\gamma)\right)$$

$$= \frac{1}{4\pi^2}\left(Li_2\left(\frac{z}{z-1}\right) + \frac{1}{2}\log\left(\frac{z}{z-1}\right)\log\left(\frac{1}{1-z}\right)\right) + \frac{1}{4}\left(\frac{1}{2\pi i}\log\left(\frac{1}{1-z}\right)\right).$$

Putting these into (6.16) combined with our choice of logarithmic branches $\log(\frac{z}{z-1}) = \log z - \log(1-z) + \pi i$, we obtain a classical functional equation from [Le]:

$$(\mathbf{6.19}) \qquad Li_2(z) + Li_2\left(\frac{z}{z-1}\right) = -\frac{1}{2}\log^2(1-z).$$

We next consider the ℓ-adic analog. The condition (iii)$_\ell$ of Theorem 5.7 reads in this case:

$$(6.20) \qquad \mathcal{L}_{nv}^{\varphi_2(f_1)_*}(z, \overrightarrow{01}; \gamma)(\sigma) + \mathcal{L}_{nv}^{\varphi_2(f_2)_*}\left(\frac{z}{z-1}, \overrightarrow{10}; f_2(\gamma)\right)(\sigma) = E(\sigma, \gamma).$$

The first term on the left-hand side is the same as (6.12). For the second term, noting that $\mathfrak{f}_\sigma^\delta = \delta \cdot \sigma(\delta)^{-1} = x^{\frac{1-\chi(\sigma)}{2}}$, we have

$$(6.21) \qquad \ell i_0(\overrightarrow{0\infty}, \delta, \vec{x}) = \begin{cases} \frac{\chi(\sigma)-1}{2}, & k = 0, \\ 0, & k \geq 1. \end{cases}$$

Hence, by Proposition 5.11(ii), it follows that

$$\mathcal{L}_{\mathrm{inv}}^{\varphi_2(f_2)_{\vec{x}}}\left(\frac{z}{z-1}, \overrightarrow{10}; f_2(\gamma)\right)(\sigma) = P_2\left(\left\{\frac{1-\chi(\sigma)}{2}, 0, 0\right\}, \left\{\ell i_i\left(\frac{z}{z-1}, \delta f_2(\gamma), \vec{x}\right)\right\}_{i=0}^2\right)$$

$$= \ell i_2\left(\frac{z}{z-1}, \delta f_2(\gamma), \vec{x}\right) + \frac{1}{2}\left(\frac{1-\chi(\sigma)}{2}\right)\ell i_1\left(\frac{z}{z-1}\right)$$

$$= -\left(\tilde{\chi}_2^{\frac{z}{z-1}}(\sigma) + \frac{1}{2}\rho_{\frac{z}{z-1}}(\sigma)\rho_{\frac{1}{1-z}}(\sigma)\right) + \frac{1}{2}\left(\frac{1-\chi(\sigma)}{2}\right)\rho_{\frac{1}{1-z}}(\sigma).$$

To evaluate the error term on the right-hand side of (6.20), we employ the same splitting of $L(\pi_{X,v})$ as (6.13). We calculate:

$$E(\sigma, \gamma) = \varphi_{2,\vec{x}}(\delta f_2([\log(\mathfrak{f}_\sigma^\gamma)^{-1}]_{<2}) \delta^{-1})$$

$$= \varphi_{2,\vec{x}}(\delta f_2(\rho_z(\sigma) \log x + \rho_{1-z}(\sigma) \log y) \delta^{-1})$$

$$= \varphi_{2,\vec{x}}\left(\rho_z(\sigma) \log x + \rho_{1-z}(\sigma) \log(y^{-1}x^{-1})\right)$$

$$= \varphi_{2,\vec{x}}\left(\frac{1}{2}\rho_{1-z}(\sigma)[\log x, \log y]\right) = \frac{1}{2}\rho_{1-z}(\sigma).$$

Taking care of the choice of paths to fix branches of the involved Kummer characters such as $\rho_{\frac{z}{z-1}} = \rho_z - \rho_{1-z} + \frac{\chi-1}{2}$, $\rho_{\frac{1}{1-z}} = -\rho_{1-z}$, we obtain from (6.20) the following functional equation:

$$\tilde{\chi}_2^z(\sigma) + \tilde{\chi}_2^{\frac{z}{z-1}}(\sigma) + \frac{1}{2}\rho_{1-z}(\sigma)^2 - \frac{\chi(\sigma)-1}{2}\rho_{1-z}(\sigma) = \frac{1}{2}\rho_{1-z}(\sigma),$$

or equivalently,

$$(6.22) \qquad \tilde{\chi}_2^z(\sigma) + \tilde{\chi}_2^{\frac{z}{z-1}}(\sigma) = \frac{\rho_{1-z}(\sigma)}{2}\left(\chi(\sigma) - \rho_{1-z}(\sigma)\right) \qquad (\sigma \in G_K).$$

Note the ℓ-integrality, i.e., $\in \mathbb{Z}_\ell$ of the right-hand side above for all $\sigma \in G_K$ even when $\ell = 2$, as should be expected from the definition of ℓ-adic polylogarithmic character appearing on the left-hand side. This shows that the error term $E(\sigma, \gamma)$ is unavoidable. Concerning the functional equation (6.22), one may also ask a question similar to what was raised just after (6.14).

6.3 Inversion formula Here, we consider two automorphisms $f_1(z) = z$ and $f_2(z) = z^{-1}$ of P_0. Take $\delta_1 : v = \overrightarrow{01} \rightsquigarrow f_1(v)$ to be the trivial path, and $\delta_2 : v \rightsquigarrow f_1(v) = \overrightarrow{\infty 1}$ to be $\langle 0, 1)[1_0^\infty]\langle \infty, 1 \rangle$ in the notation of §5.4. Then, $\delta := \delta_2$ is the same as the path δ illustrated in loc. cit. Let $n \geq 2$. In the tensor space $(\mathrm{Sym}^{n-2}O_h^{\times}) \otimes (\wedge^2 O_h^{\times})$ of $O_h^{\times} = O_h^{\times}(V^{an})$, since

$$\left(\frac{1}{z}\right)^{\odot n-2} \otimes \left(\frac{1}{z} \wedge \frac{1-z}{z}\right) = (-1)^{n-1} z^{\odot n-2} \otimes (z \wedge (1-z)),$$

we find Theorem 5.7 (ii)$_{\mathrm{C}}$ holds for $c_1 = 1$, $c_2 = (-1)^n$. Consequently, for any path $\gamma \colon \overrightarrow{01} \rightsquigarrow z$ on P_0, we have the functional equation (iii)$_{\mathrm{C}}$ in the form

$$(6.23) \qquad \mathcal{L}_{\mathrm{C}}^{\varphi_n}(z, \overrightarrow{01}; \gamma) + (-1)^n \mathcal{L}_{\mathrm{C}}^{\varphi_n}\left(\frac{1}{z}, \overrightarrow{\infty 1}; f_2(\gamma)\right) = 0 \qquad (n \geq 2).$$

The first term, $\mathcal{L}_{\mathrm{C}}^{\varphi_n}(z, \overrightarrow{01}; \gamma) = \mathrm{li}_n(z, \gamma)$, is already calculated in Proposition 5.2. For the second, applying Proposition 5.11(i) with the chain rule, we wish to compute the BCH-sum

$$(6.24) \qquad \mathcal{L}_{\mathrm{C}}^{\varphi_n}\left(\frac{1}{z}, \overrightarrow{\infty 1}; f_2(\gamma)\right) = \mathrm{P}_n(\{a_i\}_{i=0}^n, \{b_i\}_{i=0}^n),$$

where

$$\{a_i\}_{i=0}^n = \left\{\mathrm{li}_i\left(\frac{1}{z}, \delta f_2(\gamma)\right)\right\}_{i=0}^n$$

$$= \left\{(-1)^{i+1}\sum_{k=0}^{i-1}\frac{B_k}{k!}\left(\frac{-\log z}{2\pi i}\right)^k \frac{Li_{i-k}(z^{-1}, \delta f_2(\gamma))}{(2\pi i)^{i-k}}\right\}_{i=0}^n,$$

$$\{b_i\}_{i=0}^n = \left\{-\mathrm{li}_i(\overrightarrow{\infty 1}, \delta)\right\}_{i=0}^n = \left\{0, -B_1, -\frac{B_2}{2!}, \ldots, -\frac{B_n}{n!}\right\} \qquad \text{by (5.11)}.$$

It now turns out that we should work inductively on n. Let us set

$$\mathrm{L}_0 := \frac{-\log z}{2\pi i}, \quad \mathrm{L}_1 := \frac{Li_1(z)}{2\pi i} = \frac{-\log(1-z)}{2\pi i}, \qquad \mathrm{L}_k := \frac{Li_k(z)}{(2\pi i)^k} \ (k \geq 2);$$

$$\bar{\mathrm{L}}_0 := \frac{\log z}{2\pi i}, \quad \bar{\mathrm{L}}_1 := \frac{Li_1(z^{-1})}{2\pi i} = \frac{\log z - \log(z-1)}{2\pi i}, \qquad \bar{\mathrm{L}}_k := \frac{Li_k(z^{-1})}{(2\pi i)^k} \ (k \geq 2);$$

so that

$$a_0 = \bar{\mathrm{L}}_0 = -\mathrm{L}_0, \quad a_1 = \bar{\mathrm{L}}_1 = -\mathrm{L}_0 + \mathrm{L}_1 - \frac{1}{2}, \quad b_0 = 0;$$

$$a_k = (-1)^{k+1}\sum_{i=0}^{k-1}\frac{B_i}{i!}\mathrm{L}_0^i\bar{\mathrm{L}}_{k-i} \ (k \geq 2), \quad b_k = -\frac{B_k}{k!} \ (k \geq 1).$$

Consider then the generating functions:

$$(6.25) \quad \mathrm{L}(t) := \sum_{i=0}^{\infty}\mathrm{L}_{i+1}t^i, \quad \bar{\mathrm{L}}(t) := \sum_{i=0}^{\infty}\bar{\mathrm{L}}_{i+1}t^i, \quad \mathcal{B}(t) := \frac{t}{e^t - 1} = \sum_{i=0}^{\infty}\frac{B_i}{i!}t^i,$$

and define the quantities D_i, P_i ($i \geq 1$) by

$$D(t) := \sum_{i=1}^{\infty} D_i t^i = t\mathcal{B}(L_0 t)L(-t),$$

$$P(t) := \sum_{i=0}^{\infty} P_i t^i = t\mathcal{B}(-L_0 t)\bar{L}(-t) + L_0 t\mathcal{B}(t) - \mathcal{B}(t)\mathcal{B}(-L_0 t) + \mathcal{B}(-L_0 t).$$

Then, from (5.8) and (5.10) follows that this P_n coincides with $P_n(\{a_i\}_{i=0}^{n}, \{b_i\}_{i=0}^{n})$ for $n \geq 2$, and it turns out that the functional equation (6.23) is reduced to the equation $P(-t) + D(t) = 0$. After computations, we obtain

$$\bar{L}(t) - L(-t) = \frac{e^{L_0 t}}{e^{-t} - 1} + t^{-1} = \sum_{n=1}^{\infty} \frac{B_n(-L_0)}{n!}(-t)^{n-1}.$$

Comparing the coefficients, we get $\bar{L}_n + (-1)^n L_n = (-1)^{n-1}B_n(-L_0)/n!$, i.e., what is called the inversion formula of polylogarithms:

(6.26)
$$Li_n(z) + (-1)^n Li_n(\frac{1}{z}) = -\frac{(2\pi i)^n}{n!} B_n(\frac{\log z}{2\pi i}) \quad (n \geq 2).$$

Next, we consider the ℓ-adic version (iii)$_\ell$:

(6.27)
$$\mathcal{L}_{\text{inv}}^{\varphi_n(f_1)_{\check{x}}}(z, \overrightarrow{01}; \gamma)(\sigma) + (-1)^n \mathcal{L}_{\text{inv}}^{\varphi_n(f_2)_{\check{x}}}(\frac{1}{z}, \overrightarrow{\infty 1}; f_2(\gamma))(\sigma) = E(\sigma, \gamma)$$

for $\sigma \in G_K$, $n \geq 2$ ($z \in K \subset \mathbb{C}$). In the below, we shall occasionally omit σ for simplicity. The first term is $\ell i_m(z, \gamma, \vec{x})$ that is expressed by ℓ-adic polylogarithmic characters as in (5.3). For the second, applying Proposition 5.11(ii), one can write

(6.28)
$$\mathcal{L}_{\text{inv}}^{\varphi_n(f_2)_{\check{x}}}\left(\frac{1}{z}, \overrightarrow{\infty 1}; f_2(\gamma)\right) = P_n(\{a_i\}_{i=0}^{n}, \{b_i\}_{i=0}^{n}),$$

with

$$\{a_i\}_{i=0}^{n} = \{-\ell i_i(\overrightarrow{\infty 1}, \delta)\}_{i=0}^{n} = \left\{0, B_1(\chi - 1), \frac{B_2}{2!}(\chi^2 - 1), \dots, \frac{B_n}{n!}(\chi^n - 1)\right\},$$

$$\{b_i\}_{i=0}^{n} = \{\ell i_i(\frac{1}{z}, \delta f_2(\gamma))\}_{i=0}^{n} = \left\{(-1)^{i+1} \sum_{k=0}^{i-1} \frac{B_k}{k!} (\rho_z)^k \frac{\tilde{\chi}_{i-k}^{1/z}}{(i-k-1)!}\right\}_{i=0}^{n}.$$

(In the expression above for a_i, we have used the chain rule to find that $f_\sigma^\delta = f_\sigma(y, x^{-1}y^{-1})y^{\frac{1-\chi(\sigma)}{2}}f_\sigma(x, y)$. This differs from G_σ^{-1} of the first proof of

Proposition 5.13 only in the sign of $\frac{1-\chi(\sigma)}{2}$.) Let us set

$$L_0 := \rho_z, \qquad L_1 := \tilde{\chi}_1^z = \rho_{1-z}, \qquad L_k := \frac{\tilde{\chi}_k^z}{(k-1)!} \ (k \geq 2);$$

$$\bar{L}_0 := -\rho_z, \quad \bar{L}_1 := \tilde{\chi}_1^{1/z} = -\rho_z + \rho_{1-z} + \frac{\chi - 1}{2}, \quad \bar{L}_k := \frac{\tilde{\chi}_k^{1/z}}{(k-1)!} \ (k \geq 2);$$

so that

$$a_k = \frac{B_k}{k!}(\chi^k - 1) \quad (k \geq 0);$$

$$b_0 = \bar{L}_0 = -L_0, \quad b_1 = \bar{L}_1 = -L_0 + L_1 + \frac{\chi - 1}{2};$$

$$b_k = (-1)^{k+1} \sum_{i=0}^{k-1} \frac{B_i}{i!} L_0^i \bar{L}_{k-i}.$$

Keeping $L(t)$, $\bar{L}(t)$ as in (6.25), introduce the quantities D_i, P_i ($i \geq 1$) by

$$D(t) := \sum_{i=1}^{\infty} D_i t^i = t\mathcal{B}(L_0 t) L(-t),$$

$$P(t) := \sum_{i=0}^{\infty} P_i t^i = t\mathcal{B}(-L_0 t)\bar{L}(-t) + (\mathcal{B}(\chi t) - \mathcal{B}(t))\mathcal{B}(-L_0 t).$$

Then, $D_n = \ell i_n(z, \gamma, \tilde{x})$ ($n \geq 1$), and $P_n = P_n(\{a_i\}_{i=0}^n, \{b_i\}_{i=0}^n)$ ($n \geq 1$) as seen from (5.8) and (5.10). Thus, the functional equation (6.27) turns out to be in the form

(6.29) $$D_n + (-1)^n P_n = (-1)^n E_n \quad (n \geq 2),$$

where the error term on the right-hand side is evaluated by Corollary 5.8 as follows:

$$E_n := \varphi_{n,\tilde{x}}(\delta f_2([\log(\mathfrak{f}_\sigma^\gamma)^{-1}]_{<n}) \delta^{-1}).$$

Observing that $\delta f_2(x)\delta^{-1} = yzy^{-1} = x^{-1}y^{-1}$, $\delta f_2(x)\delta^{-1} = y$, we see from (5.1) that

(6.30) $$E_n = \varphi_{n,\tilde{x}}(\rho_z \log(x^{-1}y^{-1})) = -\frac{B_{n-1}}{(n-1)!}\rho_z \quad (n \geq 2).$$

If we extend the above expression of E_n also for $n = 1$, then we still have $D_1 - P_1 + E_1 = 0$. Summing up our discussions, we obtain from (6.29) the functional equation of generating functions

$$D(t) + P(-t) = \sum_{n=1}^{\infty} (-1)^n E_n t^n = \rho_z t \mathcal{B}(-t),$$

which yields

$$\mathsf{L}(-t) - \bar{\mathsf{L}}(t) = -\left(\frac{-\chi\, t}{e^{-\chi t} - 1}\right) - \left(\frac{e^{\mathsf{L}_0 t}}{e^{-t} - 1}\right).$$

Comparing the coefficients of the above, we get

$$(-1)^{n-1}\mathsf{L}_n - \bar{\mathsf{L}}_n = (-1)^n \left\{\frac{B_n(-\mathsf{L}_0)}{n!} - \frac{\chi^n B_n}{n!}\right\}$$

from which we finally conclude the ℓ-adic inversion formula:

(6.31) $\tilde{\chi}_n^z(\sigma) + (-1)^n \tilde{\chi}_n^{1/z}(\sigma) = -\frac{1}{n}\{B_n(-\rho_z(\sigma)) - B_n\chi(\sigma)^n\}$

$$(\sigma \in G_K, n \geq 1).$$

Remark   In [W9], theorem 2.6, an inversion formula without lower degree terms was obtained. Note however that, in loc. cit., one must have taken certain "suitable" ℓ-adic paths to determine those $\ell i_m(z)$ and $\ell i_m(\frac{1}{z})$. By comparison, in the functional equations (6.27) and (6.31), the ℓ-adic polylogarithms or ℓ-adic polylogarithmic characters on the left-hand sides are taken along the explicit paths composed of γ, δ and $f_2(\gamma)$, for which non-trivial lower degree terms must appear as on the right-hand sides.

Remark   Applying $z = \overrightarrow{10}$ to the above inversion formula (6.31) reproves Proposition 5.13, where our above argument specialized to this case is essentially of the same (geometric) nature as the first proof presented in §5.4. Note also that, putting $n = 2$ and $z = -1$ in (6.31) confirms the formula given in Remark 5.14, where we also apply the formula $\tilde{\chi}_2^{z=-1}(\sigma; x^{-1}p) = \tilde{\chi}_2^{z=-1}(\sigma; p) + \chi(\sigma) \cdot \rho_2(\sigma)$ ($\sigma \in G_{\mathbb{Q}}$) for $p = [0_1^\infty]\langle 0, -1\rangle$ (cf. [NW2]).

6.4 Abel's equation   In this subsection, we take for V the moduli space $M_{0,5}$ of the isomorphism classes of the projective line with ordered 5 marked points $(\mathbf{P}^1; a_1, \ldots, a_5)$. We consider $V = M_{0,5}$ to be a variety defined over a subfield $K \subset \mathbb{C}$ equipped with a standard tangential base point $\vec{v}$ determined by the $K(t)$-rational point $(\mathbf{P}^1; 0, t^2, t, 1, \infty)$. The topological fundamental group $\pi_1(V(\mathbb{C}), \vec{v})$ is known to be a quotient of pure sphere braid group with 5 strings. We fix a standard generator system $\{x_{ij} \mid i, j = 1, \ldots, 5\}$ of it as in [Ih1] or [N1] (3.1.4). Regard now $P_0 = \mathbf{P}^1 - \{0, 1, \infty\}$ as the moduli space of the $(\mathbf{P}^1; b_1, \ldots, b_4)$, i.e., of the isomorphism classes of the projective line with ordered 4 marked points, and consider, for each $i = 1, \ldots, 5$, the morphism $f_i \colon V \to P_0$ obtained by forgetting the marked point a_i and leaving the other a_j ($j \neq i$) as $b_1, \ldots, b_4$ so that the order is preserved. It is easy to check that $f_i(\vec{v}) = \overrightarrow{01}$, hence we can take all connecting paths $\delta_i\colon \overrightarrow{01} \rightsquigarrow f_i(v)$ to be trivial. We refer the reader to [N1] and [N2] for basic properties of these forgetful morphisms with respect to the generator systems of the fundamental groups. For example, one can compute $f_2(x_{24}) = 1$, $f_2(x_{14}) = (xy)^{-1}$, $f_3(x_{15}) = y$ and so on; $f_k(x_{ij})$ is equal to one of the $x, y, (xy)^{-1}$ depending on the choice of (i, j, k) with $1 \leq i, j, k \leq 5$. The graded quotient $\mathrm{gr}^2(\pi_1(V(\mathbb{C})), \vec{v})$ is a 4-dimensional vector space with a basis $[\bar{x}_{12}, \bar{x}_{23}]$, $[\bar{x}_{15}, \bar{x}_{25}]$, $[\bar{x}_{15}, \bar{x}_{35}]$, $[\bar{x}_{25}, \bar{x}_{35}]$. We summarize their images by $\mathrm{gr}^2_\Gamma(f_{i*})$ $(i = 1, \ldots, 5)$ in the following table:

♯	$\mathrm{gr}^2_\Gamma f_{1*}(\sharp)$	$\mathrm{gr}^2_\Gamma f_{2*}(\sharp)$	$\mathrm{gr}^2_\Gamma f_{3*}(\sharp)$	$\mathrm{gr}^2_\Gamma f_{4*}(\sharp)$	$\mathrm{gr}^2_\Gamma f_{5*}(\sharp)$
$[\bar{x}_{12}, \bar{x}_{23}]$	$[\bar{x}, \bar{y}]$	$-[\bar{x}, \bar{y}]$	0	0	0
$[\bar{x}_{15}, \bar{x}_{25}]$	0	$-[\bar{x}, \bar{y}]$	$[\bar{x}, \bar{y}]$	0	0
$[\bar{x}_{15}, \bar{x}_{35}]$	0	$[\bar{x}, \bar{y}]$	0	$-[\bar{x}, \bar{y}]$	0
$[\bar{x}_{25}, \bar{x}_{35}]$	0	$-[\bar{x}, \bar{y}]$	0	0	$[\bar{x}, \bar{y}]$

Consequently, we see that $\sum_{i=1}^{5} (-1)^{i-1} \mathrm{gr}^2_\Gamma(f_{i*}) = 0$ as a homomorphism of $\mathrm{gr}^2_\Gamma \pi_1(V(\mathbb{C}), v)$ to $\mathrm{gr}^2_\Gamma \pi_1(P_0(\mathbb{C}), \overrightarrow{01})) = \mathbb{C} \cdot [\bar{x}, \bar{y}]$.

Now, let us apply Theorem 5.7, and compute the functional equations $(iii)_{\mathbb{C}}$ and $(iii)_\ell$. Pick a point $z \in V(K)$ representing $(\mathbf{P}^1; 0, st, s, 1, \infty)$ with $s = \frac{\xi}{1-\eta}$, $t = \frac{\eta}{1-\xi}$. $(\xi, \eta \in K - \{0, 1\})$. Then, the images of z by the above morphisms $f_1, \ldots, f_5$ are calculated as

$$f_1(z) = \left[(\mathbf{P}^1; 0, \frac{s(1-t)}{1-st}, 1, \infty) \right] = \xi;$$

$$f_2(z) = \left[(\mathbf{P}^1; 0, s, 1, \infty) \right] = \frac{\xi}{1-\eta};$$

$$f_3(z) = \left[(\mathbf{P}^1; 0, st, 1, \infty) \right] = \frac{\xi\eta}{(1-\xi)(1-\eta)};$$

$$f_4(z) = \left[(\mathbf{P}^1; 0, t, 1, \infty) \right] = \frac{\eta}{1-\xi};$$

$$f_5(z) = \left[(\mathbf{P}^1; 0, \frac{t(1-s)}{1-st}, 1, \infty) \right] = \eta.$$

Therefore, $(iii)_{\mathbb{C}}$ leads to

$$\mathrm{li}_2(\xi, f_1(\gamma)) - \mathrm{li}_2\left(\frac{\xi}{1-\eta}, f_2(\gamma) \right) + \mathrm{li}_2\left(\frac{\xi\eta}{(1-\xi)(1-\eta)}, f_3(\gamma) \right)$$

$$- \mathrm{li}_2\left(\frac{\eta}{1-\xi}, f_4(\gamma) \right) + \mathrm{li}_2(\eta, f_5(\gamma)) = 0.$$

Applying (6.7) to each term above, we obtain what is called Abel's equation:

$$(6.32) \qquad Li_2(\frac{\xi\eta}{(1-\xi)(1-\eta)}) = Li_2(\frac{\xi}{1-\eta}) + Li_2(\frac{\eta}{1-\xi})$$
$$- Li_2(\xi) - Li_2(\eta) - \log(1-\xi)\log(1-\eta).$$

Next, we consider the ℓ-adic version. We shall state it as a theorem.

Theorem 6.1 (Abel's equation for ℓ-adic polylogarithms) *With notation as above, we have*

$$\ell i_2(\xi, f_1(\gamma), \vec{x})(\sigma) - \ell i_2\left(\frac{\xi}{1-\eta}, f_2(\gamma), \vec{x}\right)(\sigma) + \ell i_2\left(\frac{\xi\eta}{(1-\xi)(1-\eta)}, f_3(\gamma), \vec{x}\right)(\sigma)$$
$$- \ell i_2\left(\frac{\eta}{1-\xi}, f_4(\gamma), \vec{x}\right)(\sigma) + \ell i_2(\eta, f_5(\gamma), \vec{x})(\sigma) = 0 \qquad (\sigma \in G_K).$$

Remark 6.2 This functional equation seems to be nicer than the one proved in [W5], theorem 11.1.14, for $\sigma \in G_{K(\mu_{\ell^\infty})}$, because in the present approach we have no lower degree terms even for $\sigma \in G_K$.

Proof Condition (iii)$_\ell$ reads:

$$(6.33)$$
$$\ell i_2(\xi, f_1(\gamma), \vec{x})(\sigma) - \ell i_2\left(\frac{\xi}{1-\eta}, f_2(\gamma), \vec{x}\right)(\sigma) + \ell i_2\left(\frac{\xi\eta}{(1-\xi)(1-\eta)}, f_3(\gamma), \vec{x}\right)(\sigma)$$
$$- \ell i_2\left(\frac{\eta}{1-\xi}, f_4(\gamma), \vec{x}\right)(\sigma) + \ell i_2(\eta, f_5(\gamma), \vec{x})(\sigma) = E(\sigma, \gamma)$$

for all $\sigma \in G_K$. To estimate the error term $E(\sigma, \gamma)$ of (6.33), set $S = \{(1, 2), (2, 3), (1, 5), (2, 5), (3, 5)\}$ so that the $\bar{x}_{ij}$ $((i, j) \in S)$ form a basis of $\pi_1(V(\mathbb{C}))^{ab}$ and fix a splitting of the ℓ-adic Lie algebra $L(\pi_{V,v}) = L_{<2} \oplus \Gamma^2 L(\pi_{V,v})$ such that $L_{<2} = \sum_{(i,j)\in S} \mathbb{Q}_\ell X_{ij}$ where $X_{ij} = \log x_{ij}$. Write

$$[\log(f_\sigma^z)^{-1}]_{<2} = \sum_{(i,j)\in S} C_{ij}(\sigma)X_{ij}.$$

Then, by Corollary 5.8, we have

$$(6.34) \qquad E(\sigma, \gamma) = \sum_{k=1}^{5} (-1)^{i-1} \varphi_{2,\vec{x}}(f_k([\log(f_\sigma^z)^{-1}]_{<2})).$$

Noting that $f_k(X_{ij})$ $(k = 1, \ldots, 5, (i, j) \in S)$ are summarized as

$f_k(X_{ij})$	X_{12}	X_{23}	X_{15}	X_{25}	X_{35}
f_1	0	X	0	Y	$\log(y^{-1}x^{-1})$
f_2	0	0	Y	0	$\log(y^{-1}x^{-1})$
f_3	X	0	Y	$\log(y^{-1}x^{-1})$	0
f_4	X	Y	Y	$\log(y^{-1}x^{-1})$	X
f_5	X	Y	0	0	0

we find that the right-hand side of (6.34) applied to $\varphi_{2,\vec{x}} \circ f_k$ raises non-vanishing terms from $C_{ij}(\sigma)$ only when the degree 2 term of $f_k(X_{ij})$ survives, i.e., $f_k(X_{ij}) = \log(y^{-1}x^{-1}) = -X - Y - \frac{1}{2}[X, Y] + \cdots$, in which case $-\frac{1}{2}C_{ij}(\sigma)$ occurs. Summing up, we obtain

$$E(\sigma, \gamma) = -\frac{1}{2}C_{35}(\sigma) + \frac{1}{2}C_{35}(\sigma) - \frac{1}{2}C_{25}(\sigma) + \frac{1}{2}C_{25}(\sigma) = 0,$$

namely, the error term vanishes on the right-hand side of (6.33). □

Remark 6.3 In the above discussion, it is, in fact, not difficult to determine the individual coefficient characters $C_{ij} \colon G_K \to \mathbb{Q}_\ell$ as Kummer cocycles along roots of certain values (and paths from $\vec{01}$) depending on (ξ, η). This can be done only by observing Galois actions on the image of those paths in the abelianized fundamental groups after projections $f_k \colon V \to P_0$. We leave such enjoyable calculations to interested readers.

Finally, applying (6.12) allows us to interpret the left-hand side of the above theorem in terms of $\tilde{\chi}_2^z$ and $\rho_z \rho_{1-z}$. By simple calculations, we deduce the following Abel's equation for ℓ-adic polylogarithmic characters of degree 2:

$$(6.35) \quad -\tilde{\chi}_2^\xi(\sigma) + \tilde{\chi}_2^{\frac{\xi}{1-\eta}}(\sigma) - \tilde{\chi}_2^{\frac{\xi\eta}{(1-\xi)(1-\eta)}}(\sigma) + \tilde{\chi}_2^{\frac{\eta}{1-\xi}}(\sigma) - \tilde{\chi}_2^\eta(\sigma) = \rho_{1-\xi}(\sigma)\rho_{1-\eta}(\sigma),$$

which holds for all $\sigma \in G_K$.

References

[BD] A. Beilinson, P. Deligne, *Interprétation motivique de la conjecture de Zagier reliant polylogarithms et régulateurs*, Proc. Symp. in Pure Math. (AMS) **55-2** (1994), 97–121.

[Bl1] S. Bloch, *Applications of the dilogarithm function in algebraic K-theory and algebraic geometry*, Proc. Int. Symp. Alg. Geom., Kyoto, (1977), 1–14.

[Bl2] S. Bloch, *Function Theory of Polylogarithms*, in "Structural Properties of Polylogarithms", L. Lewin (ed.), Mathematical Surveys and Monographs (AMS), **37** (1991), 275–285.

[B-1]   N. Bourbaki, *Éléments de Mathématique, Algébre*, Hermann, Paris 1962.

[B-2]   N. Bourbaki, *Éléments de Mathématique, Algébre Commutative*, Hermann, Paris 1961.

[De0]   P. Deligne, *letter to Grothendieck*, November 19, 1982.

[De1]   P. Deligne, *Théorie de Hodge, II*, Publ. I.H.E.S., **40** (1971), 5–58.

[De2]   P. Deligne, *Le Groupe Fondamental de la Droite Projective Moins Trois Points*, in "Galois group over $\mathbb{Q}$" (Y. Ihara, K. Ribet, J.-P. Serre eds.), MSRI Publ. Vol. 16 (1989), 79–297.

[Dr]   V. G. Drinfeld, *On quasitriangular quasi-Hoph algebras and a group closely connected with* $\mathrm{Gal}(\overline{\mathbb{Q}}/\mathbb{Q})$, Algebra i Analiz **2** (1990), 149–181; *English translation*: Leningrad Math. J. **2** (1991), 829 – 860.

[DW1]   J.-C. Douai, Z. Wojtkowiak, *On the Galois actions on the fundamental group of* $\mathbf{P}^1_{\mathbb{Q}(\mu_n)} \setminus \{0, \mu_n, \infty\}$, Tokyo Journal of Math., **27** (2004), 21–34.

[DW2]   J.-C. Douai, Z. Wojtkowiak, *Descent for l-adic polylogarithms*, Nagoya Math. J., **192** (2008), 59–88.

[F]   H. Furusho, *The multiple zeta value algebra and the stable derivation algebra*, Publ. RIMS, Kyoto Univ. **39** (2003), 695–720.

[Ga]   H. Gangl, *Families of Functional Equations for Polylogarithms*, Comtemp. Math. (AMS) **199** (1996), 83–105.

[Gol]   K. Goldberg, *The formal power series for* $\mathrm{Log}e^x e^y$, Duke Math. J., **23** (1956), 13–21.

[Gon]   A. B. Goncharov, *Galois symmetries of fundamental groupoids and noncommutative geometry*, Duke Math. J., **128** (2005), 209–284.

[H]   R. Hain, *On a generalization of Hilbert's 21st problem*, Ann. Scient. Éc. Norm. Sup., **19** (1986), 609–627.

[HM]   R. Hain, M. Matsumoto, *Weighted completion of Galois groups and Galois actions on the fundamental group of* $\mathbb{P}^1 - \{0, 1, \infty\}$, Compositio Math. **139** (2003), 119–167.

[Ih1]   Y. Ihara, *Braids, Galois groups, and some arithmetic functions*, Proc. Intern. Congress of Math. Kyoto 1990, 99–120.

[Ih2]   Y. Ihara, *Some arithmetic aspects of Galois actions in the pro-p fundamental group of* $\mathbb{P}^1 - \{0, 1, \infty\}$, Proc. Symp. Pure Math. (AMS) **70** (2002) 247–273.

[Ii]   S. Iitaka, *Algebraic Geometry*, Springer GTM **76** 1982.

[K]   V. Kurlin, *The Baker–Campbell–Hausdorff formula in the free meta-abelian Lie algebra*, J. of Lie Theory, **17** (2007), 525–538.

[La1]   S. Lang, *Fundamentals of Diophantine Geometry*, Springer 1983.

[La2]   S. Lang, *Cyclotomic Fields I and II*, GTM **121**, Springer 1990.

[Le]   L. Lewin, *Polylogarithms and associated functions*, North Holland, 1981.

[LS]   P. Lochak, L. Schneps, *A cohomological interpretation of the Grothendieck-Teichmüller group*, Invent. math. **127** (1997), 571–600.

[MKS]   W. Magnus, A. Karrass, D. Solitar, *Combinatorial Group Theory*, Second Revised Edition, Dover 1976.

[MS]   W. G. McCallum and R. T. Sharifi, *A cup product in the Galois cohomology of number fields*, Duke Math. J., **120** (2003), 269–310.

[N0]   H. Nakamura, *Tangential base points and Eisenstein power series*, in "Aspects of Galois Theory" (H. Völklein et.al. eds.), London Math. Soc. Lect. Note Ser., **256** (1999), 202–217.

[N1] _____ , *Galois rigidity of pure sphere braid groups and profinite calculus*, J. Math. Sci. Univ. Tokyo **1** (1994), 71–136.

[N2] _____ , *Limits of Galois representations in fundamental groups along maximal degeneration of marked curves, I*, Amer. J. Math. **121** (1999) 315–358; *Part II*, Proc. Symp. Pure Math. **70** (2002) 43–78.

[NW] H. Nakamura, Z. Wojtkowiak, *On explicit formulae for l-adic polylogarithms*, Proc. Symp. Pure Math. (AMS) **70** (2002) 285–294.

[NW2] H. Nakamura, Z. Wojtkowiak, in preparation.

[Ra] M. Raynaud, *Propriétés de finitude du groupe fondamental*, SGA7, Exposé II, Lect. Notes in Math. Springer, **288** (1972), 25–31.

[Rob] A. Robert, *Elliptic Curves*, Lect. Notes in Math. **326**, Springer.

[Roq] P. Roquette, *Einheiten und Divisorkalssen in endlich erzeugbaren Körpern*, J. d. Deutschen Math. – Vereinigung, **60** (1957), 1–21.

[S1] Ch. Soulé, *On higher p-adic regulators*, Springer Lecture Notes in Math., **854** (1981), 372–401.

[S2] Ch. Soulé, *Élemens Cyclotomiques en K-Théorie*, Ast'erisque, **147/148** (1987), 225–258.

[W0] Z. Wojtkowiak, *The basic structure of polylogarithmic functional equations*, in "Structural Properties of Polylogarithms", L. Lewin (ed.), Mathematical Surveys and Monographs (AMS), **37** (1991), 205–231.

[W1] _____ , *A note on functional equations of the p-adic polylogarithms*, Bull. Soc. math. France, **119** (1991), 343–370.

[W2] _____ , *Functional equations of iterated integrals with regular singularities*, Nagoya Math. J., **142** (1996), 145–159.

[W3] _____ , *Monodromy of iterated integrals and non-abelian unipotent periods*, in "Geometric Galois Actions II", London Math. Soc. Lect. Note Ser. **243** (1997) 219–289.

[W4] _____ , *On ℓ-adic iterated integrals, I – Analog of Zagier Conjecture*, Nagoya Math. J., **176** (2004), 113–158.

[W5] _____ , *On ℓ-adic iterated integrals, II – Functional equations and ℓ-adic polylogarithms*, Nagoya Math. J., **177** (2005), 117–153.

[W6] _____ , *On ℓ-adic iterated integrals, III – Galois actions on fundamental groups*, Nagoya Math. J., **178** (2005), 1–36.

[W7] _____ , *On ℓ-adic iterated integrals, IV – ramification and generators of Galois actions on fundamental groups and torsors of paths*, Math. J. Okayama Univ., **51** (2009), 47–69.

[W8] _____ , *On the Galois actions on torsors of paths, I – Descent of Galois representations*, J. Math. Univ. Tokyo., **14** (2007), 177–259.

[W9] _____ , *A note on functional equations of ℓ-adic polylogarithms*, J. Inst. Math. Jussieu, **3** (2004), 461–471.

[W10] _____ , *On ℓ-adic Galois periods, relations between coefficients of Galois representations on fundamental groups of a projective line minus a finite number of points*, Algèbre et théorie des nombres. Années 2007–2009, Proceedings of the Conference "l-adic Cohomology and Number Theory" held at Luminy, Marseille, (December 10–14, 2007), Publ. Math. Univ. Franche-Comté Besançon Algèbr. Theor. Nr., Lab. Math. Besançon, Besançon, 2009, pp. 157–174. (available at URL:

http://www-math.univ-fcomte.fr/pp_Equipe/Algebre
TheorieDesNombres/pmb.html)

[Z] D. Zagier, *Polylogarithms, Dedekind zeta functions and the algebraic K-theory of Fields*, in "Arithmetic Algebraic Geometry", G. van der Geer et al.(eds.), Progress in Math., Birkhäuser, **89** (1991), 391–430.

Printed in the United States
by Baker & Taylor Publisher Services